Lecture Notes in Mathematics 1896

Editors:
J.-M. Morel, Cachan
F. Takens, Groningen
B. Teissier, Paris

Subseries:
Ecole d'Eté de Probabilités de Saint-Flour

T0178626

Pascal Massart

Concentration Inequalities and Model Selection

Ecole d'Eté de Probabilités
de Saint-Flour XXXIII - 2003

Editor: Jean Picard

 Springer

Author

Pascal Massart

Département de Mathématique
Université de Paris-Sud
Bât 425
91405 Orsay Cedex
France
e-mail: pascal.massart@math.u-psud.fv

Editor

Jean Picard

Laboratoire de Mathématiques Appliquées
UMR CNRS 6620
Université Blaise Pascal (Clermont-Ferrand)
63177 Aubière Cedex
France
e-mail: jean.picard@math.univ-bpclermont.fr

Cover: Blaise Pascal (1623-1662)

Library of Congress Control Number: 2007921691

Mathematics Subject Classification (2000): 60C05, 60E15, 62F10, 62B10, 62E17, 62G05, 62G07, 62G08, 62J02, 94A17

ISSN print edition: 0075-8434
ISSN electronic edition: 1617-9692
ISSN Ecole d'Eté de Probabilités de St. Flour, print edition: 0721-5363
ISBN-10 3-540-48497-3 Springer Berlin Heidelberg New York
ISBN-13 978-3-540-48497-4 Springer Berlin Heidelberg New York
DOI 10.1007/978-3-540-48503-2

Springer is a part of Springer Science+Business Media
springer.com
© Springer-Verlag Berlin Heidelberg 2007

Typesetting by the author and SPi using a Springer LaTeX macro package

Cover design: *design & production* GmbH, Heidelberg

Printed on acid-free paper   SPIN: 11917472   VA41/3100/SPi   5 4 3 2 1 0

Foreword

Three series of lectures were given at the 33rd Probability Summer School in Saint-Flour (July 6–23, 2003) by Professors Dembo, Funaki, and Massart. This volume contains the course of Professor Massart. The courses of Professors Dembo and Funaki have already appeared in volume 1869 (see below). We are grateful to the author for his important contribution.

Sixty-four participants have attended this school. Thirty-one have given a short lecture. The lists of participants and short lectures are enclosed at the end of the volume.

The Saint-Flour Probability Summer School was founded in 1971. Here are the references of Springer volumes where lectures of previous years were published. All numbers refer to the *Lecture Notes in Mathematics* series, expect S-50 which refers to volume 50 of the *Lecture Notes in Statistics* series.

1971: vol 307	1980: vol 929	1990: vol 1527	1998: vol 1738
1973: vol 390	1981: vol 976	1991: vol 1541	1999: vol 1781
1974: vol 480	1982: vol 1097	1992: vol 1581	2000: vol 1816
1975: vol 539	1983: vol 1117	1993: vol 1608	2001: vol 1837 & 1851
1976: vol 598	1984: vol 1180	1994: vol 1648	2002: vol 1840 & 1875
1977: vol 678	1985/86/87: vol 1362 & S-50	1995: vol 1690	2003: vol 1869 & 1896
1978: vol 774	1988: vol 1427	1996: vol 1665	2004: vol 1878 & 1879
1979: vol 876	1989: vol 1464	1997: vol 1717	2005: vol 1897

Further details can be found on the summer school web site
http://math.univ-bpclermont.fr/stflour/

Jean Picard
Clermont-Ferrand, April 2006

Preface

These notes would have never existed without the efforts of a number of people whom I would like to warmly thank. First of all, I would like to thank Lucien Birgé. We have spent hours working on model selection, trying to understand what was going on in depth. In these notes, I have attempted to provide a significant account of the nonasymptotic theory that we have tried to build together, year after year. Through our works we have promoted a nonasymptotic approach in statistics which consists in taking the number of observations as it is and try to evaluate the effect of all the influential parameters. At this very starting point, it seems to me that it is important to provide a first answer to the following question: why should we be interested by a nonasymptotic view for model selection at all? In my opinion, the motivation should neither be a strange interest for "small" sets of data nor a special taste for constants and inequalities rather than for limit theorems (although since mathematics is also a matter of taste, it is a possible way for getting involved in it!). On the contrary, the nonasymptotic point of view may turn out to be especially relevant when the number of observations is large. It is indeed to fit large complex sets of data that one needs to deal with possibly huge collections of models at different scales. The nonasymptotic approach for model selection precisely allows the collection of models together with their dimensions to vary freely, letting the dimensions be possibly of the same order of magnitude as the number of observations.

More than ten years ago we have been lucky enough to discover that concentration inequalities were indeed the probabilistic tools that we needed to develop a nonasymptotic theory for model selection. This offered me the opportunity to study this fascinating topic, trying first to understand the impressive works of Michel Talagrand and then taking benefits of Michel Ledoux's efforts to simplify some of Talagrand's arguments to bring my own contribution. It has been a great pleasure for me to work with Gábor Lugosi and Stéphane Boucheron on concentration inequalities. Most of the material which is presented here on this topic comes from our joint works.

Sharing my enthusiasm for these topics with young researchers and students has always been a strong motivation for me to work hard. I would like all of them to know how important they are to me, because not only seeing light in their eyes brought me happiness but also their theoretical works or their experiments have increased my level of understanding of my favorite topics. So many thanks to Sylvain Arlot, Yannick Baraud, Gilles Blanchard, Olivier Bousquet, Gwenaelle Castellan, Magalie Fromont, Jonas Kahn, Béatrice Laurent, Marc Lavarde, Emilie Lebarbier, Vincent Lepez, Frédérique Letué, Marie-Laure Martin, Bertrand Michel, Elodie Nédélec, Patricia Reynaud, Emmanuel Rio, Marie Sauvé, Christine Tuleau, Nicolas Verzelen, and Laurent Zwald.

In 2003, I had this wonderful opportunity to teach a course on concentration inequalities and model selection at the St Flour Summer school but before that I have taught a similar course in Orsay during several years. I am grateful to all the students who followed this course and whose questions have contributed to improve on the contents of my lectures.

Last but not least, I would like to thank Jean Picard for his kindness and patience and all the people who accepted to read my first draft. Of course the remaining mistakes or clumsy turns of phrase are entirely under my responsibility but (at least according to me) their comments and corrections have much improved the level of readability of these notes. You have been often too kind, sometimes pitiless and always careful and patient readers, so many thanks to all of you: Sylvain Arlot, Yannick Baraud, Lucien Birgé, Gilles Blanchard, Stéphane Boucheron, Laurent Cavalier, Gilles Celeux, Jonas Kahn, Frédérique Letué, Jean-Michel Loubes, Vincent Rivoirard, and Marie Sauvé.

Abstract

Model selection is a classical topic in statistics. The idea of selecting a model via penalizing a log-likelihood type criterion goes back to the early 1970s with the pioneering works of Mallows and Akaike. One can find many consistency results in the literature for such criteria. These results are asymptotic in the sense that one deals with a given number of models, and the number of observations tends to infinity. We shall give an overview of a nonasymptotic theory for model selection which has emerged during these last ten years. In various contexts of function estimation it is possible to design penalized log-likelihood type criteria with penalty terms depending not only on the number of parameters defining each model (as for the classical criteria) but also on the "complexity" of the whole collection of models to be considered. The performance of such a criterion is analyzed via nonasymptotic risk bounds for the corresponding penalized estimator which expresses that it performs almost as well as if the "best model" (i.e., with minimal risk) were known. For practical relevance of these methods, it is desirable to get a precise expression of the penalty terms involved in the penalized criteria on which they are based. That is why this approach heavily relies on concentration inequalities, the prototype being Talagrand's inequality for empirical processes. Our purpose is to give an account of the theory and discuss some selected applications such as variable selection or change points detection.

Contents

1   **Introduction** 1
  1.1   Model Selection 1
      1.1.1   Minimum Contrast Estimation 3
      1.1.2   The Model Choice Paradigm 5
      1.1.3   Model Selection via Penalization 7
  1.2   Concentration Inequalities 10
      1.2.1   The Gaussian Concentration Inequality 10
      1.2.2   Suprema of Empirical Processes 11
      1.2.3   The Entropy Method 12

2   **Exponential and Information Inequalities** 15
  2.1   The Cramér–Chernoff Method 15
  2.2   Sums of Independent Random Variables 21
      2.2.1   Hoeffding's Inequality 21
      2.2.2   Bennett's Inequality 23
      2.2.3   Bernstein's Inequality 24
  2.3   Basic Information Inequalities 27
      2.3.1   Duality and Variational Formulas 27
      2.3.2   Some Links Between the Moment Generating
             Function and Entropy............................ 29
      2.3.3   Pinsker's Inequality.............................. 31
      2.3.4   Birgé's Lemma 32
  2.4   Entropy on Product Spaces 35
      2.4.1   Marton's Coupling 37
      2.4.2   Tensorization Inequality for Entropy 40
  2.5 ϕ-Entropy 43
      2.5.1   Necessary Condition for the Convexity of ϕ-Entropy ... 45
      2.5.2   A Duality Formula for ϕ-Entropy.................. 46
      2.5.3   A Direct Proof of the Tensorization Inequality 49
      2.5.4   Efron–Stein's Inequality 50

3 Gaussian Processes 53
  3.1 Introduction and Basic Remarks 53
  3.2 Concentration of the Gaussian Measure on $\mathbb{R}^N$ 56
    3.2.1 The Isoperimetric Nature of the Concentration
        Phenomenon 57
    3.2.2 The Gaussian Isoperimetric Theorem 59
    3.2.3 Gross' Logarithmic Sobolev Inequality 62
    3.2.4 Application to Suprema of Gaussian
        Random Vectors 64
  3.3 Comparison Theorems for Gaussian Random Vectors 66
    3.3.1 Slepian's Lemma 66
  3.4 Metric Entropy and Gaussian Processes 70
    3.4.1 Metric Entropy 70
    3.4.2 The Chaining Argument 72
    3.4.3 Continuity of Gaussian Processes 74
  3.5 The Isonormal Process 77
    3.5.1 Definition and First Properties 77
    3.5.2 Continuity Sets with Examples 79

4 Gaussian Model Selection 83
  4.1 Introduction 83
    4.1.1 Examples of Gaussian Frameworks 83
    4.1.2 Some Model Selection Problems 86
    4.1.3 The Least Squares Procedure 87
  4.2 Selecting Linear Models 88
    4.2.1 A First Model Selection Theorem for Linear Models ... 89
    4.2.2 Lower Bounds for the Penalty Term 94
    4.2.3 Mixing Several Strategies 98
  4.3 Adaptive Estimation in the Minimax Sense 101
    4.3.1 Minimax Lower Bounds 102
    4.3.2 Adaptive Properties of Penalized Estimators for
        Gaussian Sequences 115
    4.3.3 Adaptation with Respect to Ellipsoids 116
    4.3.4 Adaptation with Respect to Arbitrary ℓ_p-Bodies 117
    4.3.5 A Special Strategy for Besov Bodies 122
  4.4 A General Model Selection Theorem 125
    4.4.1 Statement 125
    4.4.2 Selecting Ellipsoids: A Link with Regularization 131
    4.4.3 Selecting Nets Toward Adaptive Estimation for
        Arbitrary Compact Sets 139
  4.5 Appendix: From Function Spaces to Sequence Spaces 144

5 Concentration Inequalities 147
  5.1 Introduction 147
  5.2 The Bounded Difference Inequality via Marton's Coupling 148
  5.3 Concentration Inequalities via the Entropy Method 154

5.3.1 ϕ-Sobolev and Moment Inequalities 155
5.3.2   A Poissonian Inequality for Self-Bounding
        Functionals 157
5.3.3 ϕ-Sobolev Type Inequalities 162
5.3.4   From Efron–Stein to Exponential Inequalities 166
5.3.5   Moment Inequalities 172

6   **Maximal Inequalities** 183
6.1   Set-Indexed Empirical Processes 184
      6.1.1   Random Vectors and Rademacher Processes 184
      6.1.2   Vapnik–Chervonenkis Classes 186
      6.1.3 $\mathbb{L}_1$-Entropy with Bracketing 190
6.2   Function-Indexed Empirical Processes 192

7   **Density Estimation via Model Selection** 201
7.1   Introduction and Notations 201
7.2   Penalized Least Squares Model Selection 202
      7.2.1   The Nature of Penalized LSE 204
      7.2.2   Model Selection for a Polynomial Collection
              of Models 211
      7.2.3   Model Subset Selection Within a Localized Basis 219
7.3   Selecting the Best Histogram via Penalized Maximum
      Likelihood Estimation 225
      7.3.1   Some Deepest Analysis of Chi-Square Statistics 228
      7.3.2   A Model Selection Result 230
      7.3.3   Choice of the Weights $\{x_m, m \in \mathcal{M}\}$ 236
      7.3.4   Lower Bound for the Penalty Function 237
7.4   A General Model Selection Theorem for MLE 238
      7.4.1   Local Entropy with Bracketing Conditions 239
      7.4.2   Finite Dimensional Models 245
7.5   Adaptive Estimation in the Minimax Sense 251
      7.5.1   Lower Bounds for the Minimax Risk 251
      7.5.2   Adaptive Properties of Penalized LSE 263
      7.5.3   Adaptive Properties of Penalized MLE 267
7.6   Appendix 273
      7.6.1   Kullback–Leibler Information
              and Hellinger Distance 273
      7.6.2   Moments of Log-Likelihood Ratios 276
      7.6.3   An Exponential Bound for Log-Likelihood Ratios 277

8   **Statistical Learning** 279
8.1   Introduction 279
8.2   Model Selection in Statistical Learning 280
      8.2.1   A Model Selection Theorem 281

8.3   A Refined Analysis for the Risk of an Empirical
    Risk Minimizer 287
    8.3.1   The Main Theorem 288
    8.3.2   Application to Bounded Regression 293
    8.3.3   Application to Classification 296
8.4   A Refined Model Selection Theorem 301
    8.4.1   Application to Bounded Regression 303
8.5   Advanced Model Selection Problems 307
    8.5.1   Hold-Out as a Margin Adaptive Selection Procedure ... 308
    8.5.2   Data-Driven Penalties 314

References 319

Index 325

List of Participants 331

List of Short Lectures 335

1

Introduction

If one observes some random variable ξ (which can be a random vector or a random process) with unknown distribution, the basic problem of statistical inference is to take a decision about some quantity s related to the distribution of ξ, for instance estimate s or provide a confidence set for s with a given level of confidence. Usually, one starts from a genuine estimation procedure for s and tries to get some idea of how far it is from the target. Since generally speaking the exact distribution of the estimation procedure is not available, the role of probability theory is to provide relevant approximation tools to evaluate it. In the situation where $\xi = \xi^{(n)}$ depends on some parameter n (typically when $\xi = (\xi_1, \ldots, \xi_n)$, where the variables $\xi_1, \ldots, \xi_n$ are independent), *asymptotic theory* in statistics uses limit theorems (Central Limit Theorems, Large Deviation Principles, etc.) as approximation tools when n is large. One of the first examples of such a result is the use of the CLT to analyze the behavior of a *maximum likelihood estimator (MLE)* on a given regular parametric model (independent of n) as n goes to infinity. More recently, since the seminal works of Dudley in the 1970s, the theory of probability in Banach spaces has deeply influenced the development of asymptotic statistics, the main tools involved in these applications being limit theorems for empirical processes. This led to decisive advances for the theory of asymptotic efficiency in semiparametric models for instance and the interested reader will find numerous results in this direction in the books by Van der Vaart and Wellner [120] or Van der Vaart [119].

1.1 Model Selection

Designing a genuine estimation procedure requires some prior knowledge on the unknown distribution of ξ and choosing a proper model is a major problem for the statistician. The aim of model selection is to construct data-driven criteria to select a model among a given list. We shall see that in many situations motivated by applications such as signal analysis for instance, it is useful to allow the size of the models to depend on the sample size n. In these situations, classical asymptotic analysis breaks down and one needs to introduce an alternative approach that we call *nonasymptotic*. By nonasymptotic, we do not mean of course that large samples of observations are not welcome but that the size of the models as well as the size of the list of models should be allowed to be large when n is large in order to be able to warrant that the statistical model is not far from the truth. When the target quantity s to be estimated is a function, this allows in particular to consider models which have good approximation properties at different scales and use model selection criteria to choose from the data what is the best approximating model to be considered. In the past 20 years, the phenomenon of the concentration of measure has received much attention mainly due to the remarkable series of works by Talagrand which led to a variety of new powerful inequalities (see in particular [112] and [113]). The main interesting feature of concentration inequalities is that, unlike central limit theorems or large deviations inequalities, they are indeed *nonasymptotic*. The major issue of this series of Lectures is to show that these new tools of probability theory lead to a *nonasymptotic* theory for model selection and illustrate the benefits of this approach for several functional estimation problems. The basic examples of functional estimation frameworks that we have in mind are the following.

- **Density estimation**

    One observes $\xi_1, \ldots, \xi_n$ which are i.i.d. random variables with unknown density s with respect to some given measure μ.

- **Regression**

    One observes $(X_1, Y_1), \ldots, (X_n, Y_n)$ with

    $$Y_i = s(X_i) + \varepsilon_i, \ 1 \leq i \leq n.$$

    One assumes the *explanatory* variables $X_1, \ldots, X_n$ to be independent (but nonnecessarily i.i.d.) and the regression errors $\varepsilon_1, \ldots, \varepsilon_n$ to be i.i.d. with $\mathbb{E}[\varepsilon_i \mid X_i] = 0$. s is the so-called *regression function*.

- **Binary classification**

    As in the regression setting, one still observes independent pairs

    $$(X_1, Y_1), \ldots, (X_n, Y_n)$$

    but here we assume those pairs to be copies of a pair (X, Y), where the response variables Y take only two values, say: 0 or 1. The basic problem of *statistical learning* is to estimate the so-called Bayes classifier s defined by

    $$s(x) = \mathbb{1}_{\eta(x) \geq 1/2}$$

    where η denotes the regression function, $\eta(x) = \mathbb{E}[Y \mid X = x]$.

- **Gaussian white noise**

    Let $s \in \mathbb{L}_2 \left([0,1]^d \right)$. One observes the process $\xi^{(n)}$ on $[0,1]^d$ defined by

    $$d\xi^{(n)}(x) = s(x) + \frac{1}{\sqrt{n}} dB(x), \; \xi^{(n)}(0) = 0,$$

    where B denotes a Brownian sheet. The level of noise ε is here written as $\varepsilon = 1/\sqrt{n}$ for notational convenience and in order to allow an easy comparison with the other frameworks.

    In all of the above examples, one observes some random variable $\xi^{(n)}$ with unknown distribution which depends on some quantity $s \in \mathcal{S}$ to be estimated. One can typically think of s as a function belonging to some space $\mathcal{S}$ which may be infinite dimensional. For instance

- In the density framework, s is a density and $\mathcal{S}$ can be taken as the set of all probability densities with respect to μ.
- In the i.i.d. regression framework, the variables $\xi_i = (X_i, Y_i)$ are independent copies of a pair of random variables (X, Y), where X takes its values in some measurable space $\mathcal{X}$. Assuming the variable Y to be square integrable, the regression function s defined by $s(x) = \mathbb{E}[Y \mid X = x]$ for every $x \in \mathcal{X}$ belongs to $\mathcal{S} = \mathbb{L}^2(\mu)$, where μ denotes the distribution of X.

    One of the most commonly used method to estimate s is minimum contrast estimation.

1.1.1 Minimum Contrast Estimation

Let us consider some empirical criterion γ_n (based on the observation $\xi^{(n)}$) such that on the set $\mathcal{S}$

$$t \to \mathbb{E}[\gamma_n(t)]$$

achieves a minimum at point s. Such a criterion is called an *empirical contrast* for the estimation of s. Given some subset S of $\mathcal{S}$ that we call a *model*, a *minimum contrast estimator* $\hat{s}$ of s is a minimizer of γ_n over S. The idea is that, if one substitutes the empirical criterion γ_n to its expectation and minimizes γ_n on S, there is some hope to get a sensible estimator of s, at least if s belongs (or is close enough) to model S. This estimation method is widely used and has been extensively studied in the asymptotic parametric setting for which one assumes that S is a given parametric model, s belongs to S and n is large. Probably, the most popular examples are maximum likelihood and least squares estimation. Let us see what this gives in the above functional estimation frameworks. In each example given below, we shall check that a given empirical criterion is indeed an empirical contrast by showing that the associated natural loss function

$$\ell(s,t) = \mathbb{E}[\gamma_n(t)] - \mathbb{E}[\gamma_n(s)] \tag{1.1}$$

is nonnegative for all $t \in \mathcal{S}$. In the case where $\xi^{(n)} = (\xi_1, \ldots, \xi_n)$, we shall define an empirical criterion γ_n in the following way:

$$\gamma_n(t) = P_n\left[\gamma(t, .)\right] = \frac{1}{n}\sum_{i=1}^{n}\gamma(t, \xi_i),$$

so that it remains to specify for each example what is the adequate function γ to be considered.

• **Density estimation**

One observes $\xi_1, \ldots, \xi_n$ which are i.i.d. random variables with unknown density s with respect to some given measure μ. The choice

$$\gamma(t, x) = -\ln(t(x))$$

leads to the *maximum likelihood criterion* and the corresponding loss function ℓ is given by

$$\ell(s, t) = \mathbf{K}(s, t),$$

where $\mathbf{K}(s, t)$ denotes the Kullback–Leibler information number between the probabilities $s\mu$ and $t\mu$, i.e.,

$$\mathbf{K}(s, t) = \int s\ln\left(\frac{s}{t}\right)$$

if $s\mu$ is absolutely continuous with respect to $t\mu$ and $\mathbf{K}(s, t) = +\infty$ otherwise. Assuming that $s \in \mathbb{L}_2(\mu)$, it is also possible to define a *least squares criterion* for density estimation by setting this time

$$\gamma(t, x) = \|t\|^2 - 2t(x)$$

where $\|.\|$ denotes the norm in $\mathbb{L}_2(\mu)$ and the corresponding loss function ℓ is in this case given by

$$\ell(s, t) = \|s - t\|^2,$$

for every $t \in \mathbb{L}_2(\mu)$.

• **Regression**

One observes $(X_1, Y_1), \ldots, (X_n, Y_n)$ with

$$Y_i = s(X_i) + \varepsilon_i, \ 1 \leq i \leq n,$$

where $X_1, \ldots, X_n$ are independent and $\varepsilon_1, \ldots, \varepsilon_n$ are i.i.d. with $\mathbb{E}\left[\varepsilon_i \mid X_i\right] = 0$. Let μ be the arithmetic mean of the distributions of the variables $X_1, \ldots, X_n$, then least squares estimation is obtained by setting for every $t \in \mathbb{L}_2(\mu)$

$$\gamma(t, (x, y)) = (y - t(x))^2,$$

and the corresponding loss function ℓ is given by

$$\ell(s, t) = \|s - t\|^2,$$

where $\|.\|$ denotes the norm in $\mathbb{L}_2(\mu)$.

- **Binary classification**

One observes independent copies $(X_1, Y_1), \ldots, (X_n, Y_n)$ of a pair (X, Y), where Y takes its values in $\{0, 1\}$. We take the same value for γ as in the least squares regression case but this time we restrict the minimization to the set $\mathcal{S}$ of *classifiers* i.e., $\{0, 1\}$-valued measurable functions (instead of $\mathbb{L}_2(\mu)$). For a function t which takes only the two values 0 and 1, we can write

$$\frac{1}{n} \sum_{i=1}^{n} (Y_i - t(X_i))^2 = \frac{1}{n} \sum_{i=1}^{n} \mathbb{1}_{Y_i \neq t(X_i)}$$

so that minimizing the least squares criterion means minimizing the number of misclassifications on the training sample $(X_1, Y_1), \ldots, (X_n, Y_n)$. The corresponding minimization procedure can also be called *empirical risk minimization* (according to Vapnik's terminology, see [121]). Setting

$$s(x) = \mathbb{1}_{\eta(x) \geq 1/2}$$

where η denotes the regression function, $\eta(x) = \mathbb{E}[Y \mid X = x]$, the corresponding loss function ℓ is given by

$$\ell(s, t) = \mathbb{P}[Y \neq t(X)] - \mathbb{P}[Y \neq s(X)] = \mathbb{E}[|2\eta(X) - 1| \, |s(X) - t(X)|].$$

Finally, we can consider the least squares procedure in the Gaussian white noise framework too.

- **Gaussian white noise**

Recall that one observes the process $\xi^{(n)}$ on $[0, 1]^d$ defined by

$$d\xi^{(n)}(x) = s(x) + \frac{1}{\sqrt{n}} dB(x), \ \xi^{(n)}(0) = 0,$$

where W denotes a Brownian sheet. We define for every $t \in \mathbb{L}_2\left([0, 1]^d\right)$

$$\gamma_n(t) = \|t\|^2 - 2 \int_0^1 t(x) \, d\xi^{(n)}(x),$$

then the corresponding loss function ℓ is simply given by

$$\ell(s, t) = \|s - t\|^2.$$

1.1.2 The Model Choice Paradigm

The main problem which arises from minimum contrast estimation in a parametric setting is the choice of a proper model S on which the minimum contrast estimator is to be defined. In other words, it may be difficult to guess what is the right parametric model to consider in order to reflect the nature of data from the real life and one can get into problems whenever the model S is false in the sense that the true s is too far from S. One could then be tempted to choose S as big as possible. Taking S as $\mathcal{S}$ itself or as a "huge" subset of $\mathcal{S}$ is known to lead to inconsistent (see [7]) or suboptimal estimators (see [19]). We see that choosing some model S in advance leads to some difficulties:

- If S is a "small" model (think of some parametric model, defined by 1 or 2 parameters for instance) the behavior of a minimum contrast estimator on S is satisfactory as long as s is close enough to S but the model can easily turn to be false.
- On the contrary, if S is a "huge" model (think of the set of all continuous functions on $[0, 1]$ in the regression framework for instance), the minimization of the empirical criterion leads to a very poor estimator of s even if s truly belongs to S.

Illustration (White Noise)

Least squares estimators (LSE) on a linear model S (i.e., minimum contrast estimators related to the least squares criterion) can be computed explicitly. For instance, in the white noise framework, if $(\phi_j)_{1 \leq j \leq D}$ denotes some orthonormal basis of the D-dimensional linear space S, the LSE can be expressed as

$$\widehat{s} = \sum_{j=1}^{D} \left(\int_0^1 \phi_j(x) \, d\xi^{(n)}(x) \right) \phi_j.$$

Since for every $1 \leq j \leq D$

$$\int_0^1 \phi_j(x) \, d\xi^{(n)}(x) = \int_0^1 \phi_j(x) s(x) \, dx + \frac{1}{\sqrt{n}} \eta_j,$$

where the variables $\eta_1, \ldots, \eta_D$ are i.i.d. standard normal variables, the quadratic risk of $\widehat{s}$ can be easily computed. One indeed has

$$\mathbb{E}\left[\|s - \widehat{s}\|^2 \right] = d^2(s, S) + \frac{D}{n}.$$

This formula for the quadratic risk perfectly reflects the model choice paradigm since if one wants to choose a model in such a way that the risk of the resulting least square estimator is small, we have to warrant that the *bias term* $d^2(s, S)$ and the *variance term* D/n are small simultaneously. It is therefore interesting to consider a family of models instead of a single one and try to select some appropriate model among the family. More precisely, if $(S_m)_{m \in \mathcal{M}}$ is a list of finite dimensional subspaces of $\mathbb{L}_2\left([0, 1]^d\right)$ and $(\widehat{s}_m)_{m \in \mathcal{M}}$ be the corresponding list of least square estimators, an *ideal* model should minimize $\mathbb{E}\left[\|s - \widehat{s}_m\|^2 \right]$ with respect to $m \in \mathcal{M}$. Of course, since we do not know the bias term, the quadratic risk cannot be used as a model choice criterion but just as a benchmark.

More generally if we consider some empirical contrast γ_n and some (at most countable and usually finite) collection of models $(S_m)_{m \in \mathcal{M}}$, let us represent each model S_m by the minimum contrast estimator $\widehat{s}_m$ related to γ_n. The purpose is to select the "best" estimator among the collection $(\widehat{s}_m)_{m \in \mathcal{M}}$. Ideally, one would like to consider $m(s)$ minimizing the risk $\mathbb{E}\left[\ell(s, \widehat{s}_m)\right]$ with respect to $m \in \mathcal{M}$. The minimum contrast estimator $\widehat{s}_{m(s)}$ on the corresponding model $S_{m(s)}$ is called an *oracle* (according to the terminology introduced by Donoho and Johnstone, see [47] for instance). Unfortunately, since the risk depends on the unknown parameter s, so does $m(s)$ and the oracle is *not* an estimator of s. However, the risk of an oracle can serve as a benchmark which will be useful in order to evaluate the performance of any data driven selection procedure among the collection of estimators $(\widehat{s}_m)_{m \in \mathcal{M}}$. Note that this notion is different from the notion of true model. In other words, if s belongs to some model S_{m_0}, this does not necessarily imply that $\widehat{s}_{m_0}$ is an oracle. The idea is now to consider data-driven criteria to select an estimator which tends to mimic an oracle, i.e., one would like the risk of the selected estimator $\widehat{s}_{\widehat{m}}$ to be as close as possible to the risk of an oracle.

1.1.3 Model Selection via Penalization

Let us describe the method. The *model selection via penalization* procedure consists in considering some proper *penalty function* pen: $\mathcal{M} \to \mathbb{R}_+$ and take $\widehat{m}$ minimizing the *penalized criterion*

$$\gamma_n(\widehat{s}_m) + \mathrm{pen}(m)$$

over $\mathcal{M}$. We can then define the selected model $S_{\widehat{m}}$ and the selected estimator $\widehat{s}_{\widehat{m}}$.

This method is definitely not new. Penalized criteria have been proposed in the early 1970s by Akaike (see [2]) for penalized log-likelihood in the density estimation framework and Mallows for penalized least squares regression (see [41] and [84]), where the variance σ^2 of the errors of the regression framework is assumed to be known for the sake of simplicity. In both cases the penalty functions are proportional to the number of parameters D_m of the corresponding model S_m

- Akaike : D_m/n
- Mallows' C_p : $2D_m\sigma^2/n$.

Akaike's heuristics leading to the choice of the penalty function D_m/n heavily relies on the assumption that the dimensions and the number of the models are bounded with respect to n and n tends to infinity.

Let us give a simple motivating example for which those assumptions are clearly not satisfied.

A Case Example: Change Points Detection

Change points detection on the mean is indeed a typical example for which these criteria are known to fail. A noisy signal ξ_j is observed at each time j/n on $[0,1]$. We consider the fixed design regression framework

$$\xi_j = s\,(j/n) + \varepsilon_j,\, 1 \le j \le n$$

where the errors are i.i.d. centered random variables. Detecting change points on the mean amounts to select the "best" piecewise constant estimator of the true signal s on some arbitrary partition m with endpoints on the regular grid $\{j/n, 0 \le j \le n\}$. Defining S_m as the linear space of piecewise constant functions on partition m, this means that we have to select a model among the family $(S_m)_{m \in \mathcal{M}}$, where $\mathcal{M}$ denotes the collection of all possible partitions by intervals with end points on the grid. Then, the number of models with dimension D, i.e., the number of partitions with D pieces is equal to $\binom{n-1}{D-1}$ which grows polynomially with respect to n.

The Nonasymptotic Approach

The approach to model selection via penalization that we have developed (see for instance the seminal papers [20] and [12]) differs from the usual parametric asymptotic approach in the sense that:

- The number as well as the dimensions of the models may depend on n.
- One can choose a list of models because of its approximation properties: wavelet expansions, trigonometric or piecewise polynomials, artificial neural networks etc.

It may perfectly happen that many models in the list have the same dimension and in our view, the "complexity" of the list of models is typically taken into account via the choice of the penalty function of the form

$$(C_1 + C_2 L_m)\,\frac{D_m}{n}$$

where the weights L_m satisfy the restriction

$$\sum_{m \in \mathcal{M}} e^{-L_m D_m} \le 1$$

and C_1 and C_2 do not depend on n.

As we shall see, concentration inequalities are deeply involved both in the construction of the penalized criteria and in the study of the performance of the resulting *penalized estimator* $\widehat{s}_{\widehat{m}}$.

The Role of Concentration Inequalities

Our approach can be described as follows. We take as a loss function the nonnegative quantity $\ell(s, t)$ and recall that our aim is to mimic the oracle, i.e., minimize $\mathbb{E}[\ell(s, \widehat{s}_m)]$ over $m \in \mathcal{M}$.

Let us introduce the centered *empirical process*

$$\overline{\gamma}_n(t) = \gamma_n(t) - \mathbb{E}[\gamma_n(t)].$$

By definition a penalized estimator $\widehat{s}_{\widehat{m}}$ satisfies for every $m \in \mathcal{M}$ and any point $s_m \in S_m$

$$\gamma_n(\widehat{s}_{\widehat{m}}) + \text{pen}(\widehat{m}) \leq \gamma_n(\widehat{s}_m) + \text{pen}(m)$$
$$\leq \gamma_n(s_m) + \text{pen}(m)$$

or, equivalently if we substitute $\overline{\gamma}_n(t) + \mathbb{E}[\gamma_n(t)]$ to $\gamma_n(t)$

$$\overline{\gamma}_n(\widehat{s}_{\widehat{m}}) + \text{pen}(\widehat{m}) + \mathbb{E}[\gamma_n(\widehat{s}_{\widehat{m}})] \leq \overline{\gamma}_n(s_m) + \text{pen}(m) + \mathbb{E}[\gamma_n(s_m)].$$

Subtracting $\mathbb{E}[\gamma_n(s)]$ on each side of this inequality finally leads to the following important bound

$$\ell(s, \widehat{s}_{\widehat{m}}) \leq \ell(s, s_m) + \text{pen}(m)$$
$$+ \overline{\gamma}_n(s_m) - \overline{\gamma}_n(\widehat{s}_{\widehat{m}}) - \text{pen}(\widehat{m}).$$

Hence, the penalty should be

- heavy enough to annihilate the fluctuations of $\overline{\gamma}_n(s_m) - \overline{\gamma}_n(\widehat{s}_{\widehat{m}})$;
- but not too large since ideally we would like that $\ell(s, s_m) + \text{pen}(m) \leq \mathbb{E}[\ell(s, \widehat{s}_m)]$.

Therefore, we see that an accurate calibration of the penalty should rely on a sharp evaluation of the fluctuations of $\overline{\gamma}_n(s_m) - \overline{\gamma}_n(\widehat{s}_{\widehat{m}})$. This is precisely why we need local concentration inequalities in order to analyze the uniform deviation of $\overline{\gamma}_n(u) - \overline{\gamma}_n(t)$ when t is close to u and belongs to a given model. In other words, the key is to get a good control of the supremum of some conveniently weighted empirical process

$$\frac{\overline{\gamma}_n(u) - \overline{\gamma}_n(t)}{a(u, t)}, \ t \in S_{m'}.$$

The prototype of such bounds is by now the classical Gaussian concentration inequality to be proved in Chapter 3 and Talagrand's inequality for empirical processes to be proved in Chapter 5 in the non-Gaussian case.

1.2 Concentration Inequalities

More generally, the problem that we shall deal with is the following. Given independent random variables $X_1, \ldots, X_n$ taking their values in $\mathcal{X}^n$ and some functional $\zeta : \mathcal{X}^n \to \mathbb{R}$, we want to study the concentration property of $Z = \zeta(X_1, \ldots, X_n)$ around its expectation. In the applications that we have in view the useful results are sub-Gaussian inequalities. We have in mind to prove inequalities of the following type

$$\mathbb{P}\left[Z - \mathbb{E}\left[Z\right] \geq x\right] \leq \exp\left(-\frac{x^2}{2v}\right), \text{ for } 0 \leq x \leq x_0, \tag{1.2}$$

and analogous bounds on the left tail.

Ideally, one would like that $v = \operatorname{Var}(Z)$ and $x_0 = \infty$. More reasonably, we shall content ourselves with bounds for which v is a "good" upper bound for $\operatorname{Var}(Z)$ and x_0 is an explicit function of n and v.

1.2.1 The Gaussian Concentration Inequality

In the Gaussian case, this program can be fruitfully completed. We shall indeed see in Chapter 3 that whenever $\mathcal{X}^n = \mathbb{R}^n$ is equipped with the canonical Euclidean norm, $X_1, \ldots, X_n$ are i.i.d. standard normal and ζ is assumed to be Lipschitz, i.e.,

$$|\zeta(y) - \zeta(y')| \leq L \|y - y'\|, \text{ for every } y, y' \text{ in } \mathbb{R}^n$$

then, on the one hand $\operatorname{Var}(Z) \leq L^2$ and on the other hand the Cirelson–Ibragimov–Sudakov inequality ensures that

$$\mathbb{P}\left[Z - \mathbb{E}\left[Z\right] \geq x\right] \leq \exp\left(-\frac{x^2}{2L^2}\right), \text{ for all } x \geq 0.$$

The remarkable feature of this inequality is that its dependency with respect to the dimension n is entirely contained in the expectation $\mathbb{E}\left[Z\right]$. Extending this result to more general situations is not so easy. It is in particular unclear to know what kind of regularity conditions should be required on the functional ζ. A Lipschitz type condition with respect to the Hamming distance could seem to be a rather natural and attractive candidate. It indeed leads to interesting results as we shall see in Chapter 5. More precisely, if d denotes Hamming distance on $\mathcal{X}^n$ defined by

$$d(y, y') = \sum_{i=1}^{n} \mathbb{1}_{y_i \neq y'_i}, \text{ for all } y, y' \text{ in } \mathcal{X}^n$$

and ζ is assumed to be Lipschitz with respect to d

$$|\zeta(y) - \zeta(y')| \leq L d(y, y'), \text{ for all } y, y' \text{ in } \mathcal{X}^n \tag{1.3}$$

then it can be proved that

$$\mathbb{P}\left[Z - \mathbb{E}\left[Z\right] \geq x\right] \leq \exp\left(-\frac{2x^2}{nL^2}\right), \text{ for all } x \geq 0.$$

Let us now come back to the functional which naturally emerges from the study of penalized model selection criteria.

1.2.2 Suprema of Empirical Processes

Let us assume T to be countable in order to avoid any measurability problem. The supremum of an empirical process of the form

$$Z = \sup_{t \in T} \sum_{i=1}^{n} f_t\left(X_i\right)$$

provides an important example of a functional of independent variables both for theory and applications. Assuming that $\sup_{t \in T} \|f_t\|_\infty \leq 1$ ensures that the mapping

$$\zeta : y \rightarrow \sup_{t \in T} \sum_{i=1}^{n} f_t\left(y_i\right)$$

satisfies the Lipschitz condition (1.3) with respect to the Hamming distance d with $L = 2$ and therefore

$$\mathbb{P}\left[Z - \mathbb{E}\left[Z\right] \geq x\right] \leq \exp\left(-\frac{x^2}{2n}\right), \text{ for all } x \geq 0. \tag{1.4}$$

However, it may happen that the variables $f_t\left(X_i\right)$ have a "small" variance uniformly with respect to t and i. In this case, one would expect a better variance factor in the exponential bound but obviously Lipschitz's condition with respect to Hamming distance alone cannot lead to such an improvement.

In other words Lipschitz property is not sharp enough to capture the *local* behavior of empirical processes which lies at the heart of our analysis of penalized criteria for model selection. It is the merit of Talagrand's inequality for empirical processes to provide an improved version of (1.4) which will turn to be an efficient tool for analyzing the uniform increments of an empirical process as expected.

It will be one of the main goals of Chapter 5 to prove the following version of Talagrand's inequality. Under the assumption that $\sup_{t \in T} \|f_t\|_\infty \leq 1$, there exists some absolute positive constant η such that

$$\mathbb{P}\left[Z - \mathbb{E}\left[Z\right] \geq x\right] \leq \exp\left(-\eta\left(\frac{x^2}{\mathbb{E}\left[W\right]} \wedge x\right)\right), \tag{1.5}$$

where $W = \sup_{t \in T} \sum_{i=1}^{n} f_t^2\left(X_i\right)$. Note that (1.5) a fortiori implies some sub-Gaussian inequality of type (1.2) with $v = \mathbb{E}\left[W\right] / \left(2\eta\right)$ and $x_0 = \sqrt{\mathbb{E}\left[W\right]}$.

1.2.3 The Entropy Method

Building upon the pioneering works of Marton (see [87]) on the one hand
and Ledoux (see [77]) on the other hand, we shall systematically derive concentration inequalities from information theoretic arguments. The elements
of information theory that we shall need will be presented in Chapter 2 and
used in Chapter 5. One of the main tools that we shall use is the duality
formula for entropy. Interestingly, we shall see how this formula also leads to
statistical minimax lower bounds. Our goal will be to provide a simple proof
of Talagrand's inequality for empirical processes and extend it to more general functional of independent variables. The starting point for our analysis is
Efron–Stein's inequality. Let $X' = X'_1, \ldots, X'_n$ be some independent copy of
$X = X_1, \ldots, X_n$ and define

$$Z'_i = \zeta \left(X_1, \ldots, X_{i-1}, X'_i, X_{i+1}, \ldots, X_n \right).$$

Setting

$$V^+ = \sum_{i=1}^{n} \mathbb{E} \left[(Z - Z'_i)_+^2 \mid X \right],$$

Efron–Stein's inequality (see [59]) ensures that

$$\mathrm{Var}\,(Z) \leq \mathbb{E} \left[V^+ \right]. \tag{1.6}$$

Let us come back to empirical processes and focus on centered empirical
processes for the sake of simplicity. This means that we assume the vari-
ables X_i to be i.i.d. and $\mathbb{E} \left[f_t (X_1) \right] = 0$ for every $t \in T$. We also assume T to
be finite and consider the supremum of the empirical process

$$Z = \sup_{t \in T} \sum_{i=1}^{n} f_t (X_i),$$

so that for every i

$$Z'_i = \sup_{t \in T} \left[\left(\sum_{j \neq i}^{n} f_t (X_j) \right) + f_t (X'_i) \right].$$

Taking t^* such that $\sup_{t \in T} \sum_{j=1}^{n} f_t (X_j) = \sum_{j=1}^{n} f_{t^*} (X_j)$, we have for every
$i \in [1, n]$

$$Z - Z'_i \leq f_{t^*} (X_i) - f_{t^*} (X'_i)$$

which yields

$$(Z - Z'_i)_+^2 \leq (f_{t^*} (X_i) - f_{t^*} (X'_i))^2$$

and therefore by independence of X'_i from X we derive from the centering
assumption $\mathbb{E} \left[f_t \left(X'_i \right) \right] = 0$ that

$$\mathbb{E}\left[(Z - Z_i')_+^2 \mid X\right] \leq f_{t^*}^2(X_i) + \mathbb{E}\left[f_{t^*}^2(X_i')\right].$$

Hence, we deduce from Efron–Stein's inequality that

$$\mathrm{Var}\,(Z) \leq 2\mathbb{E}\,[W]\,,$$

where $W = \sup_{t \in T} \sum_{i=1}^n f_t^2(X_i)$.

The conclusion is therefore that the variance factor appearing in Talagrand's inequality turns out to be the upper bound which derives from Efron–Stein's inequality. The main guideline that we shall follow in Chapter 5 is that, more generally, the adequate variance factor v to be considered in (1.2) is (up to some absolute constant) the upper bound for the variance of Z provided by Efron–Stein's inequality.

2

Exponential and Information Inequalities

A general method for establishing exponential inequalities consists in controlling the moment generating function of a random variable and then minimize the upper probability bound resulting from Markov's inequality. Though elementary, this classical method, known as Cramér–Chernoff's method, turns out to be very powerful and surprisingly sharp. In particular it will describe how a random variable Z concentrates around its mean $\mathbb{E}[Z]$ by providing an upper bound for the probability

$$\mathbb{P}\left[|Z - \mathbb{E}[Z]| \geq x\right],$$

for every positive x. Since this probability is equal to

$$\mathbb{P}\left[Z - \mathbb{E}[Z] \geq x\right] + \mathbb{P}\left[\mathbb{E}[Z] - Z \geq x\right],$$

considering either $\widetilde{Z} = Z - \mathbb{E}[Z]$ or $\widetilde{Z} = \mathbb{E}[Z] - Z$, we can mainly focus on exponential bounds for $\mathbb{P}[Z \geq x]$ where Z is a centered random variable.

2.1 The Cramér–Chernoff Method

Let Z be a real valued random variable. We derive from Markov's inequality that for all positive λ and any real number x

$$\mathbb{P}[Z \geq x] = \mathbb{P}\left[e^{\lambda Z} \geq e^{\lambda x}\right] \leq \mathbb{E}\left[e^{\lambda Z}\right] e^{-\lambda x}.$$

So that, defining the log-moment generating function as

$$\psi_Z(\lambda) = \ln\left(\mathbb{E}\left[e^{\lambda Z}\right]\right) \qquad \text{for all } \lambda \in \mathbb{R}_+$$

and

$$\psi_Z^*(x) = \sup_{\lambda \in \mathbb{R}_+} (\lambda x - \psi_Z(\lambda)), \tag{2.1}$$

we derive Chernoff's inequality

$$\mathbb{P}\left[Z \geq x\right] \leq \exp\left(-\psi_Z^*\left(x\right)\right). \tag{2.2}$$

The function ψ_Z^* is called the *Cramér transform* of Z. Since $\psi_Z\left(0\right) = 0$, ψ_Z^* is at least a nonnegative function. Assuming Z to be integrable Jensen's inequality warrants that $\psi_Z\left(\lambda\right) \geq \lambda\mathbb{E}\left[Z\right]$ and therefore $\sup_{\lambda<0}\left(\lambda x - \psi_Z\left(\lambda\right)\right) \leq 0$ whenever $x \geq \mathbb{E}\left[Z\right]$. Hence one can also write the Cramér transform as

$$\psi_Z^*\left(x\right) = \sup_{\lambda\in\mathbb{R}}\left(\lambda x - \psi_Z\left(\lambda\right)\right).$$

This means that at every point $x \geq \mathbb{E}\left[Z\right]$, $\psi_Z^*\left(x\right)$ coincides with the Fenchel–Legendre dual function of ψ_Z.

Of course Chernoff's inequality can be trivial if $\psi_Z^*\left(x\right) = 0$. This is indeed the case whenever $\psi_Z\left(\lambda\right) = +\infty$ for all positive λ or if $x \leq \mathbb{E}\left[Z\right]$ (using again the lower bound $\psi_Z\left(\lambda\right) \geq \lambda\mathbb{E}\left[Z\right]$). Let us therefore assume the set of all positive λ such that $\mathbb{E}\left[e^{\lambda Z}\right]$ is finite to be nonvoid and consider its supremum b, with $0 < b \leq +\infty$. Note then, that Hölder's inequality warrants that this set is an interval with left end point equal to 0. Let us denote by I the set of its interior points, $I = (0, b)$. Then ψ_Z is convex (strictly convex whenever Z is not almost surely constant) and infinitely many times differentiable on I.

The case where Z is a centered random variable is of special interest. In such a case ψ_Z is continuously differentiable on $[0, b)$ with $\psi_Z'\left(0\right) = \psi_Z\left(0\right) = 0$ and we can also write the Cramér transform as $\psi_Z^*\left(x\right) = \sup_{\lambda\in I}\lambda x - \psi_Z\left(\lambda\right)$. A useful property of the Cramér transform can be derived from the following result.

Lemma 2.1 *Let ψ be some convex and continuously differentiable function on $[0, b)$ with $0 < b \leq +\infty$. Assume that $\psi\left(0\right) = \psi'\left(0\right) = 0$ and set, for every $x \geq 0$,*

$$\psi^*\left(x\right) = \sup_{\lambda\in(0,b)}\left(\lambda x - \psi\left(\lambda\right)\right).$$

Then ψ^* is a nonnegative convex and nondecreasing function on $\mathbb{R}_+$. Moreover, for every $t \geq 0$, the set $\{x \geq 0 : \psi^*\left(x\right) > t\}$ is nonvoid and the generalized inverse of ψ^* at point t, defined by

$$\psi^{*-1}\left(t\right) = \inf\{x \geq 0 : \psi^*\left(x\right) > t\},$$

can also be written as

$$\psi^{*-1}\left(t\right) = \inf_{\lambda\in(0,b)}\left[\frac{t + \psi\left(\lambda\right)}{\lambda}\right].$$

Proof. By definition, ψ^* is the supremum of convex and nondecreasing functions on $\mathbb{R}_+$ and $\psi^*\left(0\right) = 0$, hence ψ^* is a nonnegative, convex, and nondecreasing function on $\mathbb{R}_+$. Moreover, given $\lambda \in (0, b)$, since $\psi^*\left(x\right) \geq \lambda x - \psi\left(\lambda\right)$,

ψ^* is unbounded which shows that for every $t \geq 0$, the set $\{x \geq 0 : \psi^*(x) > t\}$ is nonvoid. Let

$$u = \inf_{\lambda \in (0,b)} \left[\frac{t + \psi(\lambda)}{\lambda} \right],$$

then, for every $x \geq 0$, the following property holds: $u \geq x$ if and only if for every $\lambda \in (0, b)$

$$\frac{t + \psi(\lambda)}{\lambda} \geq x.$$

Since the latter inequality also means that $t \geq \psi^*(x)$, we derive that $\{x \geq 0 : \psi^*(x) > t\} = (u, +\infty)$. This proves that $u = \psi^{*-1}(t)$ by definition of ψ^{*-1}. ∎

A Maximal Inequality

Let us derive as a first consequence of the above property some control of the expectation of the supremum of a finite family of exponentially integrable variables that will turn to be useful for developing the so-called *chaining argument* for Gaussian or empirical processes. We need to introduce a notation.

Definition 2.2 *If A is a measurable set with $\mathbb{P}[A] > 0$ and Z is integrable, we set $\mathbb{E}^A[Z] = \mathbb{E}[Z \mathbb{1}_A]/\mathbb{P}[A]$.*

The proof of Lemma 2.3 is based on an argument used by Pisier in [100] to control the expectation of the supremum of variables belonging to some Orlicz space. For exponentially integrable variables it is furthermore possible to optimize Pisier's argument with respect to the parameter involved in the definition of the moment generating function. This is exactly what is performed below.

Lemma 2.3 *Let $\{Z(t), t \in T\}$ be a finite family of real valued random variables. Let ψ be some convex and continuously differentiable function on $[0, b)$ with $0 < b \leq +\infty$, such that $\psi(0) = \psi'(0) = 0$. Assume that for every $\lambda \in (0, b)$ and $t \in T$, $\psi_{Z(t)}(\lambda) \leq \psi(\lambda)$. Then, using the notations of Lemma 2.1 and Definition 2.2, for any measurable set A with $\mathbb{P}[A] > 0$ we have*

$$\mathbb{E}^A \left[\sup_{t \in T} Z(t) \right] \leq \psi^{*-1} \left(\ln \left(\frac{|T|}{\mathbb{P}[A]} \right) \right).$$

In particular, if one assumes that for some nonnegative number σ, $\psi(\lambda) = \lambda^2 \sigma^2 / 2$ for every $\lambda \in (0, +\infty)$, then

$$\mathbb{E}^A \left[\sup_{t \in T} Z(t) \right] \leq \sigma \sqrt{2 \ln \left(\frac{|T|}{\mathbb{P}[A]} \right)} \leq \sigma \sqrt{2 \ln (|T|)} + \sigma \sqrt{2 \ln \left(\frac{1}{\mathbb{P}[A]} \right)}. \quad (2.3)$$

Proof. Setting $x = \mathbb{E}^A [\sup_{t \in T} Z(t)]$, we have by Jensen's inequality

$$\exp(\lambda x) \leq \mathbb{E}^A \left[\exp \left(\lambda \sup_{t \in T} Z(t) \right) \right] = \mathbb{E}^A \left[\sup_{t \in T} \exp(\lambda Z(t)) \right],$$

for any $\lambda \in (0, b)$. Hence, recalling that $\psi_{Z(t)}(\lambda) = \ln \mathbb{E}\left[\exp(\lambda Z(t))\right]$,

$$\exp(\lambda x) \leq \sum_{t \in T} \mathbb{E}^A\left[\exp(\lambda Z(t))\right] \leq \frac{|T|}{\mathbb{P}[A]} \exp(\psi(\lambda)).$$

Therefore for any $\lambda \in (0, b)$, we have

$$\lambda x - \psi(\lambda) \leq \ln\left(\frac{|T|}{\mathbb{P}[A]}\right),$$

which means that

$$x \leq \inf_{\lambda \in (0,b)} \left[\frac{\ln\left(|T|/\mathbb{P}[A]\right) + \psi(\lambda)}{\lambda}\right]$$

and the result follows from Lemma 2.1. ∎

The case where the variables are sub-Gaussian is of special interest and (2.3) a fortiori holds for centered Gaussian variables $\{Z(t), t \in T\}$ with $\sigma^2 = \sup_{t \in T} \mathbb{E}[Z(t)]$. It is worth noticing that Lemma 2.3 provides a control of $\sup_{t \in T} Z(t)$ in expectation by choosing $A = \Omega$. But it also implies an exponential inequality by using a device which is used repeatedly in the book by Ledoux and Talagrand [79].

Lemma 2.4 *Let Z be some real valued integrable variable and φ be some increasing function φ on $\mathbb{R}_+$ such that for every measurable set A with $\mathbb{P}[A] > 0$, $\mathbb{E}^A[Z] \leq \varphi\left(\ln\left(1/\mathbb{P}[A]\right)\right)$. Then, for any positive x*

$$\mathbb{P}[Z \geq \varphi(x)] \leq e^{-x}.$$

Proof. Just take $A = (Z \geq \varphi(x))$ and apply Markov's inequality which yields

$$\varphi(x) \leq \mathbb{E}^A[Z] \leq \varphi\left(\ln\left(\frac{1}{\mathbb{P}[A]}\right)\right).$$

Therefore $x \leq \ln\left(1/\mathbb{P}[A]\right)$, hence the result. ∎

Taking ψ as in Lemma 2.3 and assuming furthermore that ψ is strictly convex warrants that the generalized inverse ψ^{*-1} of ψ^* is a usual inverse (i.e., $\psi\left(\psi^{*-1}(x)\right) = x$). Hence, combining Lemma 2.3 with Lemma 2.4 implies that, under the assumptions of Lemma 2.3, we derive that for every positive number x

$$\mathbb{P}\left[\sup_{t \in T} Z(t) \geq \psi^{*-1}\left(\ln\left(|T|\right) + x\right)\right] \leq e^{-x},$$

or equivalently that

$$\mathbb{P}\left[\sup_{t \in T} Z(t) \geq z\right] \leq |T| e^{-\psi^*(z)}$$

for all positive z. Of course this inequality could obviously be obtained by a direct application of Chernoff's inequality and a union bound. The interesting point here is that Lemma 2.3 is sharp enough to recover it without any loss, the advantage being that the formulation in terms of conditional expectation is very convenient for forthcoming chaining arguments on Gaussian or empirical processes because of the linearity of the conditional expectation (see Chapter 3 and Chapter 6).

Computation of the Cramér Transform

Let us set $\psi'_Z(I) = (0, B)$, with $0 < B \leq +\infty$, then ψ'_Z admits an increasing inverse $(\psi'_Z)^{-1}$ on $(0, B)$ and the Cramértransform can be computed on $(0, B)$ via the following formula. For any $x \in (0, B)$

$$\psi^*_Z(x) = \lambda_x x - \psi_Z(\lambda_x) \text{ with } \lambda_x = (\psi'_Z)^{-1}(x).$$

Let us use this formula to compute the Cramér transform explicitly in three illustrative cases.

(i) Let Z be a centered Gaussian variable with variance σ^2. Then

$$\psi_Z(\lambda) = \lambda^2 \sigma^2/2, \; \lambda_x = x/\sigma^2$$

and therefore for every positive x

$$\psi^*_Z(x) = \frac{x^2}{2\sigma^2}. \tag{2.4}$$

Hence, Chernoff's inequality (2.2) yields for all positive x

$$\mathbb{P}[Z \geq x] \leq \exp\left(-\frac{x^2}{2\sigma^2}\right). \tag{2.5}$$

It is an easy exercise to prove that

$$\sup_{x>0}\left(\mathbb{P}[Z \geq x]\exp\left(\frac{x^2}{2\sigma^2}\right)\right) = \frac{1}{2}.$$

This shows that inequality (2.5) is quite sharp since it can be uniformly improved only within a factor $1/2$.

(ii) Let Y be a Poisson random variable with parameter v and $Z = Y - v$. Then

$$\psi_Z(\lambda) = v\left(e^\lambda - \lambda - 1\right), \; \lambda_x = \ln\left(1 + \frac{x}{v}\right) \tag{2.6}$$

and therefore for every positive x

$$\psi^*_Z(x) = vh\left(\frac{x}{v}\right) \tag{2.7}$$

where $h(u) = (1 + u)\ln(1 + u) - u$ for all $u \geq -1$. Similarly for every $x \leq v$

$$\psi^*_{-Z}(x) = vh\left(-\frac{x}{v}\right).$$

(iii) Let X be a Bernoulli random variable with probability of success p and $Z = X - p$. Then if $0 < x < 1 - p$

$$\psi_Z(\lambda) = \ln\left(pe^\lambda + 1 - p\right) - \lambda p, \quad \lambda_x = \ln\left(\frac{(1-p)(p+x)}{p(1-p-x)}\right)$$

and therefore for every $x \in (0, 1 - p)$

$$\psi_Z^*(x) = (1 - p - x)\ln\left(\frac{(1-p-x)}{(1-p)}\right) + (p + x)\ln\left(\frac{p+x}{p}\right)$$

or equivalently, setting $a = x + p$, for every $a \in (p, 1)$, $\psi_X^*(a) = h_p(a)$ where

$$h_p(a) = (1 - a)\ln\left(\frac{(1-a)}{(1-p)}\right) + a\ln\left(\frac{a}{p}\right). \tag{2.8}$$

From the last computation we may derive the following classical combinatorial result which will turn to be useful in the sequel.

Proposition 2.5 *For all integers D and n with $1 \le D \le n$, the following inequality holds*

$$\sum_{j=0}^{D}\binom{n}{j} \le \left(\frac{en}{D}\right)^D. \tag{2.9}$$

Proof. The right-hand side of (2.9) being increasing with respect to D, it is larger than

$$\left(\sqrt{2e}\right)^n > 2^n$$

whenever $D \ge n/2$ and therefore (2.9) is trivial in this case. Assuming now that $D < n/2$, we consider some random variable S following the binomial distribution $\text{Bin}(n, 1/2)$. Then for every $a \in (1/2, 1)$

$$\psi_S^*(a) = nh_{1/2}(a)$$

where $h_{1/2}$ is given by (2.8). We notice that for all $x \in (0, 1/2)$

$$h_{1/2}(1 - x) = \ln(2) + x - x\ln(x) + h(-x),$$

where h is the function defined in (ii) above. Hence, since h is nonnegative,

$$h_{1/2}(1 - x) \ge \ln(2) - x + x\ln(x)$$

and setting $x = D/n$, Chernoff's inequality implies that

$$\sum_{j=0}^{D}\binom{n}{j} = 2^n\mathbb{P}[S_n \ge n - D] \le \exp\left(-n\left(-\ln(2) + h_{1/2}(1-x)\right)\right)$$

$$\le \exp\left(n\left(x - x\ln(x)\right)\right)$$

which is exactly (2.9). ∎

We now turn to classical exponential bounds for sums of independent random variables.

2.2 Sums of Independent Random Variables

The Cramér-Chernoff method is especially relevant for the study of sums of independent random variables. Indeed if $X_1, \ldots, X_n$ are independent integrable random variables such that for some nonempty interval I, $e^{\lambda X_i}$ is integrable for all $i \leq n$ and all $\lambda \in I$, then defining

$$S = \sum_{i=1}^{n} (X_i - \mathbb{E}[X_i]),$$

the independence assumption implies that for all $\lambda \in I$

$$\psi_S(\lambda) = \sum_{i=1}^{n} \left(\ln \mathbb{E} \left[e^{\lambda(X_i - \mathbb{E}[X_i])} \right] \right). \tag{2.10}$$

This identity can be used under various integrability assumptions on the variables X_i to derive sub-Gaussian type inequalities. We begin with maybe the simplest one which is due to Hoeffding (see [68]).

2.2.1 Hoeffding's Inequality

Hoeffding's inequality is a straightforward consequence of the following lemma.

Lemma 2.6 (*Hoeffding's lemma*) *Let Y be some centered random variable with values in $[a, b]$. Then for every real number λ,*

$$\psi_Y(\lambda) \leq \frac{(b-a)^2 \lambda^2}{8}.$$

Proof. We first notice that whatever the distribution of Y we have

$$\left| Y - \frac{(b+a)}{2} \right| \leq \frac{(b-a)}{2}$$

and therefore

$$\mathrm{Var}(Y) = \mathrm{Var}(Y - (b+a)/2) \leq \frac{(b-a)^2}{4}. \tag{2.11}$$

Now let P denote the distribution of Y and let P_λ be the probability distribution with density

$$x \to e^{-\psi_Y(\lambda)} e^{\lambda x}$$

with respect to P. Since P_λ is concentrated on $[a, b]$, we know that inequality 2.11 holds true for a random variable Z with distribution P_λ. Hence, we have by an elementary computation

$$\psi_Y''(\lambda) = e^{-\psi_Y(\lambda)} \mathbb{E}\left[Y^2 e^{\lambda Y}\right] - e^{-2\psi_Y(\lambda)} \left(\mathbb{E}\left[Y e^{\lambda Y}\right]\right)^2$$

$$= \mathrm{Var}\,(Z) \leq \frac{(b-a)^2}{4}.$$

The result follows by integration of this inequality, noticing that $\psi_Y(0) = \psi_Y'(0) = 0$ ∎

So, if the variable X_i takes its values in $[a_i, b_i]$, for all $i \leq n$, we get from (2.10) and Lemma 2.6

$$\psi_S(\lambda) \leq \frac{\lambda^2}{8} \sum_{i=1}^n (b_i - a_i)^2$$

which via (2.4) and Chernoff's inequality implies Hoeffding's inequality that can be stated as follows.

Proposition 2.7 (*Hoeffding's inequality*) *Let* $X_1, \ldots, X_n$ *be independent random variables such that* X_i *takes its values in* $[a_i, b_i]$ *almost surely for all* $i \leq n$. *Let*

$$S = \sum_{i=1}^n (X_i - \mathbb{E}\,[X_i]),$$

then for any positive x, *we have*

$$\mathbb{P}\,[S \geq x] \leq \exp\left(-\frac{2x^2}{\sum_{i=1}^n (b_i - a_i)^2}\right). \tag{2.12}$$

It is especially interesting to apply this inequality to variables X_i of the form $X_i = \varepsilon_i \alpha_i$ where $\varepsilon_1, \ldots, \varepsilon_n$ are independent and centered random variables with $|\varepsilon_i| \leq 1$ for all $i \leq n$ and $\alpha_1, \ldots, \alpha_n$ are real numbers. Hoeffding's inequality becomes

$$\mathbb{P}\,[S \geq x] \leq \exp\left(-\frac{x^2}{2\sum_{i=1}^n \alpha_i^2}\right). \tag{2.13}$$

In particular, for Rademacher random variables, that is when the ε_i are identically distributed with $\mathbb{P}\,(\varepsilon_1 = 1) = \mathbb{P}\,(\varepsilon_1 = -1) = 1/2$, then the variance of S is exactly equal to $\sum_{i=1}^n \alpha_i^2$ and inequality (2.13) is really a sub-Gaussian inequality.

Generally speaking however, Hoeffding's inequality cannot be considered as a sub-Gaussian inequality since the variance of S may be much smaller than $\sum_{i=1}^n (b_i - a_i)^2$. This is the reason why some other bounds are needed. To establish Bennett's or Bernstein's inequality below one starts again from (2.10) and writes it as

$$\psi_S(\lambda) = \sum_{i=1}^{n} \left(\ln \mathbb{E} \left[e^{\lambda X_i} \right] - \lambda \mathbb{E} \left[X_i \right] \right).$$

Using the inequality $\ln(u) \leq u - 1$ which holds for all positive u one gets

$$\psi_S(\lambda) \leq \sum_{i=1}^{n} \mathbb{E} \left[e^{\lambda X_i} - \lambda X_i - 1 \right]. \tag{2.14}$$

We shall use this inequality under two different integrability assumptions on the variables X_i.

2.2.2 Bennett's Inequality

We begin with sums of bounded variables for which we prove Bennett's inequality (see [15]).

Proposition 2.8 (Bennett's inequality) Let $X_1, \ldots, X_n$ be independent and square integrable random variables such that for some nonnegative constant b, $X_i \leq b$ almost surely for all $i \leq n$. Let

$$S = \sum_{i=1}^{n} (X_i - \mathbb{E}[X_i])$$

and $v = \sum_{i=1}^{n} \mathbb{E}(X_i^2)$. Then for any positive x, we have

$$\mathbb{P}[S \geq x] \leq \exp\left(-\frac{v}{b^2} h\left(\frac{bx}{v} \right) \right) \tag{2.15}$$

where $h(u) = (1+u)\ln(1+u) - u$ for all positive u.

Proof. By homogeneity we can assume that $b = 1$. Setting $\phi(u) = e^u - u - 1$ for all real number u, we note that the function $u \to u^{-2}\phi(u)$ is nondecreasing. Hence for all $i \leq n$ and all positive λ

$$e^{\lambda X_i} - \lambda X_i - 1 \leq X_i^2 \left(e^\lambda - \lambda - 1 \right)$$

which, taking expectations, yields

$$\mathbb{E} \left[e^{\lambda X_i} \right] - \lambda \mathbb{E}[X_i] - 1 \leq \mathbb{E} \left[X_i^2 \right] \phi(\lambda).$$

Summing up these inequalities we get via (2.14)

$$\psi_S(\lambda) \leq v\phi(\lambda).$$

According to the above computations on Poisson variables, this means that because of identity (2.6), the moment generatingfunction of S is not larger than that of a centered Poisson variable with parameter v and therefore (2.7) yields

$$\psi_S^*(x) \geq vh\left(\frac{x}{v}\right)$$

which proves the proposition via Chernoff's inequality (2.2). ∎

Comment. It is easy to prove that

$$h(u) \geq \frac{u^2}{2(1+u/3)}$$

which immediately yields

$$\mathbb{P}[S \geq x] \leq \exp\left(-\frac{x^2}{2(v+bx/3)}\right). \tag{2.16}$$

The latter inequality is known as Bernstein's inequality. For large values of x as compared to v/b, it looses some logarithmic factor in the exponent with respect to Bennett's inequality. On the contrary, when v/b remains moderate Bennett's and Bernstein's inequality are almost equivalent and both provide a sub-Gaussian type inequality.

A natural question is then: does Bernstein's inequality hold under a weaker assumption than boundedness? Fortunately the answer is positive under appropriate moment assumptions and this refinement with respect to boundedness will be of considerable interest for the forthcoming study of MLEs.

2.2.3 Bernstein's Inequality

Bernstein's inequality for unbounded variables can be found in [117]. We begin with a statement which does not seem to be very well known (see [21]) but which is very convenient to use and implies the classical form of Bernstein's inequality.

Proposition 2.9 (Bernstein's inequality) *Let $X_1, \ldots, X_n$ be independent real valued random variables. Assume that there exists some positive numbers v and c such that*

$$\sum_{i=1}^{n} \mathbb{E}\left[X_i^2\right] \leq v \tag{2.17}$$

and for all integers $k \geq 3$

$$\sum_{i=1}^{n} \mathbb{E}\left[(X_i)_+^k\right] \leq \frac{k!}{2} vc^{k-2}. \tag{2.18}$$

Let $S = \sum_{i=1}^{n}(X_i - \mathbb{E}[X_i])$, then for every positive x

$$\psi_S^*(x) \geq \frac{v}{c^2} h_1\left(\frac{cx}{v}\right), \tag{2.19}$$

where

$$h_1(u) = 1 + u - \sqrt{1 + 2u} \text{ for all positive } u.$$

In particular for every positive x

$$\mathbb{P}\left[S \geq \sqrt{2vx} + cx\right] \leq \exp(-x). \tag{2.20}$$

Proof. We consider again the function $\phi(u) = e^u - u - 1$ and notice that

$$\phi(u) \leq \frac{u^2}{2} \text{ whenever } u \leq 0.$$

Hence for any positive λ, we have for all $i \leq n$

$$\phi(\lambda X_i) \leq \frac{\lambda^2 X_i^2}{2} + \sum_{k=3}^{\infty} \frac{\lambda^k (X_i)_+^k}{k!}$$

which implies by the monotone convergence Theorem

$$\mathbb{E}\left[\phi(\lambda X_i)\right] \leq \frac{\lambda^2 \mathbb{E}\left[X_i^2\right]}{2} + \sum_{k=3}^{\infty} \frac{\lambda^k \mathbb{E}\left[(X_i)_+^k\right]}{k!}$$

and therefore by assumptions (2.17) and (2.18)

$$\sum_{i=1}^{n} \mathbb{E}\left[\phi(\lambda X_i)\right] \leq \frac{v}{2} \sum_{k=2}^{\infty} \lambda^k c^{k-2}.$$

This proves on the one hand that for any $\lambda \in (0, 1/c)$, $e^{\lambda X_i}$ is integrable for all $i \leq n$ and on the other hand using inequality (2.14), that we have for all $\lambda \in (0, 1/c)$

$$\psi_S(\lambda) \leq \sum_{i=1}^{n} \mathbb{E}\left[\phi(\lambda X_i)\right] \leq \frac{v\lambda^2}{2(1 - c\lambda)}. \tag{2.21}$$

Therefore

$$\psi_S^*(x) \geq \sup_{\lambda \in (0, 1/c)} \left(x\lambda - \frac{\lambda^2 v}{2(1 - c\lambda)}\right).$$

Now it follows from elementary computations that

$$\sup_{\lambda \in (0, 1/c)} \left(x\lambda - \frac{\lambda^2 v}{2(1 - c\lambda)}\right) = \frac{v}{c^2} h_1\left(\frac{cx}{v}\right)$$

which yields inequality (2.19). Since h_1 is an increasing mapping from $(0, \infty)$ onto $(0, \infty)$ with inverse function $h_1^{-1}(u) = u + \sqrt{2u}$ for $u > 0$, inequality (2.20) follows easily via Chernoff's inequality (2.2). ∎

Corollary 2.10 *Let $X_1, \ldots, X_n$ be independent real valued random variables. Assume that there exist positive numbers v and c such that (2.17) and (2.18) hold for all integers $k \geq 3$. Let $S = \sum_{i=1}^{n} (X_i - \mathbb{E}[X_i])$, then for any positive x, we have*

$$\mathbb{P}[S \geq x] \leq \exp\left(-\frac{x^2}{2(v + cx)}\right) \qquad (2.22)$$

Proof. We notice that for all positive u

$$h_1(u) \geq \frac{u^2}{2(1+u)}.$$

So, it follows from (2.19) that

$$\psi_S^*(x) \geq \frac{x^2}{2(v + cx)}$$

which yields the result via Chernoff's inequality (2.2). ∎

Comment. The usual assumption for getting Bernstein's inequality involves a control of the absolute moments of the variables X_i instead of their positive part as in the above statement. This refinement has been suggested to us by Emmanuel Rio. Thanks to this refined statement we can exactly recover (2.16) from Corollary 2.10. Indeed if the variables $X_1, \ldots, X_n$ are independent and such that for all $i \leq n$, $X_i \leq b$ almost surely, then assumptions (2.17) and (2.18) hold with

$$v = \sum_{i=1}^{n} \mathbb{E}[X_i^2] \quad \text{and} \quad c = b/3$$

so that Proposition 2.9 and Corollary 2.10 apply. This means in particular that inequality (2.22) holds and it is worth noticing that, in this case (2.22) writes exactly as (2.16).

By applying Proposition 2.9 to the variables $-X_i$ one easily derives from (2.20) that under the moment condition (2.18) the following concentration inequality holds for all positive x

$$\mathbb{P}\left[|S| \geq \sqrt{2vx} + cx\right] \leq 2\exp(-x) \qquad (2.23)$$

(with $c = b/3$ whenever the variables $|X_i|$'s are bounded by b) and a fortiori by inequality (2.22) of Corollary 2.10

$$\mathbb{P}[|S| \geq x] \leq 2\exp\left(-\frac{x^2}{2(v + cx)}\right). \qquad (2.24)$$

This inequality expresses how $\sum_{i=1}^{n} X_i$ concentrates around its expectation. Of course similar concentration inequalities could be obtained from Hoeffding's inequality or Bennett's inequality as well. It is one of the main tasks of the next chapters to extend these deviation or concentration inequalities to much more general functionals of independent random variables which will include norms of sums of independent infinite dimensional random vectors. For these functionals, the moment generating function is not easily directly computable since it is no longer additive and we will rather deal with entropy which is naturally subadditive for product probability measures (see the tensorization inequality below). This property of entropy will be the key to derive most of the concentration inequalities that we shall encounter in the sequel.

2.3 Basic Information Inequalities

The purpose of this section is to establish simple but clever information inequalities. They will turn to be surprisingly powerful and we shall use them to prove concentration inequalities as well as minimax lower bounds for statistical estimation problems. Here and in the sequel we need to use some elementary properties of entropy that are recorded below.

Definition 2.11 *Let Φ denote the function defined on $\mathbb{R}_+$ by*

$$\Phi(u) = u \ln(u).$$

Let $(\Omega, \mathcal{A})$ be some measurable space. For any nonnegative random variable Y on $(\Omega, \mathcal{A})$ and any probability measure P such that Y is P-integrable, we define the entropy of Y with respect to P by

$$Ent_P[Y] = E_P[\Phi(Y)] - \Phi(E_P[Y]).$$

Moreover, if $E_P[Y] = 1$, and $Q = YP$, the Kullback–Leibler information of Q with respect to P, is defined by

$$\mathbf{K}(Q, P) = Ent_P[Y].$$

Note that since Φ is bounded from below by $-1/e$ one can always give a sense to $E_P[\Phi(Y)]$ even if $\Phi(Y)$ is not P-integrable. Hence $Ent_P[Y]$ is well-defined. Since Φ is a convex function, Jensen's inequality warrants that $Ent_P[Y]$ is a nonnegative (possibly infinite) quantity. Moreover $Ent_P[Y] < \infty$ if and only if $\Phi(Y)$ is P-integrable.

2.3.1 Duality and Variational Formulas

Some classical alternative definitions of entropy will be most helpful.

Proposition 2.12 *Let $(\Omega, \mathcal{A}, P)$ be some probability space. For any nonnegative random variable Y on $(\Omega, \mathcal{A})$ such that $\Phi(Y)$ is P-integrable, the following identities hold*

$$Ent_P[Y] = \sup\left\{E_P[UY], U : \Omega \to \overline{\mathbb{R}} \text{ with } E_P[e^U] = 1\right\} \tag{2.25}$$

and

$$Ent_P[Y] = \inf_{u>0} E_P[Y(\ln(Y) - \ln(u)) - (Y - u)]. \qquad (2.26)$$

Moreover, if U is such that $E_P[UY] \leq Ent_P[Y]$ for all nonnegative random variables Y on $(\Omega, \mathcal{A})$ such that $\Phi(Y)$ is P-integrable and $E_P[Y] = 1$, then $E_P[e^U] \leq 1$.

Comments.

- Some elementary computation shows that for any $u \in \mathbb{R}$

$$\sup_{x>0}(xu - \Phi(x)) = e^{u-1},$$

  hence, if $\Phi(Y)$ is P-integrable and $E_P[e^U] = 1$ the following inequality holds

$$UY \leq \Phi(Y) + \frac{1}{e}e^U.$$

  Therefore U^+Y is integrable and one can always define $E_P[UY]$ as $E_P[U^+Y] - E_P[U^-Y]$. This indeed gives a sense to the right hand side in identity (2.25).
- Another formulation of the duality formula is the following

$$Ent_P[Y] = \sup_T E_P[Y(\ln(T) - \ln(E_P[T]))], \qquad (2.27)$$

  where the supremum is extended to the nonnegative and integrable variables $T \neq 0$ a.s.

Proof. To prove (2.25), we note that, for any random variable U with $E_P[e^U] = 1$, the following identity is available

$$Ent_P[Y] - E_P[UY] = Ent_{e^U P}[Ye^{-U}].$$

Hence $Ent_P[Y] - E_P[UY]$ is nonnegative and is equal to 0 whenever $e^U = Y/E_P[Y]$, which means that the duality formula (2.25) holds. In order to prove the variational formula (2.26), we set

$$\Psi : u \to E_P[Y(\ln(Y) - \ln(u)) - (Y - u)]$$

and note that

$$\Psi(u) - Ent_P[Y] = E_P[Y]\left[\left(\frac{u}{E_P[Y]} - 1\right) - \ln\left(\frac{u}{E_P[Y]}\right)\right].$$

But it is easy to check that $x - 1 - \ln(x)$ is nonnegative and equal to 0 for $x = 1$ so that $\Psi(u) - Ent_P[Y]$ is nonnegative and equal to 0 for $u = E_P[Y]$. This achieves the proof of (2.26). Now if U is such that $E_P[UY] \leq Ent_P[Y]$ for all random variable Y on $(\Omega, \mathcal{A})$ such that $\Phi(Y)$ is P-integrable, then, given some integer n, one can choose $Y = e^{U \wedge n}/x_n$ with $x_n = E_P\left[e^{U \wedge n}\right]$, which leads to

$$E_P\left[UY\right] \le \mathrm{Ent}_P\left[Y\right]$$

and therefore

$$\frac{1}{x_n}E_P\left[Ue^{U \wedge n}\right] \le \frac{1}{x_n}\left[E_P\left[(U \wedge n)\,e^{U \wedge n}\right] - \ln\left(x_n\right)\right].$$

Hence

$$\ln\left(x_n\right) \le 0$$

and taking the limit when n goes to infinity, we get by monotone convergence $E_P\left[e^U\right] \le 1$, which finishes the proof of the proposition. ∎

We are now in position to provide some first connections between concentration and entropy inequalities.

2.3.2 Some Links Between the Moment Generating Function and Entropy

As a first consequence of the duality formula for entropy, we can derive a first simple but somehow subtle connection between the moment generating function and entropy that we shall use several times in the sequel.

Lemma 2.13 *Let $(\Omega, \mathcal{A}, P)$ be some probability space and Z be some real valued and P-integrable random variable. Let ψ be some convex and continuously differentiable function on $[0, b)$ with $0 < b \le +\infty$ and assume that $\psi(0) = \psi'(0) = 0$. Setting for every $x \ge 0$, $\psi^*(x) = \sup_{\lambda \in (0,b)}(\lambda x - \psi(\lambda))$, let, for every $t \ge 0$, $\psi^{*-1}(t) = \inf\{x \ge 0 : \psi^*(x) > t\}$. The following statements are equivalent:*

i) for every $\lambda \in (0, b)$, $E_P\left[\exp\left[\lambda\left(Z - E_P\left[Z\right]\right)\right]\right] \le e^{\psi(\lambda)}$,   (2.28)

ii) for any probability measure Q absolutely continuous with respect to P such that $\mathbf{K}(Q, P) < \infty$,

$$E_Q\left[Z\right] - E_P\left[Z\right] \le \psi^{*-1}\left[\mathbf{K}(Q, P)\right].   (2.29)$$

In particular, given $v > 0$, it is equivalent to state that for every positive λ

$$E_P\left[\exp\left[\lambda\left(Z - E_P\left[Z\right]\right)\right]\right] \le e^{\lambda^2 v/2}   (2.30)$$

or that for any probability measure Q absolutely continuous with respect to P and such that $\mathbf{K}(Q, P) < \infty$,

$$E_Q\left[Z\right] - E_P\left[Z\right] \le \sqrt{2v\mathbf{K}(Q, P)}.   (2.31)$$

Proof. We essentially use the duality formula for entropy. It follows from Lemma 2.1 that

$$\psi^{*-1}\left[\mathbf{K}\left(P,Q\right)\right] = \inf_{\lambda \in (0,b)} \left[\frac{\psi\left(\lambda\right) + \mathbf{K}\left(Q,P\right)}{\lambda}\right]. \tag{2.32}$$

Assuming that (2.29) holds, we derive from (2.32) that for any nonnegative random variable Y such that $E_P\left[Y\right] = 1$ and every $\lambda \in (0,b)$

$$E_P\left[Y\left(Z - E_P\left[Z\right]\right)\right] \leq \frac{\psi\left(\lambda\right) + \mathbf{K}\left(Q,P\right)}{\lambda}$$

where $Q = YP$. Hence

$$E_P\left[Y\left(\lambda\left(Z - E_P\left[Z\right]\right) - \psi\left(\lambda\right)\right)\right] \leq \mathrm{Ent}_P\left[Y\right]$$

and Proposition 2.12 implies that (2.28) holds. Conversely if (2.28) holds for any $\lambda \in (0,b)$, then, setting

$$\lambda\left(Z - E_P\left[Z\right]\right) - \psi\left(\lambda\right) - \ln\left(E_P\left(\exp\left(\lambda\left(Z - E_P\left[Z\right]\right) - \psi\left(\lambda\right)\right)\right)\right) = U,$$

(2.25) yields

$$E_P\left[Y\left(\lambda\left(Z - E_P\left[Z\right]\right) - \psi\left(\lambda\right)\right)\right] \leq E_P\left[YU\right] \leq \mathrm{Ent}_P\left[Y\right]$$

and therefore

$$E_P\left[Y\left(Z - E_P\left[Z\right]\right)\right] \leq \frac{\psi\left(\lambda\right) + \mathbf{K}\left(Q,P\right)}{\lambda}.$$

Since the latter inequality holds for any $\lambda \in (0,b)$, (2.32) leads to (2.29). Applying the previous result with $\psi\left(\lambda\right) = \lambda^2 v/2$ for every positive λ leads to the equivalence between (2.30) and (2.31) since then $\psi^{*-1}\left(t\right) = \sqrt{2vt}$. ∎

Comment. Inequality (2.31) is related to what is usually called a quadratic transportation cost inequality. If Ω is a metric space, the measure P is said to satisfy to a quadratic transportation cost inequality if (2.31) holds for every Z which is Lipschitz on Ω with Lipschitz norm not larger than 1. The link between quadratic transportation cost inequalities and Gaussian type concentration is well known (see for instance [87], [42] or [28]) and the above Lemma is indeed inspired by a related result on quadratic transportation cost inequalities in [28].

An other way of going from an entropy inequality to a control of the moment generating function is the so-called *Herbst argument* that we are presenting here for sub-Gaussian controls of the moment generating function although we shall also use in the sequel some modified version of it to derive Bennett or Bernstein type inequalities. This argument will be typically used to derive sub-Gaussian controls of the moment generating function from Logarithmic Sobolev type inequalities.

Proposition 2.14 *(**Herbst argument**) Let Z be some integrable random variable on $(\Omega, \mathcal{A}, \mathbb{P})$ such that for some positive real number v the following inequality holds for every positive number λ*

$$Ent_{\mathbb{P}}\left[e^{\lambda Z}\right] \leq \frac{\lambda^2 v}{2} \mathbb{E}\left[e^{\lambda Z}\right]. \tag{2.33}$$

Then, for every positive λ

$$\mathbb{E}\left[e^{\lambda(Z-\mathbb{E}[Z])}\right] \leq e^{\lambda^2 v/2}.$$

Proof. Let us first notice that since $Z - \mathbb{E}[Z]$ also satisfies (2.33), we can assume Z to be centered at expectation. Then, (2.33) means that

$$\lambda \mathbb{E}\left[Ze^{\lambda Z}\right] - \mathbb{E}\left[e^{\lambda Z}\right] \ln \mathbb{E}\left[e^{\lambda Z}\right] \leq \frac{\lambda^2 v}{2} \mathbb{E}\left[e^{\lambda Z}\right],$$

which yields the differential inequality

$$\frac{1}{\lambda} \frac{F'(\lambda)}{F(\lambda)} - \frac{1}{\lambda^2} \ln F(\lambda) \leq \frac{v}{2},$$

where $F(\lambda) = \mathbb{E}\left[e^{\lambda Z}\right]$. Setting $G(\lambda) = \lambda^{-1} \ln F(\lambda)$, we see that the differential inequality simply becomes $G'(\lambda) \leq v/2$, which in turn implies since $G(\lambda)$ tends to 0 as λ tends to 0, $G(\lambda) \leq \lambda v/2$ and the result follows. ∎

From a statistical point of view Pinsker's inequality below provides a lower bound on the hypothesis testing errors when one tests between two probability measures. We can derive as applications of the duality formula both Pinsker's inequality and a recent result by Birgé which extends Pinsker's inequality to multiple hypothesis testing. We shall use these results for different purposes in the sequel, namely either for establishing transportation cost inequalities or minimax lower bounds for statistical estimation.

2.3.3 Pinsker's Inequality

Pinsker's inequality relates the total variation distance to the Kullback–Leibler information number. Let us recall the definition of the total variation distance.

Definition 2.15 *We define the total variation distance between two probability distributions P and Q on $(\Omega, \mathcal{A})$ by*

$$\|P - Q\|_{TV} = \sup_{A \in \mathcal{A}} |P(A) - Q(A)|.$$

It turns out that Pinsker's inequality is a somehow unexpected consequence of Hoeffding's lemma (see Lemma 2.6).

Theorem 2.16 (*Pinsker's inequality*) *Let P and Q be probability distributions on $(\Omega, \mathcal{A})$, with Q absolutely continuous with respect to P. Then*

$$\|P - Q\|_{TV}^2 \leq \frac{1}{2} \mathbf{K}(Q, P). \tag{2.34}$$

(2.34) is known as Pinsker's inequality (see [98] where inequality (2.34) is given with constant 1, and [40] for a proof with the optimal constant $1/2$).

Proof. Let $Q = YP$ and $A = \{Y \geq 1\}$. Then, setting $Z = \mathbb{1}_A$,

$$\|P - Q\|_{TV} = Q(A) - P(A) = E_Q[Z] - E_P[Z]. \tag{2.35}$$

Now, it comes from Lemma 2.6 that for any positive λ

$$E_P\left[e^{\lambda(Z - E_P[Z])}\right] \leq e^{\lambda^2/8},$$

which by Lemma 2.13 leads to

$$E_Q[Z] - E_P[Z] \leq \sqrt{\frac{1}{2} \mathbf{K}(Q, P)}$$

and therefore to (2.34) via (2.35). ∎

We turn now to an information inequality due to Birgé [17], which will play a crucial role to establish lower bounds for the minimax risk for various estimation problems.

2.3.4 Birgé's Lemma

Let us fix some notations. For any p, let ψ_p denote the logarithm of the moment generating function of the Bernoulli distribution with parameter p, i.e.,

$$\psi_p(\lambda) = \ln\left(p\left(e^\lambda - 1\right) + 1\right), \lambda \in \mathbb{R} \tag{2.36}$$

and let h_p denote the Cramér transform of the Bernoulli distribution with parameter p, i.e., by (2.8) for any $a \in [p, 1]$

$$h_p(a) = \sup_{\lambda > 0}(\lambda a - \psi_p(\lambda)) = a \ln\left(\frac{a}{p}\right) + (1 - a) \ln\left(\frac{1 - a}{1 - p}\right). \tag{2.37}$$

Our proof of Birgé's lemma to be presented below derives from the duality formula (2.25).

Lemma 2.17 (*Birgé's lemma*) *Let $(P_i)_{0 \leq i \leq N}$ be some family of probability distributions and $(A_i)_{0 \leq i \leq N}$ be some family of disjoint events. Let $a_0 = P_0(A_0)$ and $a = \min_{1 \leq i \leq N} P_i(A_i)$, then, whenever $Na \geq 1 - a_0$*

$$h_{(1-a_0)/N}(a) \leq \frac{1}{N} \sum_{i=1}^{N} \mathbf{K}(P_i, P_0). \tag{2.38}$$

Proof. We write P instead of P_0 for short. We consider some positive λ and use (2.25) with $U = \lambda \mathbb{1}_{A_i} - \psi_{P(A_i)}(\lambda)$ and $Y = dP_i/dP$. Then for every $i \in [1, N]$

$$\lambda P_i(A_i) - \psi_{P(A_i)}(\lambda) \leq \mathbf{K}(P_i, P)$$

and thus

$$\lambda a - \frac{1}{N} \sum_{i=1}^{N} \psi_{P(A_i)}(\lambda) \leq \lambda \left(\frac{1}{N} \sum_{i=1}^{N} P_i(A_i) \right) - \frac{1}{N} \sum_{i=1}^{N} \psi_{P(A_i)}(\lambda)$$

$$\leq \frac{1}{N} \sum_{i=1}^{N} \mathbf{K}(P_i, P).$$

Now, we note that $p \to -\psi_p(\lambda)$ is a nonincreasing function. Hence, since

$$\sum_{i=1}^{N} P(A_i) \leq 1 - a_0$$

we derive that

$$\lambda a - \psi_{(1-a_0)/N}(\lambda) \leq \lambda a - \psi_{N^{-1} \sum_{i=1}^{N} P(A_i)}(\lambda)$$

Since $p \to -\psi_p(\lambda)$ is convex

$$\lambda a - \psi_{N^{-1} \sum_{i=1}^{N} P(A_i)}(\lambda) \leq \lambda a - \frac{1}{N} \sum_{i=1}^{N} \psi_{P(A_i)}(\lambda)$$

and therefore

$$\lambda a - \psi_{(1-a_0)/N}(\lambda) \leq \lambda a - \frac{1}{N} \sum_{i=1}^{N} \psi_{P(A_i)}(\lambda) \leq \frac{1}{N} \sum_{i=1}^{N} \mathbf{K}(P_i, P).$$

Taking the supremum over λ in the inequality above leads to (2.38). ∎

Comment. Note that our statement of Birgé's lemma slightly differs from the original one given in [17] (we use the two parameters a_0 and a instead of the single parameter $\min(a_0, a)$ as in [17]). The interest of this slight refinement is that, stated as above, Birgé's lemma does imply Pinsker's inequality. Indeed, taking A such that $\|P - Q\|_{TV} = Q(A) - P(A)$ and applying Birgé's lemma with $P_0 = P$, $Q = P_1$ and $A = A_1 = A_0^c$ leads to

$$h_{P(A)}(Q(A)) \leq \mathbf{K}(Q, P). \tag{2.39}$$

Since by Lemma 2.6, for every nonnegative x, $h_p(p + x) \geq 2x^2$, we readily see that (2.39) implies (2.34).

Inequality (2.38) is not that easy to invert in order to get some explicit control on a. When $N = 1$, we have just seen that a possible way to (approximately) invert it is to use Hoeffding's lemma (which in fact leads to Pinsker's inequality). The main interest of Birgé's lemma appears however when N becomes large. In order to capture the effect of possibly large values of N, it is better to use a Poisson rather than a sub-Gaussian type lower bound for the Cramér transform of a Bernoulli variable. Indeed, it comes from the proof of Bennett's inequality above that for every positive x,

$$h_p(p+x) \geq ph\left(\frac{x}{p}\right) \qquad (2.40)$$

which also means that for every $a \in [p,1]$, $h_p(a) \geq ph((a/p)-1) \geq a\ln(a/ep)$. Hence, setting $\overline{\mathbf{K}} = \frac{1}{N}\sum_{i=1}^{N}\mathbf{K}(P_i, P_0)$, (2.38) implies that

$$a\ln\left(\frac{Na}{e(1-a)}\right) \leq \overline{\mathbf{K}}.$$

Now, let $\kappa = 2e/(2e+1)$. If $a \geq \kappa$, it comes from the previous inequality that $a\ln(2N) \leq \overline{\mathbf{K}}$. Hence, whatever a

$$a \leq \kappa \vee \left(\frac{\overline{\mathbf{K}}}{\ln(1+N)}\right).$$

This means that we have proven the following useful corollary of Birgé's lemma.

Corollary 2.18 *Let* $(P_i)_{0\leq i\leq N}$ *be some family of probability distributions and* $(A_i)_{0\leq i\leq N}$ *be some family of disjoint events. Let* $a = \min_{0\leq i\leq N} P_i(A_i)$, *then, setting* $\overline{\mathbf{K}} = \frac{1}{N}\sum_{i=1}^{N}\mathbf{K}(P_i, P_0)$

$$a \leq \kappa \vee \left(\frac{\overline{\mathbf{K}}}{\ln(1+N)}\right), \qquad (2.41)$$

where κ *is some absolute constant smaller than* 1 *(* $\kappa = 2e/(2e+1)$ *works).*

This corollary will be extremely useful to derive minimax lower bounds (especially for nonparametric settings) as we shall see in Chapter 4 and Chapter 7. Let us begin with the following basic example. Let us consider some finite statistical model $\{P_\theta, \theta \in \Theta\}$ and ℓ to be the $0-1$ loss on $\Theta \times \Theta$, i.e., $\ell(\theta, \theta') = 1$ if $\theta \neq \theta'$ or $\ell(\theta, \theta') = 0$ else. Setting $\mathbf{K}_{\max} = \max_{\theta, \theta'} \mathbf{K}(P_\theta, P_{\theta'})$, whatever the estimator $\widehat{\theta}$ taking its value in Θ, the maximal risk can be expressed as

$$\max_{\theta \in \Theta} E_\theta\left[\ell\left(\theta, \widehat{\theta}\right)\right] = \max_{\theta \in \Theta} P_\theta\left[\theta \neq \widehat{\theta}\right] = 1 - \min_{\theta \in \Theta} P_\theta\left[\theta = \widehat{\theta}\right],$$

which leads via (2.41) to the lower bound

$$\max_{\theta \in \Theta} E_\theta\left[\ell\left(\theta, \widehat{\theta}\right)\right] \geq 1 - \kappa$$

as soon as $\mathbf{K}_{\max} \leq \kappa \ln(1+N)$. Some illustrations will be given in Chapter 3 which will rely on the following consequence of Corollary 2.18 which uses the previous argument in a more general context.

Corollary 2.19 *Let (S, d) be some pseudometric space, $\{\mathbb{P}_s, s \in S\}$ be some statistical model. Let κ denote the absolute constant of Corollary 2.18. Then for any estimator $\widehat{s}$ and any finite subset $\mathcal{C}$ of S, setting $\delta = \min_{s,t \in \mathcal{C}, s \neq t} d(s, t)$, provided that $\max_{s,t \in \mathcal{C}} \mathbf{K}(\mathbb{P}_s, \mathbb{P}_t) \leq \kappa \ln |\mathcal{C}|$ the following lower bound holds for every $p \geq 1$*

$$\sup_{s \in \mathcal{C}} \mathbb{E}_s \left[d^p(s, \widehat{s}) \right] \geq 2^{-p} \delta^p (1 - \kappa).$$

Proof. We define an estimator $\widetilde{s}$ taking its values in $\mathcal{C}$ such that

$$d(\widehat{s}, \widetilde{s}) = \min_{t \in \mathcal{C}} d(\widehat{s}, t).$$

Then, by definition of $\widetilde{s}$, we derive via the triangle inequality that

$$d(s, \widetilde{s}) \leq d(s, \widehat{s}) + d(\widehat{s}, \widetilde{s}) \leq 2d(s, \widehat{s}).$$

Hence

$$\sup_{s \in \mathcal{C}} \mathbb{E}_s \left[d^p(s, \widehat{s}) \right] \geq 2^{-p} \delta^p \sup_{s \in \mathcal{C}} \mathbb{P}_s \left[s \neq \widetilde{s} \right] = 2^{-p} \delta^p \left(1 - \min_{s \in \mathcal{C}} \mathbb{P}_s \left[s = \widetilde{s} \right] \right)$$

and the result immediately follows via Corollary 2.18. ∎

2.4 Entropy on Product Spaces

The basic fact to better understand the meaning of Pinsker's inequality is that the total variation distance can be interpreted in terms of optimal coupling (see [46]).

Lemma 2.20 *Let P and Q be probability distributions on $(\Omega, \mathcal{A})$ and denote by $\mathcal{P}(P, Q)$ the set of probability distributions on $(\Omega \times \Omega, \mathcal{A} \otimes \mathcal{A})$ with first marginal distribution P and second distribution Q. Then*

$$\|P - Q\|_{TV} = \min_{\mathbb{Q} \in \mathcal{P}(P,Q)} \mathbb{Q}[X \neq Y]$$

where X and Y denote the coordinate mappings $(x, y) \to x$ and $(x, y) \to y$.

Proof. Note that if $\mathbb{Q} \in \mathcal{P}(P, Q)$, then

$$\begin{aligned}
|P(A) - Q(A)| &= |\mathbb{E}_{\mathbb{Q}} \left[\mathbb{1}_A(X) - \mathbb{1}_A(Y) \right]| \\
&\leq \mathbb{E}_{\mathbb{Q}} \left[|\mathbb{1}_A(X) - \mathbb{1}_A(Y)| \, \mathbf{I}_{X \neq Y} \right] \leq \mathbb{Q}[X \neq Y]
\end{aligned}$$

which means that $\|P - Q\|_{TV} \leq \inf_{\mathbb{Q} \in \mathcal{P}(P,Q)} \mathbb{Q}[X \neq Y]$. Conversely, let us consider some probability measure μ which dominates P and Q and denote by f and g the corresponding densities of P and Q with respect to μ. Then

$$a = \|P - Q\|_{TV} = \int_\Omega [f - g]_+ \, d\mu = \int_\Omega [g - f]_+ \, d\mu = 1 - \int_\Omega (f \wedge g) \, d\mu$$

and since we can assume that $a > 0$ (otherwise the result is trivial), we define the probability measure $\mathbb{Q}$ as a mixture $\mathbb{Q} = a\mathbb{Q}_1 + (1 - a)\mathbb{Q}_2$ where $\mathbb{Q}_1$ and $\mathbb{Q}_2$ are such that, for any measurable and bounded function Ψ

$$a^2 \int_{\Omega \times \Omega} \Psi(x, y)\, d\mathbb{Q}_1[x, y]$$

equals

$$\int_{\Omega \times \Omega} [f(x) - g(x)]_+ [g(y) - f(y)]_+ \Psi(x, y)\, d\mu(x)\, d\mu(y)$$

and

$$(1 - a) \int_{\Omega \times \Omega} \Psi(x, y)\, d\mathbb{Q}_2[x, y] = \int_{\Omega} (f(x) \wedge g(x)) \Psi(x, x)\, d\mu(x).$$

It is easy to check that $\mathbb{Q} \in \mathcal{P}(P, Q)$. Moreover, since $\mathbb{Q}_2$ is concentrated on the diagonal, $\mathbb{Q}[X \neq Y] = a\mathbb{Q}_1[X \neq Y] \leq a$. ∎

As a matter of fact the problem solved by Lemma 2.20 is a special instance of the *transportation cost* problem. Given two probability distributions P and Q (both defined on $(\Omega, \mathcal{A})$) and some nonnegative measurable function w on $\Omega \times \Omega$, the idea is basically to measure how close Q is to P in terms of how much effort is required to transport a mass distributed according to P into a mass distributed according to Q relatively to w (which is called the cost function). More precisely one defines the *transportation cost* of Q to P (relatively to w) as

$$\inf_{\mathbb{Q} \in \mathcal{P}(P, Q)} \mathbb{E}_{\mathbb{Q}}[w(X, Y)].$$

The transportation problem consists in constructing an optimal coupling $\mathbb{Q} \in \mathcal{P}(P, Q)$ (i.e., a minimizer of the above infimum if it does exist) and in relating the transportation cost to some explicit distance between P and Q. Obviously, Lemma 2.20 solves the transportation cost problem for the binary cost function $w(x, y) = \mathbb{1}_{x \neq y}$. The interested reader will find in [101] much more general results, including the Kantorovich theorem which relates the transportation cost to the bounded Lipschitz distance when the cost function is a distance and several analogue coupling results for other kinds of distances between probability measures like the Prohorov distance (see also Strassen's theorem in [108]).

The interest of the interpretation of the variation distance as a transportation cost is that it leads to a very natural extension of Pinsker's inequality (2.34) to product spaces that we shall study below following an idea due to Marton (see [86]). It leads to maybe the simplest approach to concentration based on coupling, which is nowadays usually referred to as the *transportation method*. It has the advantage to be easily extendable to nonindependent frameworks (see [88] and [105]) at the price of providing sometimes suboptimal results (such as Hoeffding type inequalities instead of Bernstein type inequalities).

2.4.1 Marton's Coupling

We present below a slightly improved version of Marton's transportation cost inequality in [86]) which, because of Lemma 2.20, can be viewed as a generalization of Pinsker's inequality to a product space.

Proposition 2.21 (*Marton's inequality*) *Let* $(\Omega^n, \mathcal{A}^n, P^n)$ *be some product probability space*

$$(\Omega^n, \mathcal{A}^n, P^n) = \left(\prod_{i=1}^n \Omega_i, \bigotimes_{i=1}^n \mathcal{A}_i, \bigotimes_{i=1}^n \mu_i \right),$$

and Q *be some probability measure absolutely continuous with respect to* P^n. *Then*

$$\min_{\mathbb{Q} \in \mathcal{P}(P^n, Q)} \sum_{i=1}^n \mathbb{Q}^2 [X_i \neq Y_i] \leq \frac{1}{2} \mathbf{K}(Q, P^n), \tag{2.42}$$

where (X_i, Y_i), $1 \leq i \leq n$ *denote the coordinate mappings on* $\Omega^n \times \Omega^n$.

Note that by Cauchy–Schwarz inequality, (2.42) implies that

$$\min_{\mathbb{Q} \in \mathcal{P}(P^n, Q)} \sum_{i=1}^n \mathbb{Q}[X_i \neq Y_i] \leq \sqrt{\frac{n}{2} \mathbf{K}(Q, P^n)}$$

which is the original statement in [86]. On the other hand some other coupling result due to Marton (see [87]) ensures that

$$\min_{\mathbb{Q} \in \mathcal{P}(P^n, Q)} \sum_{i=1}^n \int_{\Omega^n} \mathbb{Q}^2 [X_i \neq Y_i \mid Y_i = y_i] \, dQ(y) \leq 2 \mathbf{K}(Q, P^n).$$

Of course this inequality implies (2.42) but with the suboptimal constant 2 instead of $1/2$.

Proof. We follow closely the presentation in [87]. Let us prove (2.42) by induction on n. For $n = 1$, (2.42) holds simply because it is equivalent to Pinsker's inequality (2.34) through Lemma 2.20. Let us now assume that for any distribution Q' on $(\Omega^{n-1}, \mathcal{A}^{n-1}, P^{n-1})$ which is absolutely continuous with respect to P^{n-1}, the following coupling inequality holds true

$$\min_{\mathbb{Q} \in \mathcal{P}(P^{n-1}, Q')} \sum_{i=1}^{n-1} \mathbb{Q}^2 \left[(x, y) \in \Omega^{n-1} \times \Omega^{n-1} : x_i \neq y_i \right] \leq \frac{1}{2} \mathbf{K}(Q', P^{n-1}).$$

$$(2.43)$$

Let $Q = gP^n$. Then

$$\mathbf{K}(Q, P^n) = E_{P^n} [\Phi(g)] = \int_{\Omega_n} \left[\int_{\Omega^{n-1}} \Phi(g(x, t)) \, dP^{n-1}(x) \right] d\mu_n(t).$$

Denoting by g_n the marginal density $g_n(t) = \int_{\Omega^{n-1}} g(x,t) dP^{n-1}(x)$ and by q_n the corresponding marginal distribution of Q, $q_n = g_n \mu_n$, we write g as $g(x,t) = g(x \mid t) g_n(t)$ and get by Fubini's Theorem

$$\mathbf{K}(Q, P^n) = \int_{\Omega_n} g_n(t) \left[\int_{\Omega^{n-1}} \Phi(g(x \mid t)) dP^{n-1}(x) \right] d\mu_n(t)$$
$$+ \int_{\Omega_n} \Phi(g_n(t)) d\mu_n(t).$$

If, for any $t \in \Omega_n$, we introduce the conditional distribution $dQ(x \mid t) = g(x \mid t) dP^{n-1}(x)$. The previous identity can be written as

$$\mathbf{K}(Q, P^n) = \int_{\Omega_n} \mathbf{K}(Q(. \mid t), P^{n-1}) dq_n(t) + \mathbf{K}(q_n, \mu_n).$$

Now (2.43) ensures that, for any $t \in \Omega_n$, there exists some probability distribution $\mathbb{Q}_t$ on $\Omega^{n-1} \times \Omega^{n-1}$ belonging to $\mathcal{P}(P^{n-1}, Q(. \mid t))$ such that

$$\sum_{i=1}^{n-1} \mathbb{Q}_t^2 \left[(x, y) \in \Omega^{n-1} \times \Omega^{n-1} : x_i \neq y_i \right] \leq \frac{1}{2} \mathbf{K}(Q(. \mid t), P^{n-1})$$

while Pinsker's inequality (2.34) implies via Lemma 2.20 that there exists a probability distribution Q_n on $\Omega_n \times \Omega_n$ belonging to $\mathcal{P}(\mu_n, q_n)$ such that

$$Q_n^2 [(u, v) : u \neq v] \leq \frac{1}{2} \mathbf{K}(q_n, \mu_n).$$

Hence

$$\frac{1}{2} \mathbf{K}(Q, P^n) \geq \int_{\Omega_n} \sum_{i=1}^{n-1} \mathbb{Q}_t^2 \left[(x, y) \in \Omega^{n-1} \times \Omega^{n-1} : x_i \neq y_i \right] dq_n(t)$$
$$+ Q_n^2 [(u, v) : u \neq v]$$

and by Jensen's inequality

$$\frac{1}{2} \mathbf{K}(Q, P^n) \geq \sum_{i=1}^{n-1} \left[\int_{\Omega_n} \mathbb{Q}_t \left[(x, y) \in \Omega^{n-1} \times \Omega^{n-1} : x_i \neq y_i \right] dq_n(t) \right]^2$$
$$+ Q_n^2 [(u, v) : u \neq v]. \tag{2.44}$$

Now, we consider the probability distribution $\mathbb{Q}$ on $\Omega^n \times \Omega^n$ with marginal distribution Q_n on $\Omega_n \times \Omega_n$ and such that the distribution of (X_i, Y_i), $1 \leq i \leq n$ conditionally to (X_n, Y_n) is equal to $\mathbb{Q}_{Y_n}$. More precisely for any measurable and bounded function Ψ on $\Omega^n \times \Omega^n$, $\int_{\Omega^n \times \Omega^n} \Psi(x, y) d\mathbb{Q}(x, y)$ is defined by

$$\int_{\Omega_n \times \Omega_n} \left[\int_{\Omega^{n-1} \times \Omega^{n-1}} \Psi[(x, x_n), (y, y_n)] d\mathbb{Q}_{y_n}(x, y) \right] dQ_n(x_n, y_n).$$

Then, by construction, $\mathbb{Q} \in \mathcal{P}(P^n, Q)$. Moreover

$$\mathbb{Q}\left[X_i \neq Y_i\right] = \int_{\Omega_n} \mathbb{Q}_{y_n}\left[(x,y) \in \Omega^{n-1} \times \Omega^{n-1} : x_i \neq y_i\right] dq_n\left(y_n\right)$$

for all $i \leq n - 1$, and

$$\mathbb{Q}\left[X_n \neq Y_n\right] = Q_n\left[(u,v) : u \neq v\right],$$

therefore we derive from (2.44) that

$$\frac{1}{2}\mathbf{K}\left(Q, P^n\right) \geq \sum_{i=1}^{n-1} \mathbb{Q}^2\left[X_i \neq Y_i\right] + \mathbb{Q}^2\left[X_n \neq Y_n\right]. \;\blacksquare$$

In order to illustrate the transportation method to derive concentration inequalities, let us study a first illustrative example.

A Case Example: Rademacher Processes

Let $Z = \sup_{t \in T} \sum_{i=1}^{n} \varepsilon_i \alpha_{i,t}$, where T is a finite set, $(\alpha_{i,t})$ are real numbers and $\varepsilon_1, \ldots, \varepsilon_n$ are independent Rademacher variables, i.e., $\mathbb{P}\left[\varepsilon_i = 1\right] = \mathbb{P}\left[\varepsilon_i = -1\right] = 1/2$, for every $i \in [1,n]$. If we apply now the coupling inequality, defining P as the distribution of $(\varepsilon_1, \ldots, \varepsilon_n)$ on the product space $\{-1, +1\}^n$ and given Q absolutely continuous with respect to P on $\{-1, +1\}^n$, we derive from Proposition 2.21 the existence of a probability distribution $\mathbb{Q}$ on $\{-1, +1\}^n \times \{-1, +1\}^n$ with first margin equal to P and second margin equal to Q such that

$$\sum_{i=1}^{n} \mathbb{Q}^2\left[X_i \neq Y_i\right] \leq \frac{1}{2}\mathbf{K}\left(Q, P\right). \tag{2.45}$$

Now, setting for every $x \in \{-1, +1\}^n$

$$\zeta\left(x\right) = \sup_{t \in T} \sum_{i=1}^{n} x_i \alpha_{i,t},$$

the marginal restrictions on $\mathbb{Q}$ imply that

$$E_Q\left[\zeta\right] - E_P\left[\zeta\right] = E_{\mathbb{Q}}\left[\zeta\left(X\right) - \zeta\left(Y\right)\right]$$

$$\leq E_{\mathbb{Q}}\left[\sum_{i=1}^{n} |X_i - Y_i| \sup_{t \in T} |\alpha_{i,t}|\right]$$

$$\leq 2 \sum_{i=1}^{n} \mathbb{Q}\left[X_i \neq Y_i\right] \sup_{t \in T} |\alpha_{i,t}|$$

which yields by Cauchy–Schwarz inequality

$$E_Q\left[\zeta\right] - E_P\left[\zeta\right] \le 2 \left(\sum_{i=1}^{n} \mathbb{Q}^2\left[X_i \neq Y_i\right]\right)^{1/2} \left(\sum_{i=1}^{n} \sup_{t \in T} \alpha_{i,t}^2\right)^{1/2}.$$

We derive from the transportation cost inequality (2.45) that

$$E_Q\left[\zeta\right] - E_P\left[\zeta\right] \le \sqrt{2\mathbf{K}\left(Q,P\right)v},$$

where $v = \sum_{i=1}^{n} \sup_{t \in T} \alpha_{i,t}^2$. Now by Lemma 2.13 this inequality means that ζ is sub-Gaussian under distribution P with variance factor v. More precisely, we derive from Lemma 2.13 that for every real number λ

$$E_P\left[\exp\left[\lambda\left(\zeta - E_P\left[\zeta\right]\right)\right]\right] \le e^{\lambda^2 v/2}$$

or equivalently

$$\mathbb{E}\left[\exp\left[\lambda\left(Z - \mathbb{E}\left[Z\right]\right)\right]\right] \le e^{\lambda^2 v/2}.$$

This clearly means that we have obtained the Hoeffding type inequality

$$\mathbb{P}\left[Z - \mathbb{E}\left[Z\right] \ge x\right] \le \exp\left(-\frac{x^2}{2v}\right), \text{ for every positive } x.$$

Since for every t, $\mathrm{Var}\left[\sum_{i=1}^{n} \varepsilon_i \alpha_{i,t}\right] = \sum_{i=1}^{n} \alpha_{i,t}^2$ one can wonder if it is possible to improve on the previous bound, replacing v by the smaller quantity $\sup_{t \in T} \sum_{i=1}^{n} \alpha_{i,t}^2$. This will indeed be possible (at least up to some absolute multiplicative constant) by using an alternative approach that we shall develop at length in the sequel. We turn to the presentation of the so-called tensorization inequality which has been promoted by Michel Ledoux in his seminal work [77] as the basic tool for this alternative approach to derive concentration inequalities for functionals of independent variables.

2.4.2 Tensorization Inequality for Entropy

The duality formula (2.25) ensures that the entropy functional Ent_P is convex. This is indeed a key property for deriving the tensorization inequality for entropy (see the end of this Chapter for more general tensorization inequalities). The proof presented below is borrowed from [5].

Proposition 2.22 (*Tensorization inequality*) *Let* $(X_1, \ldots, X_n)$ *be independent random variables on some probability space* $(\Omega, \mathcal{A}, \mathbb{P})$ *and* Y *be some nonnegative measurable function of* $(X_1, \ldots, X_n)$ *such that* $\Phi(Y)$ *is integrable. For any integer* i *with,* $1 \le i \le n$, *let*

$$X^{(i)} = (X_1, \ldots, X_{i-1}, X_{i+1}, \ldots, X_n)$$

and let us denote by $\mathrm{Ent}_{\mathbb{P}}\left(Y \mid X^{(i)}\right)$ *the entropy of* Y *conditionally to* $X^{(i)}$, *defined by*

$$Ent_{\mathbb{P}}\left[Y \mid X^{(i)}\right] = \mathbb{E}\left[\Phi(Y) \mid X^{(i)}\right] - \Phi\left(\mathbb{E}\left[Y \mid X^{(i)}\right]\right).$$

Then

$$Ent_{\mathbb{P}}[Y] \leq \mathbb{E}\left[\sum_{i=1}^{n} Ent_{\mathbb{P}}\left[Y \mid X^{(i)}\right]\right]. \tag{2.46}$$

Proof. We follow [5] and prove (2.46) by using the duality formula. We introduce the conditional operator $\mathbb{E}^i[.] = \mathbb{E}[. \mid X_i, \ldots, X_n]$ for $i = 1, \ldots, n+1$ with the convention that $\mathbb{E}^{n+1}[.] = \mathbb{E}[.]$. Then the following decomposition holds true

$$Y\left(\ln(Y) - \ln(\mathbb{E}[Y])\right) = \sum_{i=1}^{n} Y\left(\ln\left(\mathbb{E}^i[Y]\right) - \ln\left(\mathbb{E}^{i+1}[Y]\right)\right). \tag{2.47}$$

Now the duality formula (2.27) yields

$$\mathbb{E}\left[Y\left(\ln\left(\mathbb{E}^i[Y]\right) - \ln\left(\mathbb{E}\left[\mathbb{E}^i[Y] \mid X^{(i)}\right]\right)\right) \mid X^{(i)}\right] \leq Ent_{\mathbb{P}}\left[Y \mid X^{(i)}\right]$$

and since the variables $X_1, \ldots, X_n$ are independent the following identity holds $\mathbb{E}\left[\mathbb{E}^i[Y] \mid X^{(i)}\right] = \mathbb{E}^{i+1}[Y]$. Hence, taking expectation on both sides of (2.47) becomes

$$\mathbb{E}\left[Y\left(\ln(Y) - \ln(\mathbb{E}[Y])\right)\right]$$

$$= \sum_{i=1}^{n} \mathbb{E}\left[\mathbb{E}\left[Y\left(\ln\left(\mathbb{E}^i[Y]\right) - \ln\left(\mathbb{E}\left[\mathbb{E}^i[Y] \mid X^{(i)}\right]\right)\right) \mid X^{(i)}\right]\right]$$

$$\leq \sum_{i=1}^{n} \mathbb{E}\left[Ent_{\mathbb{P}}\left[Y \mid X^{(i)}\right]\right]$$

and (2.46) follows. ∎

Comment. Note that there is no measure theoretic trap here and below since by Fubini's Theorem, the conditional entropies and conditional expectations that we are dealing with can all be defined from regular versions of conditional probabilities.

It is interesting to notice that there is some close relationship between this inequality and Han's inequality for Shannon entropy (see [39], p. 491). To see this let us recall that for some random variable ξ taking its values in some finite set $\mathcal{X}$, Shannon entropy is defined by

$$H[\xi] = -\sum_{x} \mathbb{P}[\xi = x] \ln\left(\mathbb{P}[\xi = x]\right).$$

Setting, $q(x) = \mathbb{P}[\xi = x]$ for any point x in the support of ξ, Shannon entropy can also be written as $H[\xi] = \mathbb{E}[-\ln q(\xi)]$, from which one readily see that it is a nonnegative quantity. The relationship between Shannon entropy and Kullback–Leibler information is given by the following identity. Let Q be the distribution of ξ, P be the uniform distribution on $\mathcal{X}$ and N be the cardinality of $\mathcal{X}$, then

$$\mathbf{K}\left(Q, P\right) = -H\left[\xi\right] + \ln\left(N\right).$$

We derive from this equation and the nonnegativity of the Kullback–Leibler information that

$$H\left[\xi\right] \leq \ln\left(N\right) \tag{2.48}$$

with equality if and only if ξ is uniformly distributed on its support. Han's inequality can be stated as follows.

Corollary 2.23 *Let $\mathcal{X}$ be some finite set and let us consider some random variable ξ with values in $\mathcal{X}^n$ and write $\xi^{(i)} = (\xi_1, \ldots, \xi_{i-1}, \xi_{i+1}, \ldots, \xi_n)$ for every $i \in \{1, \ldots, n\}$. Then*

$$H\left[\xi\right] \leq \frac{1}{n-1} \sum_{i=1}^{n} H\left[\xi^{(i)}\right].$$

Proof. Let us consider Q to be the distribution of ξ and P^n to be the uniform distribution on $\mathcal{X}^n$. Let moreover $(X_1, \ldots, X_n)$ be the coordinate mappings on $\mathcal{X}^n$ and for every $x \in \mathcal{X}^n$, $q(x) = \mathbb{P}\left[\xi = x\right]$. Setting $Y = dQ/dP^n$ we have $Y(x) = q(x) k^n$ and

$$\text{Ent}_{P^n}\left[Y\right] = \mathbf{K}\left(Q, P^n\right) = -H\left[\xi\right] + n \ln k, \tag{2.49}$$

where k denotes the cardinality of $\mathcal{X}$. Moreover, the tensorization inequality (2.46) can be written in this case as

$$\text{Ent}_{P^n}\left[Y\right] \leq E_{P^n}\left[\sum_{i=1}^{n} \text{Ent}_{P^n}\left[Y \mid X^{(i)}\right]\right]. \tag{2.50}$$

Now

$$E_{P^n}\left[\text{Ent}_{P^n}\left[Y \mid X^{(1)}\right]\right] = \text{Ent}_{P^n}\left[Y\right]$$

$$- \sum_{x \in \mathcal{X}^{n-1}} \left[\sum_{t \in \mathcal{X}} q\left(t, x\right)\right] \ln\left[k^{n-1} \sum_{t \in \mathcal{X}} q\left(t, x\right)\right]$$

$$= \text{Ent}_{P^n}\left[Y\right] + H\left[\xi^{(1)}\right] - (n-1) \ln k.$$

and similarly for all i

$$E_{P^n}\left[\text{Ent}_{P^n}\left[Y \mid X^{(i)}\right]\right] = \text{Ent}_{P^n}\left[Y\right] - (n-1) \ln k + H\left[\xi^{(i)}\right].$$

Hence (2.50) is equivalent to

$$- (n - 1) \operatorname{Ent}_{P^n} [Y] \le \sum_{i=1}^{n} H \left[\xi^{(i)} \right] - n (n - 1) \ln k$$

and the result follows by (2.49). ∎

In the last section we investigate some possible extensions of the previous information inequalities to some more general notions of entropy.

2.5 ϕ-Entropy

Our purpose is to understand in depth the role of the function $\Phi : x \to x \ln (x)$ in the definition of entropy. If one substitutes to this function another convex function ϕ, one can wonder what kind of properties of the entropy would be preserved for the so-defined entropy functional that we will call ϕ-entropy. In particular we want to analyze the conditions on ϕ under which the tensorization inequality holds for ϕ-entropy, discussing the role of the Latala–Oleszkiewicz condition: ϕ is a continuous and convex function on $\mathbb{R}_+$ which is twice differentiable on $\mathbb{R}_+^*$ and such that either ϕ is affine or ϕ'' is strictly positive and $1/\phi''$ is concave. The main issue is to show that if ϕ is strictly convex and twice differentiable on $\mathbb{R}_+^*$, a necessary and sufficient condition for the tensorization inequality to hold for ϕ-entropy is the concavity of $1/\phi''$. We shall denote from now on by $\mathcal{LO}$ the class of functions satisfying this condition. Our target examples are the convex power functions ϕ_p, $p \in (0, +\infty)$ on $\mathbb{R}_+$ defined by $\phi_p (x) = -x^p$ if $p \in (0, 1)$ and $\phi_p (x) = x^p$ whenever $p \ge 1$. For these functions Latala and Oleszkiewicz' condition means $p \in [1, 2]$. Let us now fix some notations. For any convex function ϕ on $\mathbb{R}_+$, let us denote by $\mathbb{L}_1^+$ the convex set of nonnegative and integrable random variables Z and define the ϕ-entropy functional H_ϕ on $\mathbb{L}_1^+$ by

$$H_\phi (Z) = \mathbb{E} [\phi (Z)] - \phi (\mathbb{E} [Z])$$

for every $Z \in \mathbb{L}_1^+$. Note that here and below we use the extended notion of expectation for a (nonnecessarily integrable) random variable X defined as $\mathbb{E} [X] = \mathbb{E} [X^+] - \mathbb{E} [X^-]$ whenever either X^+ or X^- is integrable. It follows from Jensen's inequality that $H_\phi (Z)$ is nonnegative and that $H_\phi (Z) < +\infty$ if and only if $\phi (Z)$ is integrable. It is easy to prove a variational formula for ϕ-entropy.

Lemma 2.24 *Let ϕ be some continuous and convex function on $\mathbb{R}_+$, then, denoting by ϕ' the right derivative of ϕ, for every $Z \in \mathbb{L}_1^+$, the following formula holds true*

$$H_\phi (Z) = \inf_{u \ge 0} \mathbb{E} [\phi (Z) - \phi (u) - (Z - u) \phi' (u)]. \qquad (2.51)$$

Proof. Without loss of generality we assume that $\phi(0) = 0$. Let $m = \mathbb{E}[Z]$, the convexity of ϕ implies that for every positive u

$$-\phi(m) \leq -\phi(u) - (m - u)\phi'(u)$$

and therefore

$$H_\phi(Z) \leq \mathbb{E}[\phi(Z) - \phi(u) - (Z - u)\phi'(u)].$$

Since the latter inequality becomes an equality when $u = m$, the variational formula (2.51) is proven. ∎

While the convexity of the function ϕ alone is enough to imply a variational formula, it is not at all the same for the duality formula and the tensorization property which are intimately linked as we shall see below. We shall say that H_ϕ has the tensorization property if for every finite family $X_1, \ldots, X_n$ of independent random variables and every $(X_1, \ldots, X_n)$-measurable nonnegative and integrable random variable Z,

$$H_\phi(Z) \leq \sum_{i=1}^{n} \mathbb{E}\left[\mathbb{E}\left[\phi(Z) \mid X^{(i)}\right] - \phi\left(\mathbb{E}\left[Z \mid X^{(i)}\right]\right)\right], \tag{2.52}$$

where, for every integer $i \in [1, n]$, $X^{(i)}$ denotes the family of variables $\{X_1, \ldots, X_n\} \setminus \{X_i\}$. As quoted in Ledoux [77] or Latala and Oleszkiewicz [73], there is a deep relationship between the convexity of H_ϕ and the tensorization property. More precisely, it is easy to see that, for $n = 2$, setting $Z = g(X_1, X_2)$, (2.52) is exactly equivalent to the Jensen type inequality

$$H_\phi\left(\int g(x, X_2) \, d\mu_1(x)\right) \leq \int H_\phi(g(x, X_2)) \, d\mu_1(x), \tag{2.53}$$

where μ_1 denotes the distribution of X_1. Since by the induction argument of Ledoux, (2.53) leads to (2.52) for every n, we see that the tensorization property for H_ϕ is equivalent to what we could call the Jensen property, i.e., (2.53) holds for every μ_1, X_2 and g such that $\int g(x, X_2) \, d\mu_1(x)$ is integrable. We do not want to go into the details here since we shall indeed provide an explicit proof of the tensorization inequality for ϕ-entropy later on. Of course the Jensen property implies the convexity of H_ϕ. Indeed, given $\lambda \in [0, 1]$ and two elements U, V of $\mathbb{L}_1^+$, setting $g(x, U, V) = xU + (1 - x)V$ for every $x \in [0, 1]$ and taking μ_1 to be the Bernoulli distribution with parameter λ, (2.53) means that

$$H_\phi(\lambda U + (1 - \lambda)V) \leq \lambda H_\phi(U) + (1 - \lambda)H_\phi(V).$$

Hence H_ϕ is convex.

2.5.1 Necessary Condition for the Convexity of ϕ-Entropy

As proved in [30], provided that ϕ'' is strictly positive, the condition $1/\phi''$ concave is necessary for the tensorization property to hold. We can more precisely prove that $1/\phi''$ concave is a necessary condition for H_ϕ to be convex on the set $\mathbb{L}_\infty^+ (\Omega, \mathcal{A}, \mathbb{P})$ of bounded and nonnegative random variables, for suitable probability spaces $(\Omega, \mathcal{A}, \mathbb{P})$.

Proposition 2.25 Let ϕ be a strictly convex function on $\mathbb{R}_+$ which is twice differentiable on $\mathbb{R}_+^*$. Let $(\Omega, \mathcal{A}, \mathbb{P})$ be a rich enough probability space in the sense that $\mathbb{P}$ maps $\mathcal{A}$ onto $[0, 1]$. If H_ϕ is convex on $\mathbb{L}_\infty^+ (\Omega, \mathcal{A}, \mathbb{P})$, then $\phi'' (x) > 0$ for every $x > 0$ and $1/\phi''$ is concave on $\mathbb{R}_+^*$.

Proof. Let $\theta \in [0, 1]$ and x, x', y, y' be given positive real numbers. Under the assumption on the probability space we can define a pair of random variables (X, Y) to be (x, y) with probability θ and (x', y') with probability $(1 - \theta)$. Then, the convexity of H_ϕ means that

$$H_\phi (\lambda X + (1 - \lambda) Y) \leq \lambda H_\phi (X) + (1 - \lambda) H_\phi (Y) \qquad (2.54)$$

for every $\lambda \in (0, 1)$. Defining, for every $(u, v) \in \mathbb{R}_+^* \times \mathbb{R}_+^*$

$$F_\lambda (u, v) = -\phi (\lambda u + (1 - \lambda) v) + \lambda \phi (u) + (1 - \lambda) \phi (v),$$

(2.54) is equivalent to

$$F_\lambda (\theta (x, y) + (1 - \theta) (x', y')) \leq \theta F_\lambda (x, y) + (1 - \theta) F_\lambda (x', y').$$

Hence F_λ is convex on $\mathbb{R}_+^* \times \mathbb{R}_+^*$. This implies in particular that the determinant of the Hessian matrix of F_λ is nonnegative at each point (x, y). Thus, setting $x_\lambda = \lambda x + (1 - \lambda) y$

$$[\phi'' (x) - \lambda \phi'' (x_\lambda)] [\phi'' (y) - (1 - \lambda) \phi'' (x_\lambda)] \geq \lambda (1 - \lambda) [\phi'' (x_\lambda)]^2$$

which means that

$$\phi'' (x) \phi'' (y) \geq \lambda \phi'' (y) \phi'' (x_\lambda) + (1 - \lambda) \phi'' (x) \phi'' (x_\lambda). \qquad (2.55)$$

If $\phi'' (x) = 0$ for some point x, we derive from (2.55) that either $\phi'' (y) = 0$ for every y which is impossible because ϕ is assumed to be strictly convex, or there exists some y such that $\phi'' (y) > 0$ and then ϕ'' is identically equal to 0 on the nonempty open interval with extremities x and y which also leads to a contradiction with the assumption that ϕ is strictly convex. Hence ϕ'' is strictly positive at each point of $\mathbb{R}_+^*$ and (2.55) leads to

$$\frac{1}{\phi'' (\lambda x + (1 - \lambda) y)} \geq \frac{\lambda}{\phi'' (x)} + \frac{(1 - \lambda)}{\phi'' (y)}$$

which means that $1/\phi''$ is concave. ∎

Conversely, $\phi \in \mathcal{LO}$ implies the convexity of the function F_λ defined above and thus the convexity of H_ϕ as proved by Latala and Oleszkiewicz in [73]. However this does not straightforwardly leads to Jensen's property (and therefore to the tensorization property for H_ϕ) because the distribution μ_1 in (2.53) needs not be discrete and we really have to face with an infinite dimensional analogue of Jensen's inequality. The easiest way to overcome this difficulty is to follow the lines of Ledoux's proof of the tensorization property for the classical entropy (which corresponds to the case where $\phi(x) = x \ln(x)$) and mimic the duality argument used in dimension 1 to prove the usual Jensen inequality i.e., express H_ϕ as the supremum of affine functions.

2.5.2 A Duality Formula for ϕ-Entropy

Provided that $\phi \in \mathcal{LO}$, our purpose is now, following the lines of [30] to establish a duality formula for ϕ-entropy of the type

$$H_\phi(Z) = \sup_{T \in \mathcal{T}} \mathbb{E}[\psi_1(T) Z + \psi_2(T)],$$

for convenient functions ψ_1 and ψ_2 on $\mathbb{R}_+$ and a suitable class of nonnegative variables $\mathcal{T}$. Such a formula of course implies the convexity of H_ϕ but also Jensen's property (just by Fubini) and therefore the tensorization property for H_ϕ.

Lemma 2.26 (Duality formula for ϕ-entropy) Let ϕ belong to $\mathcal{LO}$ and Z belong to $\mathbb{L}_1^+$. Then if $\phi(Z)$ is integrable

$$H_\phi(Z) = \sup_{T \in \mathbb{L}_1^+, T \neq 0} \{\mathbb{E}[(\phi'(T) - \phi'(\mathbb{E}[T])) (Z - T) + \phi(T)] - \phi(\mathbb{E}[T])\}.$$

$$(2.56)$$

Note that the convexity of ϕ implies that $\phi'(T)(Z - T) + \phi(T) \leq \phi(Z)$. Hence $\mathbb{E}[\phi'(T)(Z - T) + \phi(T)]$ is well defined and is either finite or equal to $-\infty$.

Proof. The case where ϕ is affine is trivial. Otherwise, let us first assume Z and T to be bounded and bounded away from 0. For any $\lambda \in [0, 1]$, we set $T_\lambda = (1 - \lambda) Z + \lambda T$ and

$$f(\lambda) = \mathbb{E}[(\phi'(T_\lambda) - \phi'(\mathbb{E}[T_\lambda])) (Z - T_\lambda)] + H_\phi(T_\lambda).$$

Our aim is to show that f if nonincreasing on $[0, 1]$. Noticing that $Z - T_\lambda = \lambda(Z - T)$ and using our boundedness assumptions to differentiate under the expectation

$$f'(\lambda) = -\lambda \left[\mathbb{E}\left[(Z - T)^2 \phi''(T_\lambda)\right] - (\mathbb{E}[Z - T])^2 \phi''(\mathbb{E}[T_\lambda])\right]$$
$$+ \mathbb{E}[(\phi'(T_\lambda) - \phi'(\mathbb{E}[T_\lambda])) (Z - T)]$$
$$+ \mathbb{E}[\phi'(T_\lambda)(T - Z)] - \phi'(\mathbb{E}[T_\lambda]) \mathbb{E}[T - Z]$$

and so

$$f'(\lambda) = -\lambda \left[\mathbb{E}\left[(Z-T)^2 \phi''(T_\lambda)\right] - (\mathbb{E}[Z-T])^2 \phi''(\mathbb{E}[T_\lambda]) \right].$$

Now, by Cauchy–Schwarz inequality and Jensen's inequality (remember that $1/\phi''$ is assumed to be concave)

$$(\mathbb{E}[Z-T])^2 = \left(\mathbb{E}\left[(Z-T)\sqrt{\phi''(T_\lambda)} \frac{1}{\sqrt{\phi''(T_\lambda)}} \right] \right)^2$$

$$\leq \mathbb{E}\left[\frac{1}{\phi''(T_\lambda)} \right] \mathbb{E}\left[(Z-T)^2 \phi''(T_\lambda) \right]$$

and

$$\mathbb{E}\left[\frac{1}{\phi''(T_\lambda)} \right] \leq \frac{1}{\phi''(\mathbb{E}[T_\lambda])}$$

which leads to

$$(\mathbb{E}[Z-T])^2 \leq \frac{1}{\phi''(\mathbb{E}[T_\lambda])} \mathbb{E}\left[(Z-T)^2 \phi''(T_\lambda) \right].$$

Hence f' is nonpositive and therefore $f(1) \leq f(0) = H_\phi(Z)$. This means that whatever T, $\mathbb{E}[(\phi'(T) - \phi'(\mathbb{E}[T]))(Z-T)] + H_\phi(T)$ is less than $H_\phi(Z)$. Since this inequality is an equality for $T = Z$, the proof of (2.56) is complete under the extra assumption that Z and T are bounded and bounded away from 0. In the general case we consider the sequences $Z_n = (Z \vee 1/n) \wedge n$ and $T_k = (T \vee 1/k) \wedge k$ and we have in view to take the limit as k and n go to infinity in the inequality

$$H_\phi(Z_n) \geq \mathbb{E}[(\phi'(T_k) - \phi'(\mathbb{E}[T_k]))(Z_n - T_k) + \phi(T_k)] - \phi(\mathbb{E}[T_k])$$

that we can also write

$$\mathbb{E}[\psi(Z_n, T_k)] \geq -\phi'(\mathbb{E}[T_k])\mathbb{E}[Z_n - T_k] - \phi(\mathbb{E}[T_k]) + \phi(\mathbb{E}[Z_n]), \quad (2.57)$$

where $\psi(z,t) = \phi(z) - \phi(t) - (z-t)\phi'(t)$. Since we have to show that

$$\mathbb{E}[\psi(Z,T)] \geq -\phi'(\mathbb{E}[T])\mathbb{E}[Z-T] - \phi(\mathbb{E}[T]) + \phi(\mathbb{E}[Z]) \quad (2.58)$$

with $\psi \geq 0$, we can always assume $[\psi(Z,T)]$ to be integrable (otherwise (2.58) is trivially satisfied). Taking the limit when n and k go to infinity in the right hand side of (2.57) is easy while the treatment of the left hand side requires some care. Let us notice that $\psi(z,t)$ as a function of t decreases on $(0,z)$ and increases on $(z,+\infty)$. Similarly, as a function of z, $\psi(z,t)$ decreases on $(0,t)$ and increases on $(t,+\infty)$. Hence, for every t, $\psi(Z_n,t) \leq \psi(1,t) + \psi(Z,t)$ while for every z, $\psi(z,T_k) \leq \psi(z,1) + \psi(z,T)$. Hence, given k

$$\psi(Z_n, T_k) \leq \psi(1, T_k) + \psi(Z, T_k)$$

and we can apply the bounded convergence theorem and conclude that $\mathbb{E}\left[\psi\left(Z_n, T_k\right)\right]$ converges to $\mathbb{E}\left[\psi\left(Z, T_k\right)\right]$ as n goes to infinity. Hence the following inequality holds

$$\mathbb{E}\left[\psi\left(Z, T_k\right)\right] \geq -\phi'\left(\mathbb{E}\left[T_k\right]\right) \mathbb{E}\left[Z - T_k\right] - \phi\left(\mathbb{E}\left[T_k\right]\right) + \phi\left(\mathbb{E}\left[Z\right]\right). \quad (2.59)$$

Now we also have $\psi\left(Z, T_k\right) \leq \psi\left(Z, 1\right) + \psi\left(Z, T\right)$ and we can apply the bounded convergence theorem again to ensure that $\mathbb{E}\left[\psi\left(Z, T_k\right)\right]$ converges to $\mathbb{E}\left[\psi\left(Z, T\right)\right]$ as k goes to infinity. Taking the limit as k goes to infinity in (2.59) implies that (2.58) holds for every $T, Z \in \mathbb{L}_1^+$ such that $\phi\left(Z\right)$ is integrable and $\mathbb{E}\left[T\right] > 0$. If $Z \neq 0$ a.s., (2.58) is achieved for $T = Z$ while if $Z = 0$ a.s., it is achieved for $T = 1$ and the proof of the Lemma is now complete in its full generality. ∎

Comments.

- First note that since the supremum in (2.56) is achieved for $T = Z$ (or $T = 1$ if $Z = 0$), the duality formula remains true if the supremum is restricted to the class $\mathcal{T}_\phi$ of variables T such that $\phi\left(T\right)$ is integrable. Hence the following alternative formula also holds

$$H_\phi\left(Z\right) = \sup_{T \in \mathcal{T}_\phi} \left\{\mathbb{E}\left[\left(\phi'\left(T\right) - \phi'\left(\mathbb{E}\left[T\right]\right)\right)\left(Z - T\right)\right] + H_\phi\left(T\right)\right\}. \quad (2.60)$$

- Formula (2.56) takes the following form for the "usual" entropy (which corresponds to $\phi\left(x\right) = x \ln\left(x\right)$)

$$\text{Ent}\left(Z\right) = \sup_T \left\{\mathbb{E}\left[\left(\ln\left(T\right) - \ln\left(\mathbb{E}\left[T\right]\right)\right) Z\right]\right\}$$

where the supremum is extended to the set of nonnegative and integrable random variables T with $\mathbb{E}\left[T\right] > 0$. We recover the duality formula of Proposition 2.12.

- Another case of interest is $\phi\left(x\right) = x^p$, where $p \in (1, 2]$. In this case, the duality formula (2.60) becomes

$$H_\phi\left(Z\right) = \sup_T \left\{p \mathbb{E}\left[Z\left(T^{p-1} - \left(\mathbb{E}\left[T\right]\right)^{p-1}\right)\right] - \left(p - 1\right) H_\phi\left(T\right)\right\},$$

where the supremum is extended to the set of nonnegative variables in $\mathbb{L}_p$.

ϕ-Entropy for Real Valued Random Variables

For the sake of simplicity we have focused on nonnegative variables and restrict ourselves to convex functions ϕ on $\mathbb{R}_+$. Of course, this restriction can be avoided and one can consider the case where ϕ is a convex function on $\mathbb{R}$ and define the ϕ-entropy of a real valued integrable random variable Z by the same formula as in the nonnegative case. Assuming this time that ϕ is differentiable on $\mathbb{R}$ and twice differentiable on $\mathbb{R}^*$, the proof of the duality formula which is presented above can be easily adapted to cover this case provided that $1/\phi''$ can be extended to a concave function on $\mathbb{R}$. In particular if $\phi(x) = |x|^p$, where $p \in (1, 2]$ one gets

$$H_\phi(Z) = \sup_T \left\{ p\mathbb{E}\left[Z\left(\frac{|T|^p}{T} - \frac{|\mathbb{E}[T]|^p}{\mathbb{E}[T]} \right) \right] - (p-1)H_\phi(T) \right\}$$

where the supremum is extended to $\mathbb{L}_p$. Note that this formula reduces for $p = 2$ to the classical one for the variance

$$\mathrm{Var}(Z) = \sup_T \left\{ 2\,\mathrm{Cov}(Z, T) - \mathrm{Var}(T) \right\},$$

where the supremum is extended to the set of square integrable variables. This means that the tensorization inequality for ϕ-entropy also holds for convex functions ϕ on $\mathbb{R}$ under the condition that $1/\phi''$ is the restriction to $\mathbb{R}^*$ of a concave function on $\mathbb{R}$.

2.5.3 A Direct Proof of the Tensorization Inequality

Starting from the duality formula, it is possible to design a direct proof of the tensorization inequality which does not involve any induction argument. This means that the proof of Proposition 2.22 nicely extends to ϕ-entropy.

Theorem 2.27 (*Tensorization inequality for ϕ-entropy*) *Assume that $\phi \in \mathcal{LO}$, then for every finite family $X_1, \ldots, X_n$ of independent random variables and every $(X_1, \ldots, X_n)$-measurable nonnegative and integrable random variable Z,*

$$H_\phi(Z) \leq \sum_{i=1}^n \mathbb{E}\left[\mathbb{E}\left[\phi(Z) \mid X^{(i)} \right] - \phi\left(\mathbb{E}\left[Z \mid X^{(i)} \right] \right) \right], \qquad (2.61)$$

where, for every integer $i \in [1, n]$, $X^{(i)}$ denotes the family of variables $\{X_1, \ldots, X_n\} \setminus \{X_i\}$.

Proof. Of course we may assume $\phi(Z)$ to be integrable, otherwise (2.61) is trivial. We introduce the conditional operator $\mathbb{E}^i[.] = \mathbb{E}[. \mid X_i, \ldots, X_n]$ for $i = 1, \ldots, n+1$ with the convention that $\mathbb{E}^{n+1}[.] = \mathbb{E}[.]$. Note also that $\mathbb{E}^1[.]$ is just identity when restricted to the set of $(X_1, \ldots, X_n)$-measurable and integrable random variables. Let us introduce the notation $H_\phi(Z \mid X^{(i)}) = \mathbb{E}\left[\phi(Z) \mid X^{(i)} \right] - \phi\left(\mathbb{E}\left[Z \mid X^{(i)} \right] \right)$. By (2.60) we know that $H_\phi(Z \mid X^{(i)})$ is bounded from below by

$$\mathbb{E}\left[\left(\phi'\left(\mathbb{E}^i[Z] \right) - \phi'\left(\mathbb{E}\left[\mathbb{E}^i[Z] \mid X^{(i)} \right] \right) \right) \left(Z - \mathbb{E}^i[Z] \right) \mid X^{(i)} \right]$$
$$+ \mathbb{E}\left[\phi\left(\mathbb{E}^i[Z] \right) \mid X^{(i)} \right] - \phi\left(\mathbb{E}\left[\mathbb{E}^i[Z] \mid X^{(i)} \right] \right).$$

Now, by independence, for every i, we note that $\mathbb{E}\left[\mathbb{E}^i\left[Z\right] \mid X^{(i)}\right] = \mathbb{E}^{i+1}\left[Z\right]$. Hence, using in particular the identity

$$\mathbb{E}\left[\phi'\left(\mathbb{E}^{i+1}\left[Z\right]\right)\mathbb{E}^i\left[Z\right] \mid X^{(i)}\right] = \phi'\left(\mathbb{E}^{i+1}\left[Z\right]\right)\mathbb{E}^{i+1}\left[Z\right],$$

we derive from the previous inequality that

$$\mathbb{E}\left[H_\phi\left(Z \mid X^{(i)}\right)\right] \geq \mathbb{E}\left[Z\left(\phi'\left(\mathbb{E}^i\left[Z\right]\right) - \phi'\left(\mathbb{E}^{i+1}\left[Z\right]\right)\right)\right]$$
$$+ \mathbb{E}\left[\phi'\left(\mathbb{E}^{i+1}\left[Z\right]\right)\mathbb{E}^{i+1}\left[Z\right] - \phi'\left(\mathbb{E}^i\left[Z\right]\right)\mathbb{E}^i\left[Z\right]\right]$$
$$+ \mathbb{E}\left[\phi\left(\mathbb{E}^i\left[Z\right]\right) - \phi\left(\mathbb{E}^{i+1}\left[Z\right]\right)\right].$$

Summing up these inequalities leads to

$$\sum_{i=1}^n \mathbb{E}\left[H_\phi\left(Z \mid X^{(i)}\right)\right] \geq \mathbb{E}\left[Z\left(\phi'\left(Z\right) - \phi'\left(\mathbb{E}\left[Z\right]\right)\right)\right]$$
$$+ \mathbb{E}\left[\phi'\left(\mathbb{E}\left[Z\right]\right)\mathbb{E}\left[Z\right] - \phi'\left(Z\right)Z\right]$$
$$+ \mathbb{E}\left[\phi\left(Z\right) - \phi\left(\mathbb{E}\left[Z\right]\right)\right]$$

and the result follows. ■

To see the link between the tensorization inequality and concentration, let us consider the case of the variance as a training example.

2.5.4 Efron–Stein's Inequality

Let us show how to derive an inequality due to Efron and Stein (see [59]) from the tensorization inequality of the variance, i.e., the ϕ-entropy when ϕ is defined on the whole real line as $\phi\left(x\right) = x^2$. In this case the tensorization inequality can be written as

$$\text{Var}\left(Z\right) \leq \mathbb{E}\left[\sum_{i=1}^n \mathbb{E}\left[\left(Z - \mathbb{E}\left[Z \mid X^{(i)}\right]\right)^2 \mid X^{(i)}\right]\right].$$

Now let X' be a copy of X and define

$$Z_i' = \zeta\left(X_1, \ldots, X_{i-1}, X_i', X_{i+1}, \ldots, X_n\right).$$

Since conditionally to $X^{(i)}$, Z_i' is an independent copy of Z, we can write

$$\mathbb{E}\left[\left(Z - \mathbb{E}\left[Z \mid X^{(i)}\right]\right)^2 \mid X^{(i)}\right] = \frac{1}{2}\mathbb{E}\left[\left(Z - Z_i'\right)^2 \mid X^{(i)}\right]$$
$$= \mathbb{E}\left[\left(Z - Z_i'\right)_+^2 \mid X^{(i)}\right],$$

which leads to Efron–Stein's inequality

$$\mathrm{Var}\,(Z) \leq \frac{1}{2} \sum_{i=1}^{n} \mathbb{E}\left[(Z - Z_i')^2\right] = \sum_{i=1}^{n} \mathbb{E}\left[(Z - Z_i')_+^2\right]. \qquad (2.62)$$

To be more concrete let us consider again the example of Rademacher processes and see what Efron–Stein's inequality is telling us and compare it with what we can derive from the coupling inequality.

Rademacher Processes Revisited

As an exercise we can apply the tensorization technique to suprema of Rademacher processes. Let $Z = \sup_{t \in T} \sum_{i=1}^{n} \varepsilon_i \alpha_{i,t}$, where T is a finite set, $(\alpha_{i,t})$ are real numbers and $\varepsilon_1, \ldots, \varepsilon_n$ are independent random signs. Since we are not ready yet to derive exponential bounds from the tensorization inequality for entropy, for this first approach, by sake of keeping the calculations as simple as possible we just bound the second moment and not the moment generating function of $Z - \mathbb{E}[Z]$. Then, taking $\varepsilon_1', \ldots, \varepsilon_n'$ as an independent copy of $\varepsilon_1, \ldots, \varepsilon_n$ we set for every $i \in [1, n]$

$$Z_i' = \sup_{t \in T}\left[\left(\sum_{j \neq i}^{n} \varepsilon_j \alpha_{j,t}\right) + \varepsilon_i' \alpha_{i,t}\right].$$

Considering t^* such that $\sup_{t \in T} \sum_{j=1}^{n} \varepsilon_j \alpha_{j,t} = \sum_{j=1}^{n} \varepsilon_j \alpha_{j,t^*}$ we have for every $i \in [1, n]$

$$Z - Z_i' \leq (\varepsilon_i - \varepsilon_i') \alpha_{i,t^*}$$

which yields

$$(Z - Z_i')_+^2 \leq (\varepsilon_i - \varepsilon_i')^2 \alpha_{i,t^*}^2$$

and therefore by independence of ε_i' from $\varepsilon_1, \ldots, \varepsilon_n$

$$\mathbb{E}\left[(Z - Z_i')_+^2\right] \leq \mathbb{E}\left[(1 + \varepsilon_i^2) \alpha_{i,t^*}^2\right] \leq 2\mathbb{E}\left[\alpha_{i,t^*}^2\right].$$

Hence, we derive from Efron–Stein's inequality that

$$\mathrm{Var}\,(Z) \leq 2\mathbb{E}\left[\sum_{i=1}^{n} \alpha_{i,t^*}^2\right] \leq 2\sigma^2, \qquad (2.63)$$

where $\sigma^2 = \sup_{t \in T} \sum_{i=1}^{n} \alpha_{i,t}^2$. As compared to what we had derived from the coupling approach, we see that this time we have got the expected order for the variance i.e., $\sup_{t \in T} \sum_{i=1}^{n} \alpha_{i,t}^2$ instead of $\sum_{i=1}^{n} \sup_{t \in T} \alpha_{i,t}^2$.

3

Gaussian Processes

3.1 Introduction and Basic Remarks

The aim of this chapter is to treat two closely related problems for a Gaussian process: the question of the sample path regularity and the derivation of tail bounds for proper functionals such as its supremum over a given set in its parameter space. A *stochastic process* $X = (X(t))_{t \in T}$ indexed by T is a collection of random variables $X(t)$, $t \in T$. Then X is a random variable as a map into $\mathbb{R}^T$ equipped with the σ-field generated by the cylinder sets (i.e., the product of a collection of Borel sets which are trivial except for a finite subcollection). The law of X is determined by the collection of all marginal distributions of the finite dimensional random vectors $(X(t_1), \ldots, X(t_N))$ when $\{t_1, \ldots, t_N\}$ varies. It is generally not possible to show (except when T is countable) that any *version* of X, i.e., any stochastic process Y with same law as X is almost surely continuous on T (equipped with some given distance). Instead one has to deal with a special version of X which can typically be constructed as a *modification* $\widetilde{X}$ of X i.e., $\widetilde{X}(t) = X(t)$ a.s. for all $t \in T$. A (centered) Gaussian process is a stochastic process $X = (X(t))_{t \in T}$, such that each finite linear combination $\sum \alpha_t X(t)$ is (centered) Gaussian (in other words, each finite dimensional random vector $(X(t_1), \ldots, X(t_N))$ is Gaussian). Since the transition from centered to noncentered Gaussian variables is via the addition of a constant, in this chapter, we shall deal exclusively with centered Gaussian processes. The parameter space T can be equipped with the intrinsic $\mathbb{L}_2$-pseudodistance

$$d(s,t) = \mathbb{E}\left[(X(s) - X(t))^2\right]^{1/2}.$$

Note that d is a distance which does not necessarily separate points ($d(s,t) = 0$ does not always imply that $s = t$) and therefore (T, d) is only a pseudometric space. One of the major issues will be to derive tail bounds for $\sup_{t \in T} X(t)$. Possibly applying these bounds to the supremum of the process of increments $X(s) - X(t)$ over the sets $\{(s,t) ; d(s,t) \leq \eta\}$ and letting η go to 0, we shall deal with the question of the sample boundedness and uniform continuity (with respect to d) at the same time. Of course there exists some unavoidable measurability questions that we would like to briefly discuss once and for all. When T is countable there is no problem, the (possibly infinite) quantity $\sup_{t \in T} X(t)$ is always measurable and its integrability is equivalent to that of $(\sup_{t \in T} X(t))_+$ so that one can always speak of $\mathbb{E}[\sup_{t \in T} X(t)]$ which is either finite or equals $+\infty$. Otherwise we shall assume (T,d) to be *separable*, i.e., there exists some countable set $D \subseteq T$ which is dense into (T,d). Hence, if $(X(t))_{t \in T}$ is a.s. continuous $\sup_{t \in T} X(t) = \sup_{t \in D} X(t)$ a.s. and again we can speak of $\mathbb{E}[\sup_{t \in T} X(t)] = \mathbb{E}[\sup_{t \in D} X(t)]$ and one also has

$$\mathbb{E}\left[\sup_{t \in T} X(t)\right] = \sup\left\{\mathbb{E}\left[\sup_{t \in F} X(t)\right] ; \ F \subseteq T \text{ and } F \text{ finite}\right\}. \tag{3.1}$$

Note that the same conclusions would hold true if d were replaced by another distance τ. As we shall see in the sequel, the special role of d arises from the fact that it is possible to completely characterize the existence of a continuous and bounded version of $(X(t))_{t \in T}$ in terms of the "geometry" of T with respect to d. Moreover, we shall also discuss the continuity question with respect to another distance than d and show that it is essentially enough to solve the problem for the intrinsic pseudometric. At a more superficial level, let us also notice that since the use of d as the referent pseudometric on T warrants the continuity of $(X(t))_{t \in T}$ in $\mathbb{L}_2$, the separability of (T,d) by itself tells something *without* requiring the a.s. continuity. Indeed the following elementary but useful remark holds true.

Lemma 3.1 *Assume that $D \subseteq T$ is countable and dense into (T,d), then for any $t \in T$, there exists some sequence (t_n) of elements of D such that (t_n) converges to t as n goes to infinity and $(X(t_n))$ converges to $X(t)$ a.s.*

Proof. Just choose t_n such that $d^2(t_n, t) \leq n^{-2}$, then the result follows by Bienaymé–Chebycheff's inequality and the Borel–Cantelli lemma. ∎

Hence, if one wants to build a version $(\widetilde{X}(t))_{t \in T}$ of $(X(t))_{t \in T}$ with a.s. uniformly continuous sample paths on T, it is enough to show that (under some appropriate condition to be studied below) $(X(t))_{t \in D}$ has this property on D and define $(\widetilde{X}(t))_{t \in T}$ by extending each uniformly continuous sample path of $(X(t))_{t \in D}$ on D to a uniformly continuous sample path on T. Then, Lemma 3.1 ensures that $\widetilde{X}$ is a modification (thus a version) of X which is of course a.s. uniformly continuous by construction. Another consequence of Lemma 3.1 is that even if $\sup_{t \in T} X(t)$ is not necessarily measurable, $(X(t))_{t \in T}$ admits an *essential measurable supremum*.

Proposition 3.2 *Assume that (T,d) is separable, then there exists an almost surely unique random variable Z such that $X(t) \leq Z$ a.s. for all $t \in T$ and* *such that if U is a random variable sharing the same domination property, then $Z \leq U$ a.s. The random variable Z is called the essential supremum of $(X(t))_{t \in T}$. Moreover if $\sup_{t \in T} X(t)$ happens to be finite a.s., so is Z.*

Proof. Indeed, in view of Lemma 3.1, $Z = \sup_{t \in D} X(t)$ is the a.s. unique candidate and is of course a.s. finite when so is $\sup_{t \in T} X(t)$. ∎

One derives from the existence of a finite essential supremum, the following easy necessary condition for the sample boundedness of $(X(t))_{t \in T}$.

Corollary 3.3 *Assume that (T, d) is separable, if $\sup_{t \in T} X(t)$ is almost surely finite then $\sup_{t \in T} \mathbb{E}\left[X^2(t)\right] < \infty$.*

Proof. Let t be given such that $\sigma_t = \left(\mathbb{E}\left[X^2(t)\right]\right)^{1/2} > 0$. Since $\sup_{t \in T} X(t)$ is almost surely finite we can take Z as in Proposition 3.2 above and choosing z such that $\mathbb{P}[Z > z] \leq 1/4$, if we denote by Φ the cumulative distribution function of the standard normal distribution, we derive from the inequalities

$$\frac{3}{4} \leq \mathbb{P}[Z \leq z] \leq \mathbb{P}[X(t) \leq z] = \Phi\left(\frac{z}{\sigma_t}\right),$$

that $\sigma_t \leq z/\Phi^{-1}(3/4)$. ∎

As a conclusion, in view of the considerations about the process of increments and of (3.1), it appears that the hard work to study the continuity question is to get dimension free bounds on the supremum of the components of a Gaussian random vector. One reason for focusing on $\sup_{t \in T} X(t)$ rather than $\sup_{t \in T} |X(t)|$ is that one can freely use the formula

$$\mathbb{E}\left[\sup_{t \in T}(X(t) - Y)\right] = \mathbb{E}\left[\sup_{t \in T} X(t)\right]$$

for any centered random variable Y. Another deeper reason is that the comparison results (such as Slepian's lemma below) hold for one-sided suprema only. At last let us notice that by symmetry, since for every $t_0 \in T$

$$\mathbb{E}\left[\sup_{t \in T} |X(t)|\right] \leq \mathbb{E}[|X(t_0)|] + \mathbb{E}\left[\sup_{s,t \in T}(X(t) - X(s))\right]$$

$$\leq \mathbb{E}[|X(t_0)|] + 2\mathbb{E}\left[\sup_{t \in T} X(t)\right],$$

$\mathbb{E}[\sup_{t \in T} |X(t)|]$ cannot be essentially larger than $\mathbb{E}[\sup_{t \in T} X(t)]$. Chapter 3 is organized as follows. Focusing on the case where T is finite, i.e., on Gaussian random vectors,

- we study the concentration of the random variables $\sup_{t \in T} |X(t)|$ or $\sup_{t \in T} X(t)$ around their expectations and more generally the concentration of any Lipschitz function on the Euclidean space equipped with the standard Gaussian measure,

- we prove Slepian's comparison Lemma which leads to the Sudakov minoration for $\mathbb{E}\left[\sup_{t \in T} X(t)\right]$.

Then, turning to the general case we provide necessary and sufficient conditions for the sample boundedness and continuity in terms of metric entropy (we shall also state without proof the Fernique–Talagrand characterization in terms of majorizing measures). Finally we introduce the isonormal process which will be used in the next chapter devoted to Gaussian model selection. We do not pretend at all to provide here a complete overview on the topic of Gaussian processes. Our purpose is just to present some of the main ideas of the theory with a biased view towards the finite dimensional aspects. Our main sources of inspiration have been [79], [57] and [1], where the interested reader will find much more detailed and exhaustive treatments of the subject.

3.2 Concentration of the Gaussian Measure on $\mathbb{R}^N$

The concentration of measure phenomenon for product measures has been investigated in depth by M. Talagrand in a most remarkable series of works (see in particular [112] for an overview and [113] for recent advances). One of the first striking results illustrating this phenomenon has been obtained in the seventies. It is the concentration of the standard Gaussian measure on $\mathbb{R}^N$.

Theorem 3.4 *Consider some Lipschitz function ζ on the Euclidean space $\mathbb{R}^N$ with Lipschitz constant L, if P denotes the canonical Gaussian measure on $\mathbb{R}^N$, then, for every $x \geq 0$*

$$P\left[|\zeta - M| \geq x\right] \leq 2 \exp\left(-\frac{x^2}{2L^2}\right) \tag{3.2}$$

and

$$P\left[\zeta \geq M + x\right] \leq \exp\left(-\frac{x^2}{2L^2}\right), \tag{3.3}$$

where M denotes either the mean or the median of ζ with respect to P.

Usually the first inequality is called a concentration inequality while the latter is called a deviation inequality. These inequalities are due to Borell [29] when M is a median (the same result has been published independently by Cirelson and Sudakov [37]) and to Cirelson, Ibragimov and Sudakov [36] when M is the mean. We refer to [76] for various proofs and numerous applications of these statements, we shall content ourselves here to give a complete proof of (3.3) when M is the mean via the Gaussian Logarithmic Sobolev inequality. As a matter of fact this proof leads to the slightly stronger result that the moment generating function of $\zeta - E_P\left[\zeta\right]$ is sub-Gaussian. Namely we shall prove the following result.

Proposition 3.5 *Consider some Lipschitz function ζ on the Euclidean space $\mathbb{R}^N$ with Lipschitz constant L, if P denotes the canonical Gaussian measure on $\mathbb{R}^N$, then*

$$\ln E_P \left[\exp\left(\lambda\left(\zeta - E_P\left[\zeta\right]\right)\right)\right] \leq \frac{\lambda^2 L^2}{2}, \text{ for every } \lambda \in \mathbb{R}. \qquad (3.4)$$

In particular one also has

$$Var_P\left[\zeta\right] \leq L^2 \qquad (3.5)$$

One could think to derive the Poincaré type inequality (3.5) directly from Theorem 3.4 by integration. This technique leads to the same inequality but with a worse constant. More generally, integrating (3.2) one easily gets for every integer q

$$\left(E_P \left|\zeta - E_P\left[\zeta\right]\right|^q\right)^{1/q} \leq C\sqrt{q}L,$$

where C is some absolute constant (see [78] for details about that and a "converse" assertion).

3.2.1 The Isoperimetric Nature of the Concentration Phenomenon

The isoperimetric inequality means something to any mathematician. Something different of course depending on his speciality and culture. For sure, whatever he knows about the topic, he has in mind a statement which is close to the following: "among all compact sets A in the N-dimensional Euclidean space with smooth boundary and with fixed volume, Euclidean balls are the one with minimal surface."

The History of the Concept

As it stands this statement is neither clearly connected to the concentration of any measure nor easy to generalize to abstract metric spaces say. Fortunately the Minkowski content formula (see [60]) allows to interpret the surface of the boundary of A as

$$\liminf_{\varepsilon \to 0} \frac{1}{\varepsilon}\left(\lambda\left(A^\varepsilon\right) - \lambda\left(A\right)\right)$$

where λ denotes the Lebesgue measure and A^ε the ε-neighborhood (or enlargement) of A with respect to the Euclidean distance d,

$$A^\varepsilon = \left\{x \in \mathbb{R}^N : d\left(x, A\right) < \varepsilon\right\}.$$

This leads (see Federer [60] for instance) to the equivalent formulation for the classical isoperimetric statement: "Given a compact set A and a Euclidean ball B with the same volume, $\lambda\left(A^\varepsilon\right) \geq \lambda\left(B^\varepsilon\right)$ for all $\varepsilon > 0$." In this version the measure λ and the Euclidean distance play a fundamental role. Of course the Lebesgue measure is not bounded and we shall get closer to the heart of the matter by considering the somehow less universally known but more "probabilistic" isoperimetric theorem for the sphere which is usually referred as Lévy's isoperimetric theorem although it has apparently been proved by Lévy and Schmidt independently (see [82], [106] and [64] for extensions of Lévy's proof to Riemannian manifolds with positive curvature). Again this theorem can be stated in two equivalent ways. We just present the statement which enlightens the role of the distance and the measure (see [63]): "Let S_ρ^N be the sphere of radius ρ in $\mathbb{R}^{N+1}$, P be the rotation invariant probability measure on S_ρ^N. For any measurable subset A of S_ρ^N, if B is a geodesic ball (i.e., a cap) with the same measure as A then, for every positive ε

$$P\left(A^\varepsilon\right) \geq P\left(B^\varepsilon\right),$$

where the ε-neighborhoods A^ε and B^ε are taken with respect to the geodesic distance on the sphere." The concentration of measure principle precisely arises from this statement. Indeed from an explicit computation of the measure of B^ε for a cap B with measure $1/2$ one derives from the isoperimetric theorem that for any set A with $P\left(A\right) \geq 1/2$

$$P\left(S_\rho^N \backslash A^\varepsilon\right) \leq \exp\left(-\frac{(N-1)}{\rho^2} \frac{\varepsilon^2}{2}\right).$$

In other words, as soon as $P\left(A\right) \geq 1/2$, the measure of A^ε grows very fast as a function of ε. This is the *concentration of measure phenomenon* which is analyzed and studied in details by Michel Ledoux in his recent book [78].

A General Formulation

Following [79], let us define the notion of concentration function and explain how it is linked to concentration inequalities for Lipschitz functions on a metric space.

Let us consider some metric space $(\mathbb{X}, d)$ and a continuous functional $\zeta : \mathbb{X} \rightarrow \mathbb{R}$. Given some probability measure P on $(\mathbb{X}, d)$, one is interested in controlling the deviation probabilities

$$P\left(\zeta \geq M + x\right) \quad \text{or} \quad P\left(|\zeta - M| \geq x\right)$$

where M is a median of ζ with respect to P. If the latter probability is small it expresses a concentration of ζ around the median M. Given some Borel set A, let for all positive ε, A^ε denote as usual the ε-neighborhood of A

$$A^\varepsilon = \{x \in \mathbb{X} : d\left(x, A\right) < \varepsilon\}.$$

If ζ is a Lipschitz function we define its Lipschitz seminorm

$$\|\zeta\|_L = \sup_{x \neq y} \frac{|\zeta\left(x\right) - \zeta\left(y\right)|}{d\left(x, y\right)},$$

and choose $A = \{\zeta \leq M\}$ so that for all $x \in A^\varepsilon$

$$\zeta(x) < M + \|\zeta\|_L \varepsilon.$$

Therefore

$$P(\zeta \geq M + \|\zeta\|_L \varepsilon) \leq P(\mathbb{X} \backslash A^\varepsilon) = P[d(.,A) \geq \varepsilon].$$

We can now forget about what is exactly the set A and just retain the fact that it is of probability at least $1/2$. Indeed, denoting by $d(.,A)$ the function $x \to d(x,A)$ and defining

$$\gamma(\varepsilon) = \sup \left\{ P[d(.,A) \geq \varepsilon] \mid A \text{ is a Borel set with } P(A) \geq \frac{1}{2} \right\}, \qquad (3.6)$$

we have

$$P(\zeta \geq M + \|\zeta\|_L \varepsilon) \leq \gamma(\varepsilon).$$

Setting $\varepsilon = x/\|\zeta\|_L$ we derive that for all positive x

$$P(\zeta \geq M + x) \leq \gamma \left(\frac{x}{\|\zeta\|_L} \right) \qquad (3.7)$$

and changing ζ into $-\zeta$,

$$P(\zeta \leq M - x) \leq \gamma \left(\frac{x}{\|\zeta\|_L} \right). \qquad (3.8)$$

Combining these inequalities of course implies the concentration inequality

$$P(|\zeta - M| \geq x) \leq 2\gamma \left(\frac{x}{\|\zeta\|_L} \right) \qquad (3.9)$$

The conclusion is that if one is able to control the "concentration function" γ as in the "historical case" of the sphere described above, then one immediately gets a concentration inequality for any Lipschitz function through (3.9).

An important idea brought by Talagrand is that the concentration function γ can happen to be controlled without an exact determination of the extremal sets as it was the case for the sphere. This approach allowed him to control the concentration function of general product measures. For Gaussian measures however, one can completely solve the isoperimetric problem.

3.2.2 The Gaussian Isoperimetric Theorem

If we take $(\mathbb{X}, d)$ to be the Euclidean space $\mathbb{R}^N$ and P to be the standard Gaussian probability measure we indeed get a spectacular example for which the above program can be successfully applied.

Indeed in that case, the isoperimetric problem is connected to that of the sphere via the Poincaré limit procedure. It is completely solved by the following theorem due to Borell, Cirel'son and Sudakov (see [29] and also [37] where the same result has been published independently).

Theorem 3.6 (*Gaussian isoperimetric Theorem*) *Let P be the standard Gaussian measure on the Euclidean space $\left(\mathbb{R}^N, d\right)$. For any Borel set A, if H is a half-space with $P[A] = P[H]$, then $P[d(., A) \geq \varepsilon] \leq P[d(., H) \geq \varepsilon]$ for all positive ε.*

We refer to [76] for a proof of this theorem (see also [78] for complements and a discussion about Ehrhardt's direct approach to the Gaussian isoperimetric Theorem). Theorem 3.6 readily implies an exact computation for the concentration function as defined by (3.6). Indeed if we define the standard normal tail function $\overline{\Phi}$ by

$$\overline{\Phi}(x) = \frac{1}{\sqrt{2\pi}} \int_x^{+\infty} e^{-u^2/2} du,$$

we can consider for a given Borel set A the point x_A such that $1 - \overline{\Phi}(x_A) = P[A]$. Then, taking H to be the half-space $\mathbb{R}^{N-1} \times (-\infty, x_A)$ we see that

$$P[A] = P[H] \quad \text{and} \quad P[d(., H) \geq \varepsilon] = \overline{\Phi}(x_A + \varepsilon).$$

Now, if $P(A) \geq 1/2$, then $x_A \geq 0$ and therefore $P[d(., H) \geq \varepsilon] \leq \overline{\Phi}(\varepsilon)$. Hence Theorem 3.6 implies that the concentration function γ of the standard Gaussian measure P (as defined by 3.6) is exactly equal to the standard Gaussian tail function $\overline{\Phi}$ so that one gets the following Corollary via (3.7) and (3.9).

Corollary 3.7 *Let P be the standard Gaussian measure on the Euclidean space $\mathbb{R}^N$ and ζ be any real valued Lipschitz function on $\mathbb{R}^N$. If M denotes a median of ζ, then*

$$P[\zeta \geq M + x] \leq \overline{\Phi}\left(\frac{x}{\|\zeta\|_L}\right)$$

and

$$P[|\zeta - M| \geq x] \leq 2\overline{\Phi}\left(\frac{x}{\|\zeta\|_L}\right)$$

for all positive x.

Remark. These inequalities are sharp and unimprovable since if we take $\zeta(x_1, \ldots, x_N) = x_1$, they become in fact identities. Moreover, they a fortiori imply Theorem 3.4 when M is the median by using the standard upper bound for the Gaussian tail

$$\overline{\Phi}(u) \leq \frac{1}{2} \exp\left(-\frac{u^2}{2}\right) \quad \text{for all positive } u. \tag{3.10}$$

The inequalities of Corollary 3.7 are often more convenient to use with M being the mean rather than the median. Thus let us try to derive concentration inequalities around the mean from Corollary 3.7. We have

$$E_P\left[\zeta - M\right]_+ = \int_0^{+\infty} P\left[\zeta - M > x\right] dx \leq \|\zeta\|_L \int_0^{+\infty} \overline{\Phi}(u)\, du.$$

Integrating by parts, we get

$$E_P\left[\zeta\right] - M \leq E_P\left[\zeta - M\right]_+ \leq \|\zeta\|_L \left(\frac{1}{\sqrt{2\pi}} \int_0^{+\infty} u e^{-u^2/2} du\right) \leq \frac{1}{\sqrt{2\pi}} \|\zeta\|_L,$$

hence, by symmetry

$$\left|E_P\left[\zeta\right] - M\right| \leq \frac{1}{\sqrt{2\pi}} \|\zeta\|_L.$$

Thus, for any positive x

$$P\left[|\zeta - E_P\left[\zeta\right]| > x\right] \leq 2\overline{\Phi}\left(\frac{x}{\|\zeta\|_L} - \frac{1}{\sqrt{2\pi}}\right)$$

As a matter of fact one can do better by a direct attack of the problem. Cirelson, Ibragimov and Sudakov (see [36]) use a very subtle imbedding argument in the Brownian motion. More precisely they prove that $\zeta - E_P(\zeta)$ has the same distribution as B_τ, where $(B_t)_{t\geq 0}$ is a standard Brownian motion and τ is a stopping time which is almost surely bounded by $\|\zeta\|_L$. Now a classical result of Paul Lévy ensures that for all positive x

$$P\left[\sup_{t\leq\|\zeta\|_L} B_t \geq x\right] = 2\overline{\Phi}\left(\frac{x}{\|\zeta\|_L}\right).$$

and by symmetry the imbedding argument implies

Theorem 3.8 *Let P be the standard Gaussian measure on the Euclidean space $\mathbb{R}^N$ and ζ be any real valued Lipschitz function on $\mathbb{R}^N$. Then*

$$P\left[\zeta \geq E_P\left[\zeta\right] + x\right] \leq 2\overline{\Phi}\left(\frac{x}{\|\zeta\|_L}\right)$$

and

$$P\left[|\zeta - E_P(\zeta)| \geq x\right] \leq 4\overline{\Phi}\left(\frac{x}{\|\zeta\|_L}\right)$$

for all positive x.

By using again the classical upper bound (3.10) one derives easily from Theorem 3.8 Inequality (3.3) of Theorem 3.4 when M is the mean. We do not provide the proof of this theorem because of its too specifically Gaussian flavor. Instead we focus on its consequence Inequality (3.3). As a matter of fact an alternative way of proving this deviation inequality consists in solving the differential inequality for the moment generating function of $\zeta - E_P\left[\zeta\right]$ that derives from a Logarithmic Sobolev inequality. This proof is given in details in the next section and is much more illustrative of what will be achievable in much more general frameworks than the Gaussian one.

3.2.3 Gross' Logarithmic Sobolev Inequality

The connection between the concentration of measure phenomenon and Logarithmic Sobolev inequalities relies on Herbst argument (see Proposition 2.14). Let us state Gross' Logarithmic Sobolev inequality (see [65]) for the standard Gaussian measure on $\mathbb{R}^N$, show how it implies (3.3) and then prove it.

Theorem 3.9 (*Gross' inequality*) Let P be the standard Gaussian measure on the Euclidean space $\mathbb{R}^N$ and u be any continuously differentiable function on $\mathbb{R}^N$. Then

$$Ent_P\left[u^2\right] \leq 2E_P\left(\|\nabla u\|^2\right). \tag{3.11}$$

If we now consider some Lipschitz function ζ on the Euclidean space $\mathbb{R}^N$ with Lipschitz constant L and if we furthermore assume ζ to be continuously differentiable, we have for all x in $\mathbb{R}^N$, $\|\nabla\zeta(x)\| \leq L$ and given $\lambda > 0$, we can apply (3.11) to $u = e^{\lambda\zeta/2}$. Since for all x in $\mathbb{R}^N$ we have

$$\|\nabla u(x)\|^2 = \frac{\lambda^2}{4}\|\nabla\zeta(x)\|^2 e^{\lambda\zeta(x)} \leq \frac{\lambda^2 L^2}{4}e^{\lambda\zeta(x)},$$

we derive from (3.11) that

$$\mathrm{Ent}_P\left[e^{\lambda\zeta}\right] \leq \frac{\lambda^2 L^2}{2}E_P\left[e^{\lambda\zeta}\right]. \tag{3.12}$$

This inequality holds for all positive λ and therefore Herbst argument (see Proposition 2.14) yields for any positive λ

$$E_P\left[e^{\lambda(\zeta - E_P(\zeta))}\right] \leq \exp\left(\frac{\lambda^2 L^2}{2}\right). \tag{3.13}$$

Using a regularization argument (by convolution), this inequality remains valid when ζ is only assumed to be Lipschitz and (3.3) follows by Chernoff's inequality.

We turn now to the proof of Gross' Logarithmic Sobolev inequality. There exists several proofs of this subtle inequality. The original proof of Gross is rather long and intricate while the proof that one can find for instance in [76] is much shorter but uses the stochastic calculus machinery. The proof that we present here is borrowed from [5]. It relies on the tensorization principle for entropy and uses only elementary arguments. Indeed we recall that one derives from the tensorization inequality for entropy (2.46) that it is enough to prove (3.11) when the dimension N is equal to 1. So that the problem reduces to show that

$$\mathrm{Ent}_P\left[u^2\right] \leq 2E_P\left(u'^2\right). \tag{3.14}$$

when P denotes the standard Gaussian measure on the real line. We start by proving a related result for the symmetric Bernoulli distribution which will rely on the following elementary inequality.

Lemma 3.10 *Let h be the convex function $x \longrightarrow (1+x)\ln(1+x) - x$ on $[-1, +\infty)$, then, for any $x \in [-1, +\infty)$ the following inequality holds*

$$h(x) - 2h\left(\frac{x}{2}\right) \leq \left(\sqrt{1+x} - 1\right)^2. \tag{3.15}$$

Proof. We simply consider the difference between the right hand side of this inequality and the left hand side

$$\psi(x) = \left(\sqrt{1+x} - 1\right)^2 - h(x) + 2h\left(\frac{x}{2}\right).$$

Then $\psi(0) = \psi'(0) = 0$ and for any $x \in [-1, +\infty)$

$$\psi''(x) = \frac{1}{2}(1+x)^{-3/2} - h''(x) + \frac{1}{2}h''\left(\frac{x}{2}\right)$$

$$= \frac{1}{2}(1+x)^{-3/2}(2+x)^{-1}\left(2 + x - 2\sqrt{1+x}\right) \geq 0.$$

Hence ψ is a convex and nonnegative function. ∎

We can now prove the announced result for the symmetric Bernoulli distribution.

Proposition 3.11 *Let ε be a Rademacher random variable, i.e., $\mathbb{P}[\varepsilon = +1] = \mathbb{P}[\varepsilon = -1] = 1/2$, then for any real valued function f on $\{-1, +1\}$,*

$$Ent_{\mathbb{P}}\left[f^2(\varepsilon)\right] \leq \frac{1}{2}(f(1) - f(-1))^2. \tag{3.16}$$

Proof. Since the inequality is trivial when f is constant and since

$$|f(1) - f(-1)| \geq ||f(1)| - |f(-1)||,$$

we can always assume that f is nonnegative and that either $f(1)$ or $f(-1)$ is positive. Assuming for instance that $f(-1) > 0$ we derive that (3.16) is equivalent to

$$2Ent_{\mathbb{P}}\left[\frac{f^2(\varepsilon)}{f^2(-1)}\right] \leq \left(\frac{f(1)}{f(-1)} - 1\right)^2. \tag{3.17}$$

Now setting $f(1)/f(-1) = \sqrt{1+x}$, we notice that the left hand side of (3.17) equals $h(x) - 2h(x/2)$ while the right-hand side equals $\left(\sqrt{1+x} - 1\right)^2$. Hence (3.15) implies that (3.17) is valid and the result follows. ∎

We turn now to the proof of (3.14). We first notice that it is enough to prove (3.14) when u has a compact support and is twice continuously differentiable. Let $\varepsilon_1, \ldots, \varepsilon_n$ be independent Rademacher random variables. Applying the tensorization inequality (2.46) again and setting

$$S_n = n^{-1/2}\sum_{j=1}^{n}\varepsilon_j,$$

we derive from Proposition 3.11 that

$$\operatorname{Ent}_{\mathbb{P}}\left[u^2\left(S_n\right)\right] \leq \frac{1}{2}\sum_{i=1}^{n}\mathbb{E}\left[\left(u\left(S_n+\frac{1-\varepsilon_i}{\sqrt{n}}\right)-u\left(S_n-\frac{1+\varepsilon_i}{\sqrt{n}}\right)\right)^2\right].$$

(3.18)

The Central Limit Theorem implies that S_n converges in distribution to X, where X has the standard normal law. Hence the left hand side of (3.18) converges to $\operatorname{Ent}_{\mathbb{P}}\left[u^2\left(X\right)\right]$. Let K denote the supremum of the absolute value of the second derivative of u. We derive from Taylor's formula that for every i

$$\left|u\left(S_n+\frac{1-\varepsilon_i}{\sqrt{n}}\right)-u\left(S_n-\frac{1+\varepsilon_i}{\sqrt{n}}\right)\right| \leq \frac{2}{\sqrt{n}}\left|u'\left(S_n\right)\right|+\frac{2K}{n}$$

and therefore

$$\frac{n}{4}\left(u\left(S_n+\frac{1-\varepsilon_i}{\sqrt{n}}\right)-u\left(S_n-\frac{1+\varepsilon_i}{\sqrt{n}}\right)\right)^2 \leq u'^2\left(S_n\right)+\frac{2K}{\sqrt{n}}\left|u'\left(S_n\right)\right|$$
$$+\frac{K^2}{n}. \qquad (3.19)$$

One derives from (3.19), and the Central Limit Theorem that

$$\overline{\lim_{n}}\frac{1}{2}\sum_{i=1}^{n}\mathbb{E}\left[\left(u\left(S_n+\frac{1-\varepsilon_i}{\sqrt{n}}\right)-u\left(S_n-\frac{1+\varepsilon_i}{\sqrt{n}}\right)\right)^2\right] \leq 2\mathbb{E}\left[u'^2\left(X\right)\right],$$

which means that (3.18) leads to (3.14) by letting n go to infinity. ∎

3.2.4 Application to Suprema of Gaussian Random Vectors

A very remarkable feature of these inequalities for the standard Gaussian measure on $\mathbb{R}^N$ is that they do not depend on the dimension N. This allows to extend them easily to an infinite dimensional setting (see [76] for various results about Gaussian measures on separable Banach spaces). We are mainly interested in controlling suprema of Gaussian processes. Under appropriate separability assumptions this problem reduces to a finite dimensional one and therefore the main task is to deal with Gaussian random vectors rather than Gaussian processes.

Theorem 3.12 *Let X be some centered Gaussian random vector on $\mathbb{R}^N$. Let $\sigma \geq 0$ be defined as*

$$\sigma^2 = \sup_{1\leq i\leq N}\left(\mathbb{E}\left[X_i^2\right]\right)$$

and Z denote either $\sup_{1\leq i\leq N}X_i$ or $\sup_{1\leq i\leq N}|X_i|$. Then,

$$\ln\mathbb{E}\left[\exp\left(\lambda\left(Z-\mathbb{E}\left[Z\right]\right)\right)\right] \leq \frac{\lambda^2\sigma^2}{2}, \text{ for every } \lambda \in \mathbb{R}, \qquad (3.20)$$

leading to

$$\mathbb{P}\left[Z - \mathbb{E}\left[Z\right] \geq \sigma\sqrt{2x}\right] \leq \exp\left(-x\right) \tag{3.21}$$

and

$$\mathbb{P}\left[\mathbb{E}\left[Z\right] - Z \geq \sigma\sqrt{2x}\right] \leq \exp\left(-x\right) \tag{3.22}$$

for all positive x. Moreover, the following bound for the variance of Z is available

$$\mathrm{Var}\left[Z\right] \leq \sigma^2. \tag{3.23}$$

Proof. Let Γ be the covariance matrix of the centered Gaussian vector X. Denoting by A the square root of the nonnegative matrix Γ, we define for all $u \in \mathbb{R}^N$

$$\zeta\left(u\right) = \sup_{i \leq N}\left(Au\right)_i.$$

As it is well known the distribution of $\sup_{1 \leq i \leq N} X_i$ is the same as that of ζ under the standard Gaussian law on $\mathbb{R}^N$. Hence we can apply Proposition 3.5 by computing the Lipschitz constant of ζ on the Euclidean space $\mathbb{R}^N$. We simply write that, by Cauchy–Schwarz inequality, we have for all $i \leq N$

$$\left|\left(Au\right)_i - \left(Av\right)_i\right| = \left|\sum_j A_{i,j}\left(u_j - v_j\right)\right| \leq \left(\sum_j A_{i,j}^2\right)^{1/2} \|u - v\|,$$

hence, since $\sum_j A_{i,j}^2 = \mathrm{Var}(X_i)$ we get

$$\left|\zeta\left(u\right) - \zeta\left(v\right)\right| \leq \sup_{i \leq N}\left|\left(Au\right)_i - \left(Av\right)_i\right| \leq \sigma\|u - v\|.$$

Therefore ζ is Lipschitz with $\|\zeta\|_L \leq \sigma$. The case where

$$\zeta\left(u\right) = \sup_{i \leq N}\left|\left(Au\right)_i\right|$$

for all $u \in \mathbb{R}^N$ leads to the same conclusion. Hence (3.20) follows from (3.4) yielding (3.21) and (3.22). Similarly, (3.23) derives from (3.5). ■

It should be noticed that inequalities (3.21) and (3.22) would remain true for the median instead of the mean (simply use Corollary 3.7 instead of (3.4) at the end of the proof). This tells us that median and mean of Z must be close to each other. Indeed, denoting by $\mathrm{Med}[Z]$ a median of Z and taking $x = \ln\left(2\right)$ in (3.21) and (3.22) implies that

$$\left|\mathrm{Med}\left[Z\right] - \mathbb{E}\left[Z\right]\right| \leq \sigma\sqrt{2\ln\left(2\right)}. \tag{3.24}$$

Although we have put the emphasis on concentration inequalities around the mean rather than the median, arguing that the mean is usually easier to manage than the median, there exists a situation where the median is exactly computable while getting an explicit expression for the mean is more delicate. We think here to the case where $X_1, \ldots, X_N$ are i.i.d. standard normal variables. In this case

$$\operatorname{Med}[Z] = \overline{\Phi}^{-1}\left(1 - 2^{-1/N}\right) \geq \overline{\Phi}^{-1}\left(\frac{\ln(2)}{N}\right)$$

and therefore by (3.24)

$$\mathbb{E}[Z] \geq \overline{\Phi}^{-1}\left(\frac{\ln(2)}{N}\right) - \sqrt{2\ln(2)}. \tag{3.25}$$

In order to check that this lower bound is sharp, let us see what Lemma 2.3 gives when applied to our case. By (2.3) we get $\mathbb{E}[Z] \leq \sqrt{2\ln(N)}$. Since $\overline{\Phi}^{-1}(\varepsilon) \sim \sqrt{2|\ln(\varepsilon)|}$ when ε goes to 0, one derives that

$$\mathbb{E}[Z] \sim \sqrt{2\ln(N)} \text{ as } N \text{ goes to infinity}, \tag{3.26}$$

which shows that (3.25) and also (2.3) are reasonably sharp.

3.3 Comparison Theorems for Gaussian Random Vectors

We present here some of the classical comparison theorems for Gaussian random vectors. Our aim is to be able to prove Sudakov's minoration on the expectation of the supremum of the components of a Gaussian random vector by comparison with the extremal i.i.d. case.

3.3.1 Slepian's Lemma

The proof is adapted from [79].

Lemma 3.13 (*Slepian's lemma*) *Let X and Y be some centered Gaussian random vectors in $\mathbb{R}^N$. Assume that*

$$\mathbb{E}[X_i X_j] \leq \mathbb{E}[Y_i Y_j] \text{ for all } i \neq j \tag{3.27}$$

$$\mathbb{E}[X_i^2] = \mathbb{E}[Y_i^2] \text{ for all } i \tag{3.28}$$

then, for every x

$$\mathbb{P}\left[\sup_{1 \leq i \leq N} Y_i > x\right] \leq \mathbb{P}\left[\sup_{1 \leq i \leq N} X_i > x\right]. \tag{3.29}$$

In particular, the following comparison is available

$$\mathbb{E}\left[\sup_{1 \leq i \leq N} Y_i\right] \leq \mathbb{E}\left[\sup_{1 \leq i \leq N} X_i\right]. \tag{3.30}$$

Proof. We shall prove a stronger result than (3.29). Namely we intend to show that

$$\mathbb{E}\left[\prod_{i=1}^{N} f(X_i)\right] \leq \mathbb{E}\left[\prod_{i=1}^{N} f(Y_i)\right]. \tag{3.31}$$

for every nonnegative and nonincreasing differentiable f such that f and f' are bounded on $\mathbb{R}$. Since one can design a sequence (f_n) of such functions which are uniformly bounded by 1 and such that (f_n) converges pointwise to $\mathbb{1}_{(-\infty, x]}$, by dominated convergence (3.31) leads to

$$\mathbb{P}\left[\sup_{1 \leq i \leq N} X_i \leq x\right] \leq \mathbb{P}\left[\sup_{1 \leq i \leq N} Y_i \leq x\right]$$

and therefore to (3.29). In order to prove (3.31) we may assume X and Y to be independent. Let, for $t \in [0,1]$, $Z(t) = (1-t)^{1/2} X + t^{1/2} Y$ and $\rho(t) = \mathbb{E}\left[\prod_{i=1}^{N} f(Z_i(t))\right]$. Then, for every $t \in (0,1)$

$$\rho'(t) = \sum_{i=1}^{N} \mathbb{E}\left[Z_i'(t) f'(Z_i(t)) \prod_{j \neq i} f(Z_j(t))\right]. \tag{3.32}$$

Moreover, for every $i, j \in [1, N]$ and $t \in (0, 1)$

$$\mathbb{E}[Z_j(t) Z_i'(t)] = \frac{1}{2} \left(\mathbb{E}[Y_i Y_j] - \mathbb{E}[X_i X_j]\right) \geq 0$$

so that $\mathbb{E}[Z_j(t) \mid Z_i'(t)] = \theta_{i,j}(t) Z_i'(t)$ with $\theta_{i,j}(t) \geq 0$. Now, setting $Z_{i,j}(t) = Z_j(t) - \theta_{i,j}(t) Z_i'(t)$, we note that for every $\alpha \in \mathbb{R}^N$ and every $j \neq i$

$$\frac{\partial}{\partial \alpha_j} \mathbb{E}\left[Z_i'(t) f'(Z_i(t)) \prod_{k \neq i} f(Z_{i,k}(t) + \alpha_k Z_i'(t))\right]$$

$$= \mathbb{E}\left[Z_i'^2(t) f'(Z_i(t)) f'(Z_{i,j}(t) + \alpha_j Z_i'(t)) \prod_{k \neq i, k \neq j} f(Z_{i,k}(t) + \alpha_k Z_i'(t))\right]$$

$$\geq 0$$

and therefore

$$\mathbb{E}\left[Z_i'(t) f'(Z_i(t)) \prod_{j \neq i} f(Z_j(t))\right] \geq \mathbb{E}\left[Z_i'(t) f'(Z_i(t)) \prod_{j \neq i} f(Z_{i,j}(t))\right].$$

But $(Z_{i,j}(t))_{1\leq j\leq N}$ is independent from $Z_i'(t)$, with $Z_{i,i}(t) = Z_i(t)$ because $\theta_{i,i}(t) = 0$, thus

$$\mathbb{E}\left[Z_i'(t)\, f'(Z_i(t)) \prod_{j\neq i} f(Z_{i,j}(t))\right] = \mathbb{E}\left[Z_i'(t)\right]\mathbb{E}\left[f'(Z_i(t)) \prod_{j\neq i} f(Z_{i,j}(t))\right]$$
$$= 0,$$

which implies that each summand in (3.32) is nonnegative. Hence ρ is nondecreasing on $[0,1]$ and the inequality $\rho(0) \leq \rho(1)$ means that (3.31) holds. Of course (3.30) easily derives from (3.29) via the integration by parts formula

$$\mathbb{E}[Z] = \int_0^\infty (\mathbb{P}[Z > x] - \mathbb{P}[Z < -x])\, dx$$

which achieves the proof of Slepian's lemma. ∎

Slepian's lemma implies another comparison theorem which holds without the requirement that the components of X and Y have the same variances.

Theorem 3.14 *Let X and Y be some centered Gaussian random vectors in $\mathbb{R}^N$. Assume that*

$$\mathbb{E}\left[(Y_i - Y_j)^2\right] \leq \mathbb{E}\left[(X_i - X_j)^2\right] \text{ for all } i \neq j \tag{3.33}$$

then,

$$\mathbb{E}\left[\sup_{1\leq i\leq N} Y_i\right] \leq 2\mathbb{E}\left[\sup_{1\leq i\leq N} X_i\right]. \tag{3.34}$$

Proof. Note that $\mathbb{E}\left[\sup_{1\leq i\leq N} X_i\right] = \mathbb{E}\left[\sup_{1\leq i\leq N}(X_i - X_1)\right]$ and similarly $\mathbb{E}\left[\sup_{1\leq i\leq N} Y_i\right] = \mathbb{E}\left[\sup_{1\leq i\leq N}(Y_i - Y_1)\right]$. Hence, possibly replacing X and Y respectively by $(X_i - X_1)_{1\leq i\leq N}$ and $(Y_i - Y_1)_{1\leq i\leq N}$ which also satisfy to (3.33), we may assume that $X_1 = Y_1 = 0$. Assuming from now that this assumption holds, let us apply Lemma 3.13 to convenient modifications of X and Y. Setting $\sigma = \left(\sup_{1\leq i\leq N}\mathbb{E}\left[X_i^2\right]\right)^{1/2}$, let $\widetilde{X}$ and $\widetilde{Y}$ be defined by

$$\widetilde{X}_i = X_i + Z\sigma_i$$
$$\widetilde{Y}_i = Y_i + Z\sigma,$$

where $\sigma_i = \left(\sigma^2 + \mathbb{E}[Y_i^2] - \mathbb{E}[X_i^2]\right)^{1/2}$ and Z is a standard normal variable independent from X and Y. Then one has $\mathbb{E}\left[\widetilde{X}_i^2\right] = \sigma^2 + \mathbb{E}[Y_i^2] = \mathbb{E}\left[\widetilde{Y}_i^2\right]$ for every i and for every $i \neq j$, $\mathbb{E}\left[\left(\widetilde{Y}_i - \widetilde{Y}_j\right)^2\right] = \mathbb{E}\left[(Y_i - Y_j)^2\right]$ and $\mathbb{E}\left[\left(\widetilde{X}_i - \widetilde{X}_j\right)^2\right] \geq \mathbb{E}\left[(X_i - X_j)^2\right]$. Hence, the Gaussian random vectors $\widetilde{X}$

and $\widetilde{Y}$ satisfy to assumptions (3.28) and (3.33) and thus also to assumption (3.27). So, the hypothesis of Slepian's lemma are fulfilled and therefore

$$\mathbb{E}\left[\sup_{1\leq i\leq N} Y_i\right] = \mathbb{E}\left[\sup_{1\leq i\leq N} \widetilde{Y}_i\right] \leq \mathbb{E}\left[\sup_{1\leq i\leq N} \widetilde{X}_i\right].$$

It remains to relate $\mathbb{E}\left[\sup_{1\leq i\leq N} \widetilde{X}_i\right]$ to $\mathbb{E}\left[\sup_{1\leq i\leq N} X_i\right]$. By assumption (3.33) we know that $\mathbb{E}\left[Y_i^2\right] \leq \mathbb{E}\left[X_i^2\right]$ (remember that $X_1 = Y_1 = 0$) and therefore $\sigma_i \leq \sigma$, hence

$$\mathbb{E}\left[\sup_{1\leq i\leq N} \widetilde{X}_i\right] \leq \mathbb{E}\left[\sup_{1\leq i\leq N} X_i\right] + \sigma\mathbb{E}\left[Z^+\right]. \tag{3.35}$$

Now let i_0 be such that $\sigma = \left(\mathbb{E}\left[X_{i_0}^2\right]\right)^{1/2}$, then on the one hand $\mathbb{E}\left[X_{i_0}^+\right] = \sigma\mathbb{E}\left[Z^+\right]$ and on the other hand since $X_1 = 0$, $\mathbb{E}\left[X_{i_0}^+\right] \leq \mathbb{E}\left[\sup_{1\leq i\leq N} X_i\right]$. Hence (3.35) leads to (3.34). ■

We are now in position to prove a result known as the Sudakov minoration which completes the upper bounds for the suprema of Gaussian random variables established in Chapter 2.

Proposition 3.15 (*Sudakov's minoration*) *There exists some absolute positive constant C such that the following inequality holds for any centered Gaussian random vector X on $\mathbb{R}^N$*

$$\min_{i\neq j} \sqrt{\mathbb{E}\left[(X_i - X_j)^2\right]\ln(N)} \leq C\mathbb{E}\left[\sup_{1\leq i\leq N} X_i\right]. \tag{3.36}$$

Proof. Let us consider N i.i.d. standard normal random variables $Z_1,\ldots,Z_N$. Let

$$\delta = \min_{i\neq j}\left(\mathbb{E}\left[(X_i - X_j)^2\right]\right)^{1/2}$$

and

$$Y_i = \frac{\delta}{\sqrt{2}}Z_i, \text{ for every } i.$$

Since for every $i \neq j$, $\mathbb{E}\left[(Y_i - Y_j)^2\right] = \delta^2 \leq \mathbb{E}\left[(X_i - X_j)^2\right]$, we may apply Theorem 3.14. Hence

$$\delta\mathbb{E}\left[\sup_{1\leq i\leq N} Z_i\right] \leq 2\sqrt{2}\mathbb{E}\left[\sup_{1\leq i\leq N} X_i\right]$$

and it remains to show that for some absolute constant κ, one has (whatever N)

$$\mathbb{E}\left[\sup_{1\leq i\leq N} Z_i\right] \geq \kappa\sqrt{\ln(N)}. \tag{3.37}$$

We may assume that $N \geq 2$ (otherwise the inequality trivially holds) and therefore

$$\mathbb{E}\left[\sup_{1 \leq i \leq N} Z_i\right] = \mathbb{E}\left[\sup_{1 \leq i \leq N} (Z_i - Z_1)\right] \geq \mathbb{E}\left[(Z_2 - Z_1)^+\right] = \frac{1}{\sqrt{\pi}},$$

which in turn shows that we may assume N to be large enough. But we know from (3.25) that

$$\mathbb{E}\left[\sup_{1 \leq i \leq N} Z_i\right] \sim \sqrt{2 \ln (N)} \text{ as } N \text{ goes to infinity,}$$

thus (3.37) holds, completing the proof of the proposition. ∎

The analysis of the boundedness of the sample paths of a Gaussian process lies at the heart of the next section. Sudakov's minoration is essential for understanding in depth the role of the intrinsic metric structure of the set of parameter in this matter.

3.4 Metric Entropy and Gaussian Processes

Our purpose is here to investigate the conditions which warrant the sample boundedness or uniform continuity of a centered Gaussian process $(X(t))_{t \in T}$ with respect to the intrinsic pseudodistance defined by its covariance structure. Recall that this pseudodistance is defined by

$$d(s,t) = \left(\mathbb{E}\left[(X(t) - X(s))^2\right]\right)^{1/2}, \text{ for every } t \in T.$$

Since the uniform continuity means the study of the behavior of the supremum of the increment process $\{X(t) - X(s), d(s,t) \leq \sigma\}$, the problems of boundedness and uniform continuity are much more correlated than it should look at a first glance. The hard work will consist in controlling the supremum of a Gaussian process with given diameter. The role of the size of T as a metric space will be essential.

3.4.1 Metric Entropy

Metric entropy allows to quantify the "size" of a metric space. In the context of Gaussian processes, it has been introduced by Dudley (see [56]) in order to provide a sufficient condition for the existence of an almost surely continuous version of a Gaussian process. A completely general approach would consist in using majorizing measures as introduced by Fernique (see [61]) rather than metric entropy (we refer to [79] or [1] for an extensive study of this topic). However the metric entropy framework turns out to be sufficient to cover the examples that we have in view. This is the reason why we choose to pay the price of slightly loosing in generality in order to gain in simplicity for the proofs. If (S, d) is a totally bounded pseudometric space and $\delta > 0$ is given, a finite subset $\mathcal{S}_\delta$ of S with maximal cardinality such that for every distinct points s and t in $\mathcal{S}_\delta$ one has $d(s, t) > \delta$ is an δ-net, which means that the closed balls with radius δ which are centered on the points of $\mathcal{S}_\delta$ are covering (S, d). The cardinality $N(\delta, S)$ of $\mathcal{S}_\delta$ is called the δ-packing number (since it is measuring the maximal number of disjoint closed balls with radius $\delta/2$ that can be "packed" into S), while the minimal cardinality $N'(\delta, S)$ of an δ-net is called the δ-covering number (since it is measuring the minimal number of closed balls with radius δ which is necessary to cover S). Both quantities are measuring the *massiveness* of the totally bounded metric space (S, d). They really behave in the same way as δ goes to 0, since we have seen already that $N'(\delta, S) \leq N(\delta, S)$ and conversely $N'(\delta/2, S) \geq N(\delta, S)$ because if S is covered by a family of closed balls with radius $\delta/2$, each ball in this family contains at most a point of $\mathcal{S}_\delta$. Although, in the "Probability in Banach spaces" literature, δ-covering numbers are maybe more commonly used than their twin brothers δ-packing numbers, we prefer to work with packing numbers just because they increase when S increases. More precisely, while obviously $N(\delta, S) \leq N(\delta, T)$ whenever $S \subseteq T$, the same property does not necessarily hold for N' instead of N (think here to the position of the centers of the balls). Therefore we are defining below the metric entropy in a slightly unusual way but once again this is completely harmless for what follows because of the tight inequalities relating N to N'

$$N'(\delta, S) \leq N(\delta, S) \leq N'(\delta/2, S).$$

Definition 3.16 *Let (S, d) be a totally bounded pseudometric space. For any positive δ, let $N(\delta, S)$ denote the δ-packing number of (S, d). We define the δ-entropy number $H(\delta, S)$ by*

$$H(\delta, S) = \ln(N(\delta, S)).$$

We call $H(., S)$ the metric entropy of (S, d).

The role of metric entropy is clear from Sudakov's minoration. Indeed, assuming (T, d) to be separable, whenever $\sup_{t \in T} X(t)$ is a.s. finite, it comes from the preliminary remarks made in Section 3.1 that $(X(t))_{t \in T}$ admits an essential supremum Z and that $\sigma = \left(\sup_{t \in T} \mathbb{E}\left[X^2(t)\right]\right)^{1/2} < \infty$. One readily derives from (3.24) that for every finite subset F of T, one has $\mathbb{E}\left[\sup_{t \in F} X(t)\right] \leq \text{Med}[Z] + \sigma\sqrt{2\ln(2)}$. Hence Sudakov's minoration implies that (T, d) must be totally bounded with

$$\delta\sqrt{H(\delta, T)} \leq C \sup\left\{\mathbb{E}\left[\sup_{t \in F} X(t)\right]; F \subseteq T \text{ and } F \text{ finite}\right\} < \infty.$$

Our purpose is to build explicit exponential bounds for Gaussian processes and to prove Dudley's regularity criterion (which holds under a slightly stronger entropy condition than the one just above) simultaneously. This program is achieved through an explicit control of expectation of the supremum of a Gaussian process.

3.4.2 The Chaining Argument

To take into account the "size" of (T, d) it is useful to use the so-called chaining argument which goes back to Kolmogorov. Once again the main issue is to deal with the case where T is finite and get some estimates which are free from the cardinality of T. This can be performed if we take into account the covariance metric structure to use Lemma 2.3 in a clever way. We prove the following inequality, first for Gaussian random vectors and then extend it to separable Gaussian processes later on.

Theorem 3.17 *Let T be some finite set and $(X(t))_{t \in T}$ be some centered Gaussian process. Let $\sigma = \left(\sup_{t \in T} \mathbb{E}\left[X^2(t)\right]\right)^{1/2}$ and consider, for any positive δ, $H(\delta, T)$ to be the δ-entropy number of T equipped with the intrinsic covariance pseudometric d of $(X(t))_{t \in T}$. Then, for any $\varepsilon \in \left]0, 1\right]$ and any measurable set A with $\mathbb{P}[A] > 0$*

$$\mathbb{E}^A\left[\sup_{t \in T} X(t)\right] \leq \frac{12}{\varepsilon} \int_0^{\varepsilon \sigma} \sqrt{H(x, T)} dx + (1 + 3\varepsilon) \sigma \sqrt{2 \ln\left(\frac{1}{\mathbb{P}[A]}\right)}.$$

Proof. We write p for $\mathbb{P}[A]$ for short. For any integer j, we set $\delta_j = \varepsilon \sigma 2^{-j}$. We write H instead of $H(., T)$ for short. By definition of H, for any integer j we can define some mapping Π_j from T to T such that

$$\ln|\Pi_j(T)| \leq H(\delta_j)$$

and

$$d(t, \Pi_j t) \leq \delta_j \quad \text{for all } t \in T.$$

Since T is finite, there exists some integer J such that for all $t \in T$

$$X(t) = X(\Pi_0 t) + \sum_{j=0}^J X(\Pi_{j+1} t) - X(\Pi_j t),$$

from which we deduce that

$$\mathbb{E}^A\left[\sup_{t \in T} X(t)\right] \leq \mathbb{E}^A\left[\sup_{t \in T} X(\Pi_0 t)\right] + \sum_{j=0}^J \mathbb{E}^A\left[\sup_{t \in T} X(\Pi_{j+1} t) - X(\Pi_j t)\right].$$

Since $\ln|\Pi_0(T)| \leq H(\delta_0)$, we derive from (2.3) that

$$\mathbb{E}^A\left[\sup_{t \in T} X(\Pi_0 t)\right] \leq \frac{\sqrt{2}}{\varepsilon} \delta_0 \sqrt{H(\delta_0)} + \sigma \sqrt{2 \ln\left(\frac{1}{p}\right)}.$$

Moreover, for any integer j, $(\Pi_j t, \Pi_{j+1} t)$ ranges in a set with cardinality not larger than $\exp\left(2H\left(\delta_{j+1}\right)\right)$ when t varies and

$$d\left(\Pi_j t, \Pi_{j+1} t\right) \leq 3\delta_{j+1}$$

for all $t \in T$. Hence, by (2.3), we get

$$\sum_{j=0}^{J} \mathbb{E}^A \left[\sup_{t \in T} X\left(\Pi_{j+1} t\right) - X\left(\Pi_j t\right)\right] \leq 6 \sum_{j=0}^{J} \delta_{j+1} \sqrt{H\left(\delta_{j+1}\right)}$$

$$+ 3\sqrt{2\ln\left(\frac{1}{p}\right)} \sum_{j=0}^{J} \delta_{j+1}$$

and therefore

$$\mathbb{E}^A \left[\sup_{t \in T} X\left(t\right)\right] \leq \frac{6}{\varepsilon} \sum_{j=0}^{J} \delta_j \sqrt{H\left(\delta_j\right)} + (1 + 3\varepsilon)\,\sigma \sqrt{2\ln\left(\frac{1}{p}\right)}.$$

We can easily complete the proof by using the monotonicity of H. ∎

We have now at our disposal two different ways for deriving an exponential inequality from Theorem 3.17.

- First, setting $E_\varepsilon = \frac{12}{\varepsilon} \int_0^{\varepsilon\sigma} \sqrt{H\left(u\right)}du$, we can use Lemma 2.4 with $\varphi\left(x\right) = E_\varepsilon + (1 + 3\varepsilon)\,\sigma\sqrt{2x}$ and get for any positive x

$$\mathbb{P}\left[\sup_{t \in T} X\left(t\right) \geq E_\varepsilon + (1 + 3\varepsilon)\,\sigma\sqrt{2x}\right] \leq e^{-x}.$$

- Second, we can retain the conclusion of Theorem 3.17 only for $A = \Omega$ and $\varepsilon = 1$ and use inequality (3.21), which yields

$$\mathbb{P}\left[\sup_{t \in T} X\left(t\right) \geq E_1 + \sigma\sqrt{2x}\right] \leq e^{-x}.$$

Since E_ε is a nonincreasing function of ε, we see that the second method always produces a better result than the first one and this demonstrates the power of the Gaussian concentration inequality (3.21). The first method however misses the target from rather short and we can see that at the price of increasing E_ε by taking a small ε, we can recover the optimal term $\sigma\sqrt{2x}$ up to some factor which can be chosen arbitrary close to 1.

While the conclusion of this study is clearly not in favor of the conditional formulation of Theorem 3.17 since some control of the unconditional expectation is enough for building a sharp exponential probability bound via the

Gaussian concentration argument. However the conditional statement that we present here has the merit to introduce an alternative way to the concentration approach for establishing exponential inequalities. The comparison above shows that this alternative approach is not ridiculous and produces exponential bounds which up to constants have the right structure. This is an important fact since for unbounded empirical processes for instance, no concentration inequality is yet available and we are forced to use this alternative approach to study such processes.

3.4.3 Continuity of Gaussian Processes

Applying Theorem 3.17 to the increments of a Gaussian process allows to recover Dudley's regularity criterion without any additional effort.

Dudley's Criterion

In his landmark paper [56], Dudley has established the following metric entropy criterion for the sample continuity of some version of a Gaussian process.

Theorem 3.18 *Let $(X(t))_{t \in T}$ be some centered Gaussian process and d be the covariance pseudometric of $(X(t))_{t \in T}$. Assume that (T, d) is totally bounded and denote by $H(\delta, T)$ the δ-entropy number of (T, d), for all positive δ. If $\sqrt{H(., T)}$ is integrable at 0, then $(X(t))_{t \in T}$ admits a version which is almost surely uniformly continuous on (T, d). Moreover, if $(X(t))_{t \in T}$ is almost surely continuous on (T, d), then*

$$\mathbb{E}\left[\sup_{t \in T} X(t)\right] \leq 12 \int_0^\sigma \sqrt{H(x, T)} dx,$$

where $\sigma = \left(\sup_{t \in T} \mathbb{E}\left[X^2(t)\right]\right)^{1/2}$.

Proof. We note that since $S \subset T$ implies that $H(\delta, S) \leq H(\delta, T)$ for all positive δ, by monotone convergence Theorem 3.17 still holds whenever T is countable. We first assume T to be at most countable and introduce the process of increments $(X(t) - X(t'))_{(t, t') \in T^2}$. Since the covariance pseudometric of the process of increments is not larger than $2d$, we have for all positive δ

$$H\left(\delta, T^2\right) \leq 2H\left(\delta/2, T\right).$$

Hence $\sqrt{H(., T^2)}$ is integrable at zero and applying Theorem 3.17 to the Gaussian process $(X(t) - X(t'))_{(t, t') \in T_\delta^2}$, where

$$T_\delta^2 = \left\{(t, t') \in T^2 : d(t, t') \leq \delta\right\}$$

we get

$$\mathbb{E}\left[\sup_{(t,t')\in T_\delta^2}|X(t)-X(t')|\right]=\mathbb{E}\left[\sup_{(t,t')\in T_\delta^2}X(t)-X(t')\right]$$

$$\leq 12\sqrt{2}\int_0^\delta\sqrt{H(x/2,T)}dx$$

$$\leq 24\sqrt{2}\int_0^{\delta/2}\sqrt{H(u,T)}du.$$

This means that the nondecreasing function ψ defined for all positive δ by

$$\psi(\delta)=\mathbb{E}\left[\sup_{(t,t')\in T_\delta^2}|X(t)-X(t')|\right],$$

tends to 0 as δ tends to 0. Defining some positive sequence $(\delta_j)_{j\geq 0}$ tending to 0 such that the series $\sum_j\psi(\delta_j)$ converges, we deduce from Markov's inequality and the Borell–Cantelli Lemma that the process $(X(t))_{t\in T}$ is almost surely uniformly continuous on (T,d). If T is no longer assumed to be at most countable, we can argue as follows. Since (T,d) is totally bounded it is separable and we can apply the previous arguments to (D,d) where D is a countable and dense subset of T. Hence $(X(t))_{t\in D}$ is almost surely uniformly continuous and we can construct an almost surely continuous modification of $(X(t))_{t\in T}$ by using the standard extension argument described in Section 3.1. Of course for such a version $\widetilde{X}$, one has almost surely

$$\sup_{t\in T}\widetilde{X}(t)=\sup_{t\in D}\widetilde{X}(t)$$

which completes the proof since as previously noticed, Theorem 3.17 can be applied to $\left(\widetilde{X}(t)\right)_{t\in D}$. ∎

Now, if (T,d) is separable, we have at our disposal two entropy conditions. On the one hand Sudakov's minoration ensures that a *necessary* condition for the a.s. sample boundedness of $(X(t))_{t\in T}$ is that $\delta\to\delta\sqrt{H(\delta,T)}$ is bounded while on the other hand Dudley's *sufficient* condition for the existence of an almost surely uniformly continuous and bounded version of $(X(t))_{t\in T}$ says that $\delta\to\sqrt{H(\delta,T)}$ is integrable at 0 (which of course implies by monotonicity of $H(.,T)$ that $\delta\sqrt{H(\delta,T)}$ tends to 0 as δ goes to 0). Hence there is some little gap between these necessary and sufficient conditions which unfortunately cannot be filled if one only considers metric entropy conditions.

Majorizing Measures

In order to characterize the almost sure sample boundedness and uniform continuity of some version of X, one has to consider the sharper notion of *majorizing measure.* By definition, a majorizing probability measure μ on the metric space (T, d) satisfies

$$\sup_{t \in T} \int_0^\infty \sqrt{|\ln [\mu (\mathcal{B} (t, \delta))]|} d\delta < \infty, \tag{3.38}$$

where $\mathcal{B} (t, \delta)$ denotes the closed ball with radius δ centered at t. The existence of a majorizing measure is indeed a necessary and sufficient condition for the a.s. boundedness of some version of X, while the slightly stronger condition that there exists some majorizing measure μ satisfying

$$\lim_{\eta \to 0} \sup_{t \in T} \int_0^\eta \sqrt{|\ln [\mu (\mathcal{B} (t, \delta))]|} d\delta = 0 \tag{3.39}$$

is necessary and sufficient for the a.s. uniform continuity of some version of X. These striking definitive results are due to Fernique for the sufficient part (see [61]) and Talagrand for the necessary part ([109]). Of course Dudley's integrability condition of the square root of the metric entropy implies (3.39) thus a fortiori (3.38). Indeed it suffices to consider for every integer j a 2^{-j}-net and the corresponding uniform distribution μ_j on it. Then, defining the discrete probability measure μ as

$$\mu = \sum_{j \geq 1} 2^{-j} \mu_j,$$

we see that whatever t

$$\int_0^\eta \sqrt{|\ln [\mu (\mathcal{B} (t, \delta))]|} d\delta \leq \sum_{2^{-j} \leq \eta} 2^{-j} \sqrt{|\ln [\mu (\mathcal{B} (t, 2^{-j}))]|}$$

$$\leq \sum_{2^{-j} \leq \eta} 2^{-j} \left(\sqrt{\ln (2^j)} + \sqrt{H (2^{-j}, T)} \right)$$

which tends to 0 (uniformly with respect to t) when η goes to 0 under Dudley's integrability assumption for $\sqrt{H (., T)}$. As it will become clear in the examples studied below, we shall however dispense ourselves from using this more general concept since either entropy calculations will be a sharp enough tool or we shall use a direct approach based on finite dimensional approximations (as for the continuity of the isonormal process on a Hilbert–Schmidt ellipsoid). Let us finish this section devoted to the sample paths continuity of Gaussian processes by a simple remark concerning other possible distances than the intrinsic pseudometric d. If we take τ to be a pseudometric such that (T, τ) is compact, then, provided that $(X (t))_{t \in T}$ is continuous on (T, τ) in $\mathbb{L}_2$, the identity map from (T, τ) to (T, d) is continuous thus bi-continuous since (T, τ) is compact. This means that τ and d are defining equivalent topologies in the sense that a real valued mapping f is continuous on (T, τ) if and only if it is continuous on (T, d) (of course here continuity also means uniform continuity because of the compactness of (T, τ) and (T, d)). Hence under the mild requirements that (T, τ) is compact and that $(X(t))_{t \in T}$ is continuous on (T, τ) in $\mathbb{L}_2$, studying the sample paths continuity on (T, τ) amounts to study the sample paths continuity on (T, d). This means that although the above criteria involve the specific pseudometric d, we also have at our disposal continuity conditions for more general distances than the intrinsic pseudometric.

Concentration Inequalities for Gaussian Processes

As announced, for an almost surely continuous version of a Gaussian process on the totally bounded set of parameters (T, d), one can derive for free concentration inequalities for the suprema from the corresponding result for Gaussian random vectors.

Proposition 3.19 *If $(X(t))_{t \in T}$ is some almost surely continuous centered Gaussian process on the totally bounded set (T, d), letting $\sigma \geq 0$ be defined as*

$$\sigma^2 = \sup_{t \in T} \left(\mathbb{E} \left[X^2(t) \right] \right)$$

and Z denote

$$\text{either} \quad \sup_{t \in T} X(t) \quad \text{or} \quad \sup_{t \in T} |X(t)|.$$

Then,

$$\ln \mathbb{E} \left[\exp \left(\lambda \left(Z - \mathbb{E}[Z] \right) \right) \right] \leq \frac{\lambda^2 \sigma^2}{2}, \text{ for every } \lambda \in \mathbb{R},$$

leading to

$$\mathbb{P} \left[Z - \mathbb{E}[Z] \geq \sigma \sqrt{2x} \right] \leq \exp(-x)$$

and

$$\mathbb{P} \left[\mathbb{E}[Z] - Z \geq \sigma \sqrt{2x} \right] \leq \exp(-x)$$

for all positive x. Moreover the following bound is available

$$\mathbb{E} \left[Z^2 \right] - \left(\mathbb{E}[Z] \right)^2 \leq \sigma^2.$$

Proof. The proof is trivial from Theorem 3.12, using a separability argument and monotone convergence, the integrability of Z being a consequence of (3.24). ∎

We turn now to the introduction and the study of a generic example of a Gaussian process.

3.5 The Isonormal Process

3.5.1 Definition and First Properties

The *isonormal process* is the natural extension of the notion of standard normal random vector to an infinite dimensional setting.

Definition 3.20 *Let* $\mathbb{H}$ *be some separable Hilbert space, a Gaussian process* $(W(t))_{t \in \mathbb{H}}$ *is said to be isonormal if it is centered with covariance given by* $\mathbb{E}[W(t) W(u)] = \langle t, u \rangle$, *for every* t, u *in* $\mathbb{H}$.

If $\mathbb{H} = \mathbb{R}^N$, then, starting from a standard random vector $(\xi_1, \ldots, \xi_N)$, one can easily define the isonormal process as

$$W(t) = \langle t, \xi \rangle \text{ for every } t \in \mathbb{R}^N.$$

Note then that W is a linear (and thus continuous!) random map. Assuming now $\mathbb{H}$ to be infinite dimensional, the same kind of linear representation of W is available except that we have to be careful with negligible sets when we speak of linearity. Indeed, if $(W(t))_{t \in \mathbb{H}}$ is an isonormal process, given some finite linear combination $t = \sum_{i=1}^{k} \lambda_i t_i$ with $t_1, \ldots, t_k$ in $\mathbb{H}$, since

$$\mathbb{E}\left[\left(W(t) - \sum_{i=1}^{k} \lambda_i W(t_i)\right)^2\right] = \left\|t - \sum_{i=1}^{k} \lambda_i t_i\right\|^2 = 0,$$

$W(t) = \sum_{i=1}^{k} \lambda_i W(t_i)$ except on a set with probability 0 which does depend on (λ_i) and (t_i). So that W can be interpreted as a linear map (as a matter of fact as an isometry) from $\mathbb{H}$ into $\mathbb{L}_2(\Omega)$ but not as an almost sure linear map on $\mathbb{H}$ as it was the case in the previous finite dimensional setting. Conversely any isometry W from $\mathbb{H}$ into some Gaussian linear subspace ("Gaussian" meaning that every element of this subspace has a centered normal distribution) of $\mathbb{L}_2(\Omega)$ is a version of the isonormal process, which makes sense since remember that one has to deal with finite dimensional marginal distributions only. Now, let us take some Hilbertian basis $(\phi_j)_{j \geq 1}$ of $\mathbb{H}$. Then, $(\xi_j = W(\phi_j))_{j \geq 1}$ is a sequence of i.i.d. standard normal random variables and given $t \in \mathbb{H}$, since W is an isometry, one has by the three series Theorem

$$W(t) = \sum_{j \geq 1} \langle t, \phi_j \rangle \xi_j \text{ a.s. (and in } \mathbb{L}_2(\Omega)). \tag{3.40}$$

Conversely, given a sequence of i.i.d. standard normal random variables $(\xi_j)_{j \geq 1}$, by the three series Theorem again, given $t \in \mathbb{H}$, one can define $W(t)$ as the sum of the series $\sum_{j \geq 1} \langle t, \phi_j \rangle \xi_j$ in $\mathbb{L}_2(\Omega)$. It is easy to verify that the so-defined map W from $\mathbb{H}$ into $\mathbb{L}_2(\Omega)$ is an isometry. Moreover the image of $\mathbb{H}$ through this isometry is a Gaussian subspace (because $\sum_{j \geq 1} \langle t, \phi_j \rangle \xi_j$ is centered normal as the limit in $\mathbb{L}_2(\Omega)$ of centered normal variables). In order to avoid any ambiguity concerning the role of negligible sets in a statement like (3.40), we would like to emphasize the following two important points which both derive from the fact that by the three series Theorem $\sum_{j \geq 1} \xi_j^2 = +\infty$ a.s.

• First, if for some ω, the series $\sum_{j \geq 1} \langle t, \phi_j \rangle \xi_j(\omega)$ converges for every t, then this implies that $(\xi_j(\omega))_{j \geq 1}$ belongs to ℓ_2, which is absurd except on a null probability set. In other words, while for every t, $W(t) = \sum_{j \geq 1} \langle t, \phi_j \rangle \xi_j$ a.s., the set of ω such that $W(t)(\omega) = \sum_{j \geq 1} \langle t, \phi_j \rangle \xi_j(\omega)$ holds for every t is negligible.

- In the same way there is no almost surely continuous version of the isonormal process on the hole Hilbert space $\mathbb{H}$. More than that, any version $(W(t))_{t \in \mathbb{H}}$ of the isonormal process is a.s. discontinuous at some point of $\mathbb{H}$. Indeed, we can use a separability argument to derive that, except if ω belongs to some null probability set, as soon as $t \to W(t)(\omega)$ is continuous on $\mathbb{H}$ then $t \to W(t)(\omega)$ is a linear (and continuous!) map on $\mathbb{H}$ and therefore, by the Riesz representation Theorem, there exists some element $\xi(\omega) \in \mathbb{H}$ such that $W(t)(\omega) = \langle t, \xi(\omega) \rangle$. Hence, setting $\xi_j(\omega) = \langle \xi(\omega), \phi_j \rangle$ for every integer j, one has $\sum_{j \geq 1} \xi_j^2(\omega) < \infty$, which is possible only if ω belongs to some null probability set.

So, on the one hand we must keep in mind that while W is linear from $\mathbb{H}$ to $\mathbb{L}_2(\Omega)$, it is not true that a.s. W is a linear map on $\mathbb{H}$. On the other hand, the question of finding continuity sets C, i.e., subsets of $\mathbb{H}$ for which there exists some version of $(W(t))_{t \in \mathbb{H}}$ which is continuous on C is relevant. Of course, we do know that concerning a closed linear subspace of $\mathbb{H}$, it is a continuity set if and only if it is finite dimensional and we would like to turn now to the study of bounded (and even compact) continuity sets to which the results of Section 3.4 can be applied.

3.5.2 Continuity Sets with Examples

By Theorem 3.18, denoting by $H(\delta, C)$ the δ-entropy number of C equipped by the Hilbertian distance of $\mathbb{H}$, a sufficient condition for C to be a continuity set for the isonormal process is that $\sqrt{H(., C)}$ is integrable at 0.

Hölder Smooth Functions and the Gaussian White Noise

If we take $\mathbb{H}$ to be $\mathbb{L}_2[0, 1]$ and C to be $\{\mathbb{1}_{[0,x]}, x \in [0, 1]\}$, then the restriction of the isonormal process to C is nothing else than the Wiener process $B(x) = W(\mathbb{1}_{[0,x]})$, $x \in [0, 1]$. Obviously, since $\left\| \mathbb{1}_{[0,x]} - \mathbb{1}_{[0,x']} \right\|^2 = |x - x'|$, one has $H(\delta, C) \leq \ln(1/\delta)$, so that $\sqrt{H(., C)}$ is integrable. Hence, we recover the existence of the Brownian motion, i.e., an almost surely continuous version the Wiener process on $[0, 1]$. Furthermore, the isonormal process itself can be interpreted as a version of the stochastic integral, i.e., for every t

$$W(t) = \int_0^1 t(x) \, dB(x) \quad \text{a.s.}$$

This process is commonly called the *Gaussian white noise* (this is the reason why more generally the isonormal process on some abstract infinite dimensional and separable Hilbert space is also often called Gaussian white noise).

Typical examples of continuity sets for this process are proper compact subsets of $\mathcal{C}[0,1]$. Given $R > 0$ and $\alpha \in (0,1]$, a classical result from approximation theory (see for instance [57]) ensures that if C denotes the set of α-Hölder smooth functions t on $[0,1]$ such that $t(0) = 0$ and

$$|t(x) - t(y)| \leq R|x - y|^{\alpha},$$

then for some absolute positive constants κ_1 and κ_2, one has for every $\delta \in (0, R)$

$$\kappa_1 \left(\frac{R}{\delta}\right)^{1/\alpha} \leq H(\delta, C) \leq \kappa_2 \left(\frac{R}{\delta}\right)^{1/\alpha}.$$

Hence, according to Dudley's criterion, C is a continuity set whenever $\alpha > 1/2$, while by the Sudakov minoration, C is not a continuity set whenever $\alpha < 1/2$ (the case where $\alpha = 1/2$ would require a sharper analysis but as a matter of fact C is not a continuity set for the isonormal process in that case too). From Theorem 3.18, we also derive a more precise result, namely, provided that $\alpha > 1/2$, if $(W(t))_{t \in C}$ denotes an almost continuous version on C of the isonormal process, one has for every $\sigma \in (0, 1)$

$$\mathbb{E}\left[\sup_{\|t-u\| \leq \sigma} (W(t) - W(u))\right] \leq \kappa(\alpha) R^{1/2\alpha} \sigma^{1-(1/2\alpha)},$$

where the supremum in the above inequality is taken over all t and u belonging to C and $\kappa(\alpha)$ is a positive constant depending only on α. Of course, the same analysis would hold for the Gaussian white noise on $[0,1]^d$, simply replacing α in the entropy computations above by α/d.

Hilbert–Schmidt ellipsoids are "generic" examples of continuity sets for the isonormal process. As we shall see later on, choosing a proper basis of $\mathbb{L}_2[0,1]$ such as the Fourier or the Haar basis, the restriction that t belongs to some ellipsoid is closely linked to a regularity restriction of the same type as the one considered in the previous example.

Ellipsoids

Given an orthonormal basis $(\phi_j)_{j \geq 1}$ of $\mathbb{H}$ and a nonincreasing sequence $(c_j)_{j \geq 1}$ tending to 0 at infinity, one defines the ellipsoid

$$\mathcal{E}_2(c) = \left\{ t \in \mathbb{H}, \sum_{j \geq 1} \frac{\langle t, \phi_j \rangle^2}{c_j^2} \leq 1 \right\}. \tag{3.41}$$

the ellipsoid is said to be Hilbert–Schmidt if $(c_j)_{j \geq 1} \in \ell_2$. It is an easy exercise to show that $\mathcal{E}_2(c)$ is a compact subset of $\mathbb{H}$. Unless c_j has a well known behavior when j goes to infinity such as an arithmetical decay (i.e., $c_j \sim \kappa j^{-\alpha}$), the metric entropy of $\mathcal{E}_2(c)$ is not the right tool to determine whether

$\mathcal{E}_2(c)$ is a continuity set or not. Of course one could think to use majorizing measures and it indeed can be done (see [79]). But in this case, the solution of the continuity problem is due to Dudley (see [56]) and easily derives from a simpler direct approach essentially based on Cauchy–Schwarz inequality.

Theorem 3.21 *Let $(c_j)_{j\geq 1}$ be a nonincreasing sequence tending to 0 at infinity and $\mathcal{E}_2(c)$ be the ellipsoid defined by (3.41). Then $\mathcal{E}_2(c)$ is a continuity set for the isonormal process if and only if it is Hilbert–Schmidt. Moreover, if the ellipsoid $\mathcal{E}_2(c)$ is Hilbert–Schmidt and if $(W(t))_{t\in\mathbb{H}}$ is a version of the isonormal process which is almost surely continuous on $\mathcal{E}_2(c)$, one has*

$$\mathbb{E}\left[\left(\sup_{t\in\mathcal{E}_2(c)} W(t)\right)^2\right] = \sum_{j\geq 1} c_j^2 \tag{3.42}$$

and

$$\mathbb{E}\left[\left(\sup_{w\in\mathcal{E}_{2,\sigma}(c)} W(w)\right)^2\right] \leq 8\sum_{j\geq 1}\left(c_j^2\wedge\sigma^2\right), \tag{3.43}$$

where $\mathcal{E}_{2,\sigma}(c) = \{t-u;\ (t,u)\in\mathcal{E}_2(c)\times\mathcal{E}_2(c)\ \text{with}\ \|t-u\|\leq\sigma\}$.

Proof. If $\mathcal{E}_2(c)$ is a continuity set, then let $(W(t))_{t\in\mathbb{H}}$ be a version of the isonormal process which is almost surely continuous on $\mathcal{E}_2(c)$. Then Proposition 3.19 implies in particular that $Z = \sup_{t\in\mathcal{E}_2(c)} W(t)$ is square integrable. Moreover, since the a.s. continuity of $(W(t))$ implies that a.s., the supremum of $W(t)$ over $\mathcal{E}_2(c)$ equals the supremum of $W(t)$ on a countable and dense subset of $\mathcal{E}_2(c)$, one derives from (3.40) that almost surely

$$Z = \sup_{t\in\mathcal{E}_2(c)}\sum_{j\geq 1}\langle t,\phi_j\rangle\,\xi_j,$$

where $(\xi_j)_{j\geq 1}$ is a sequence of i.i.d. standard normal random variables. Hence $Z^2 = \sum_{j\geq 1} c_j^2\xi_j^2$ a.s. and therefore the integrability of Z^2 implies the summability of $(c_j^2)_{j\geq 1}$. Conversely, assume that $(c_j)_{j\geq 1}\in\ell_2$. Let $\mathcal{E}_2'(c)$ be a countable subset of $\mathcal{E}_2(c)$. Then, almost surely, for every t belonging to $\mathcal{E}_2'(c)$

$$W(t) = \sum_{j\geq 1}\langle t,\phi_j\rangle\,\xi_j$$

and Cauchy–Schwarz inequality implies that

$$\left(\sum_{j\geq 1}\langle t,\phi_j\rangle\,\xi_j\right)^2 \leq \left(\sum_{j\geq 1}\frac{\langle t,\phi_j\rangle^2}{c_j^2}\right)\left(\sum_{j\geq 1} c_j^2\xi_j^2\right) \leq \sum_{j\geq 1} c_j^2\xi_j^2.$$

Hence, defining

$$Z' = \sup_{t \in \mathcal{E}_2'(c)} \sum_{j \geq 1} \langle t, \phi_j \rangle \xi_j,$$

one has

$$\mathbb{E}\left[Z'^2\right] \leq \sum_{j \geq 1} c_j^2. \tag{3.44}$$

The point now is that if we set

$$\mathcal{E}_{2,\sigma}'(c) = \{t - u; \; (t, u) \in \mathcal{E}'(c) \times \mathcal{E}'(c) \text{ with } \|t - u\| \leq \sigma\},$$

then $\mathcal{E}_{2,\sigma}'(c)$ is a countable subset of the ellipsoid $\mathcal{E}_2(\gamma)$, where

$$\gamma_j = 2\sqrt{2}(c_j \wedge \sigma)$$

for every $j \geq 1$. We may therefore use (3.44) with γ instead of c and derive that

$$\mathbb{E}\left[\left(\sup_{w \in \mathcal{E}_{2,\sigma}'(c)} W(w)\right)^2\right] \leq 8 \sum_{j \geq 1} (c_j^2 \wedge \sigma^2). \tag{3.45}$$

Since the right-hand side of this inequality tends to 0 as σ tends to 0, Borel–Cantelli lemma ensures the almost sure continuity of W on $\mathcal{E}_2'(c)$. Therefore, by the extension principle explained in Section 3.1, there exists some modification of $(W(t))_{t \in \mathbb{H}}$ which is almost surely continuous on $\mathcal{E}_2(c)$. For such a version the supremum in (3.45) can be taken over $\mathcal{E}_{2,\sigma}'(c)$ or $\mathcal{E}_{2,\sigma}(c)$ indifferently, which leads to (3.43). The same argument also implies that $Z = Z'$ a.s. hence (3.44) means that $\mathbb{E}\left[Z^2\right] \leq \sum_{j \geq 1} c_j^2$. Finally since

$$\sum_{j \geq 1} \left(\frac{c_j^2 \xi_j}{\sqrt{\sum_{k \geq 1} c_k^2 \xi_k^2}}\right) \phi_j$$

belongs to $\mathcal{E}_2(c)$ we derive that a.s.

$$Z^2 \geq \sum_{j \geq 1} c_j^2 \xi_j^2$$

which leads to (3.42). ∎

Gaussian Model Selection

4.1 Introduction

We consider the generalized linear Gaussian model as introduced in [23]. This means that, given some separable Hilbert space $\mathbb{H}$, one observes

$$Y_\varepsilon(t) = \langle s, t \rangle + \varepsilon W(t), \text{ for all } t \in \mathbb{H}, \tag{4.1}$$

where W is some isonormal process (according to Definition 3.20), i.e., W maps isometrically $\mathbb{H}$ onto some Gaussian subspace of $\mathbb{L}_2(\Omega)$. This framework is convenient to cover both the infinite dimensional white noise model for which $\mathbb{H} = \mathbb{L}_2([0,1])$ and $W(t) = \int_0^1 t(x) dB(x)$, where B is a standard Brownian motion, and the finite dimensional linear model for which $\mathbb{H} = \mathbb{R}^n$ and $W(t) = \langle \xi, t \rangle$, where ξ is a standard n-dimensional Gaussian vector.

4.1.1 Examples of Gaussian Frameworks

Let us see in more details what are the main statistical Gaussian frameworks which can be covered by the above general model.

The Classical Linear Gaussian Regression Model

In this case one observes a random vector

$$Y_j = s_j + \sigma \xi_j, \ 1 \leq j \leq n \tag{4.2}$$

where the random variables are i.i.d. standard normal. Considering the scalar product

$$\langle u, v \rangle = \frac{1}{n} \sum_{j=1}^n u_j v_j,$$

and setting

$$W(t) = \sqrt{n} \langle \xi, t \rangle,$$

we readily see that W is an isonormal process on $\mathbb{R}^n$ and that $Y_\varepsilon(t) = \langle Y, t \rangle$ satisfies to (4.1) with $\varepsilon = \sigma/\sqrt{n}$.

The White Noise Framework

In this case one observes $(\zeta_\varepsilon (x), x \in [0,1])$ given by the stochastic differential equation

$$d\zeta_\varepsilon (x) = s(x)\, dx + \varepsilon dB(x) \text{ with } \zeta_\varepsilon (0) = 0.$$

Hence, setting for every $t \in \mathbb{L}_2 ([0,1])$, $W(t) = \int_0^1 t(x)\, dB(x)$, W is indeed an isonormal process on $\mathbb{L}_2 ([0,1])$ and $Y_\varepsilon (t) = \int_0^1 t(x)\, d\zeta_\varepsilon (x)$ obeys to (4.1), provided that $\mathbb{L}_2 [0,1]$ is equipped with its usual scalar product $\langle s, t \rangle = \int_0^1 s(x) t(x)\, dx$. Typically, s is a signal and $d\zeta_\varepsilon (x)$ represents the noisy signal received at time x. This framework easily extends to a d-dimensional setting if one considers some multivariate Brownian sheet B on $[0,1]^d$ and takes $\mathbb{H} = \mathbb{L}_2 \left([0,1]^d\right)$.

The Fixed Design Gaussian Regression Framework

We consider the special case of (4.2) for which $s_j = s(j/n)$, $j = 1, \ldots, n$, where s denotes some function on $[0,1]$ (in this case we denote abusively by the same symbol s the function on $[0,1]$ and the vector with coordinates $s(j/n)$ in $\mathbb{R}^n$). If s is a signal, for every j, Y_j represents the noisy signal at time j/n. It is some discrete version of the white noise model. Indeed if one observes $(\zeta_\varepsilon (x), x \in [0,1])$ such that

$$d\zeta_\varepsilon (x) = s(x)\, dx + \varepsilon dB(x)$$

only at the discrete points $\{j/n, 1 \leq j \leq n\}$, setting $\sigma = \varepsilon \sqrt{n}$ and

$$\xi_j = \sqrt{n}\, (B(j/n) - B((j-1)/n)) \text{ for all } j \in [1,n],$$

the noisy signal received at time j/n is given by

$$Y_j = n\,(\xi_\varepsilon (j/n) - \xi_\varepsilon ((j-1)/n)) = n \int_{(j-1)/n}^{j/n} s(x)\, dx + \sigma \xi_j.$$

Since the variables $\{\xi_j, 1 \leq j \leq n\}$ are i.i.d. standard normal, we are indeed back to the fixed design Gaussian regression model with $s_j = s^{(n)}(j/n)$ if one sets $s^{(n)}(x) = n \int_{(j-1)/n}^{j/n} s(y)\, dy$ whenever $x \in [(j-1)/n, j/n)$. If s is a smooth enough function, $s^{(n)}$ represents a good piecewise constant approximation of s and this shows that there is a link between the fixed design Gaussian regression setting and the white noise model.

The Gaussian Sequence Framework

The process given by (4.1) is not connected to any specific orthonormal basis of $\mathbb{H}$ but assuming $\mathbb{H}$ to be infinite dimensional, once we have chosen such a basis $\{\varphi_j\}_{j\geq 1}$, it can be transformed into the so-called *Gaussian sequence framework*. This means that we observe the filtered version of Y_ε through the basis, i.e., the sequence $\widehat{\beta}_j = Y_\varepsilon(\varphi_j)$, where Y_ε is the process defined by (4.1). This is a Gaussian sequence of the form

$$\widehat{\beta}_j = \beta_j + \varepsilon\,\xi_j, \quad j \in \mathbb{N}^*, \quad (\beta_j)_{j\geq 1} \in \ell_2. \tag{4.3}$$

Here $\beta_j = \langle s, \varphi_j \rangle$ and the random variables $\xi_j = W(\varphi_j)$ are i.i.d. standard normal. One can identify s with the sequence $\beta = (\beta_j)_{j\geq 1}$ and estimate it by some $\widetilde{\beta} \in \ell_2$. Since $\mathbb{E}[\widehat{\beta}_j] = \beta_j$ the problem of estimating β within the framework described by (4.3) can also be considered as an infinite-dimensional extension of the Gaussian linear regression framework (4.2). The study of minimax and adaptive estimation in the Gaussian sequence framework has been mainly developed by Pinsker (see [99]) and Efroimovich and Pinsker (see [58]) for ellipsoids and by Donoho and Johnstone (see [47], [48], [50], [51] and [52]) for ℓ_p-bodies. Let us now recall that given $p \in (0,2]$ and $c = (c_j)_{j\geq 1}$ be a nonincreasing sequence of numbers in $[0,+\infty]$, converging to 0 when $j \to +\infty$, the ℓ_p-*body* $\mathcal{E}_p(c)$ is the subset of $\mathbb{R}^{\mathbb{N}^*}$ given by

$$\mathcal{E}_p(c) = \left\{ (\beta_j)_{j\geq 1} \,\middle|\, \sum_{j\geq 1} \left| \frac{\beta_j}{c_j} \right|^p \leq 1 \right\},$$

with the convention that $0/0 = 0$ and $x/(+\infty) = 0$ whatever $x \in \mathbb{R}$. An ℓ_2-body is called an *ellipsoid* and it follows from classical inequalities between the norms in ℓ_2 and ℓ_p that $\mathcal{E}_p(c) \subset \ell_2$.

The interest and importance of the Gaussian sequence framework and the geometrical objects such as ℓ_p-bodies come from curve estimation. Indeed, if $\mathbb{H} = \mathbb{L}_2([0,1])$ for instance, for proper choices of the basis $\{\varphi_j\}_{j\geq 1}$, smoothness properties of s can be translated into geometric properties of $\beta \in \ell_2$. One should look at [94] for the basic ideas, Section 2 of [52] for a review and the Appendix of this Chapter below. Many classical functional classes in some $\mathbb{L}_2$ space $\mathbb{H}$ can therefore be identified with specific geometric objects in ℓ_2 via the natural isometry between $\mathbb{H}$ and ℓ_2 given by $s \leftrightarrow (\beta_j)_{j\geq 1}$ if $s = \sum_{j\geq 1} \beta_j \varphi_j$. Let us illustrate this fact by the following classical example. For α some positive integer and $R > 0$, the Sobolev ball $W^\alpha(R)$ on the torus $\mathbb{R}/\mathbb{Z}$ is defined as the set of functions s on $[0,1]$ which are the restriction to $[0,1]$ of periodic functions on the line with period 1 satisfying $\|s^{(\alpha)}\| \leq R$. Given the trigonometric basis $\varphi_1 = 1$ and, for $k \geq 1$, $\varphi_{2k}(z) = \sqrt{2}\cos(2\pi k z)$ and $\varphi_{2k+1}(z) = \sqrt{2}\sin(2\pi k z)$, it follows from Plancherel's formula that $s = \sum_{j\geq 1} \beta_j \varphi_j$ belongs to $W^\alpha(R)$ if and only if $\sum_{k=1}^{\infty}(2\pi k)^{2\alpha}\left[\beta_{2k}^2 + \beta_{2k+1}^2\right] \leq R^2$ or equivalently if the sequence $(\beta_j)_{j\geq 1}$ belongs to the ellipsoid

$$\left\{ (\beta_j)_{j\geq 1} \,\middle|\, \sum_{j\geq 1} \left(\frac{\beta_j}{c_j} \right)^2 \leq 1 \right\}, \tag{4.4}$$

with

$$c_1 = +\infty \text{ and } c_{2k} = c_{2k+1} = R(2\pi k)^{-\alpha} \quad \text{for } k \geq 1.$$

This means that, via the identification between $s \in \mathbb{L}_2([0,1])$ and its coordinates vector $(\langle s, \varphi_j \rangle)_{j \geq 1} \in \ell_2(\mathbb{N}^*)$, one can view a Sobolev ball as a geometric object which is an infinite dimensional ellipsoid in ℓ_2.

More generally, balls in Besov spaces can be identified with special types of ℓ_p-bodies when expanded on suitable wavelet bases (the interested reader will find some more details in the Appendix of this Chapter). It is important here to notice that the ordering, induced by $\mathbb{N}^*$, that we have chosen on $\{\varphi_j\}_{j \geq 1}$, plays an important role since ℓ_p-bodies are not invariant under permutations of $\mathbb{N}^*$.

4.1.2 Some Model Selection Problems

Let us give some examples of model selection problems which appear naturally in the above frameworks.

Variable Selection

If we consider the finite dimensional Gaussian regression framework, the usual requirement for the mean vector s is that it belongs to some N-dimensional subspace of $\mathbb{R}^n$ or equivalently

$$s = \sum_{\lambda=1}^{N} \beta_\lambda \varphi_\lambda,$$

where the vectors $\varphi_\lambda, \lambda = 1, \ldots, N$ are linearly independent. It may happen that the vectors $\varphi_\lambda, \lambda = 1, \ldots, N$ represent explanatory variables and an interesting question (especially if N is large) is to select among the initial collection, the "most significant" variables $\{\varphi_\lambda, \lambda \in m\}$.

Interestingly, this variable selection problem also makes sense within the white noise model. In order to reconstruct the signal s, one indeed can consider some family $\{\varphi_\lambda, \lambda \in \Lambda\}$ of linearly independent functions, where Λ is either some finite set as before $\Lambda = 1, \ldots, N$ or may be infinite $\Lambda = \mathbb{N}^*$. If we think of the situation where $\varphi_\lambda, \lambda = 1, \ldots, N$ denotes some rather large subset of a wavelet basis, one would like to select some "ideal" subset m of Λ to represent the signal s on $\{\varphi_\lambda, \lambda \in m\}$. This means that the *complete variable selection* problem is of interest in a variety of situations for which the variables can be provided by Nature or created by the statistician in order to solve some estimation problem. Another interesting question is *ordered variable selection*. If we consider the case where $\Lambda = \mathbb{N}^*$ and $\{\varphi_\lambda, \lambda \in \Lambda\}$ denotes the trigonometric basis taken according to its natural ordering, then we could think to restrict ourselves to ordered subsets $\{[1, D], D \in \mathbb{N}^*\}$ of $\mathbb{N}^*$ when searching some convenient finite dimensional expansion approximating the signal s. Note that if $\{\varphi_\lambda, \lambda \in \Lambda\}$ denotes some orthonormal basis of a Hilbert space $\mathbb{H}$, the variable selection problem can be considered within the Gaussian sequence framework (4.3) where it amounts to select a proper subset m of significant components of β.

Multiple Change-Points Detection

This problem appears naturally in the fixed design Gaussian regression setting. The question here is to select the "best" partition m of

$$\left\{0, \frac{1}{n}, \ldots, \frac{(n-1)}{n}, 1\right\}$$

by intervals $\{I, I \in m\}$ on which the unknown signal s can be represented by some piecewise constant function. The motivations come from the analysis of seismic signals for which the change points (i.e., the extremities of the intervals of the partition) correspond to different geological materials.

As illustrated by the previous examples, a major problem in estimation is connected with the choice of a suitable set of "significant" parameters to be estimated. In the classical finite dimensional Gaussian regression case, one should select some subset $\{\varphi_\lambda\}_{\lambda \in m}$ of the explanatory variables; for the Gaussian sequence problem we just considered, ordered variable selection means selecting a value of D and only estimate the D parameters $\beta_1, \ldots, \beta_D$. In any case, this amounts to pretend that the unknown target s belongs to some *model* S_m and estimate it as if this were actually true, although we know this is not necessarily the case. In this approach, a model should therefore always be viewed as an *approximate* model.

4.1.3 The Least Squares Procedure

Of course we have used several times some notions which are not well-defined like "best" partition or "most significant" variables. It is our purpose now to provide some precise framework allowing to give a mathematical sense to these intuitive questions or notions. Basically we take the squared distance as our loss function. Given some closed subset S_m of $\mathbb{H}$, the best approximating point of s belonging to S_m minimizes $\|t - s\|^2$ (or equivalently $-2\langle s, t\rangle + \|t\|^2$) over S_m. It is known to be the orthogonal projection of s onto S_m whenever S_m is a finite dimensional subspace of $\mathbb{H}$. The basic idea is to consider a minimizer of $-2Y_\varepsilon(t) + \|t\|^2$ over S_m as an estimator of s representing model S_m.

Definition 4.1 *Let S be some subset of $\mathbb{H}$ and let us set $\gamma_\varepsilon(t) = \|t\|^2 - 2Y_\varepsilon(t)$. One defines a* least squares estimator (LSE) *on S as a minimizer of the least squares criterion $\gamma_\varepsilon(t)$ with respect to $t \in S$.*

We do not pretend that such an estimator does exist without any restriction on the model S_m to be considered, but if it is the case we denote it by

$\widehat{s}_m$ and take as a measure of quality for model S_m the *quadratic risk* $\mathbb{E}_s\left[\|\widehat{s}_m - s\|^2\right]$. *Model selection* actually proceeds in two steps: first consider some family of models S_m with $m \in \mathcal{M}$ together with the LSE $\widehat{s}_m$ with values in S_m. Then use the data to select a value $\widehat{m}$ of m and take $\widehat{s}_{\widehat{m}}$ as the final estimator. A "good" model selection procedure is one for which the risk of the resulting estimator is as close as possible to the minimal risk of the estimators $\widehat{s}_m, m \in \mathcal{M}$.

4.2 Selecting Linear Models

Restricting ourselves to linear models could appear as a very severe restriction. However, as suggested by the examples given above, selecting among a collection of linear models is already an important issue for a broad range of applications. Surprisingly at first glance, this includes curve estimation problems since we should keep in mind that approximation theory provides many examples of finite dimensional linear subspaces of functions which approximate a given regular function with an accuracy depending on the dimension of the subspace involved in the approximation and the regularity of the function. Therefore, even though we shall deal with much more general models in a subsequent section, it seems to us that linear models are so important and useful that it is relevant to devote a specific section to this case, the results presented here being borrowed from [23].

If one takes S_m as a linear space with dimension D_m, one can compute the LSE explicitly. Indeed, if $(\varphi_j)_{1 \leq j \leq D_m}$ denotes some orthonormal basis of S_m, one has

$$\widehat{s}_m = \sum_{j=1}^{D_m} Y_\varepsilon\left(\varphi_j\right)\varphi_j.$$

Since for every $1 \leq j \leq D_m$,

$$Y_\varepsilon\left(\varphi_j\right) = \langle s, \varphi_j \rangle + \varepsilon\eta_j,$$

where the variables $\eta_1, \ldots, \eta_D$ are i.i.d. standard normal variables, $\widehat{s}_m$ appears as some kind of empirical projection on S_m which is indeed an unbiased estimator of the orthogonal projection

$$s_m = \sum_{j=1}^{D_m} \langle s, \varphi_j \rangle \varphi_j$$

of s onto S_m. Its quadratic risk as an estimator of s can be easily computed:

$$\mathbb{E}_s\left[\|s - \widehat{s}_m\|^2\right] = \|s - s_m\|^2 + \varepsilon^2 D_m.$$

This formula for the quadratic risk perfectly reflects the model choice paradigm since if one wants to choose a model in such a way that the risk of the resulting LSE is small, we have to warrant that the bias term $\|s - s_m\|^2$ and the variance term $\varepsilon^2 D$ are small simultaneously. In other words if $\{S_m\}_{m \in \mathcal{M}}$ is a list of finite dimensional subspaces of $\mathbb{H}$ and $(\widehat{s}_m)_{m \in \mathcal{M}}$ denotes the corresponding list of LSEs, an "ideal" model should minimize $\mathbb{E}_s\left[\|s - \widehat{s}_m\|^2\right]$ with respect to $m \in \mathcal{M}$. Of course, since we do not know the bias term, the quadratic risk cannot be used as a model choice criterion.

Considering $m(s)$ minimizing the risk $\mathbb{E}_s\left[\|s - \widehat{s}_m\|^2\right]$ with respect to $m \in \mathcal{M}$, the LSE $\widehat{s}_{m(s)}$ on the corresponding model $S_{m(s)}$ is called an *oracle* (according to the terminology introduced by Donoho and Johnstone, see [47] for instance). Unfortunately, since the risk depends on the unknown parameter s, so does $m(s)$ and the oracle is definitely not an estimator of s. However, the risk of an oracle is a benchmark which will be useful in order to evaluate the performance of any data driven selection procedure among the collection of estimators $(\widehat{s}_m)_{m \in \mathcal{M}}$. Note that this notion is different from the notion of "true model." In other words if s belongs to some model S_{m_0}, this does not necessarily imply that $\widehat{s}_{m_0}$ is an oracle. The idea is now to consider data-driven criteria to select an estimator which tends to mimic an oracle, i.e., one would like the risk of the selected estimator $\widehat{s}_{\widehat{m}}$ to be as close as possible to the risk of an oracle.

4.2.1 A First Model Selection Theorem for Linear Models

Since the aim is to mimic the oracle, a natural approach consists in estimating the risk and then minimizing the corresponding criterion. Historically, this is exactly the first path which has been followed in the pioneering works of Akaike and Mallows (see [41], [2] and [84]), the main point being: how to estimate the risk?

Mallows' Heuristics

The classical answer given by Mallows' C_p heuristics is as follows. An "ideal" model should minimize the quadratic risk

$$\|s_m - s\|^2 + \varepsilon^2 D_m = \|s\|^2 - \|s_m\|^2 + \varepsilon^2 D_m,$$

or equivalently

$$- \|s_m\|^2 + \varepsilon^2 D_m.$$

Substituting to $\|s_m\|^2$ its natural unbiased estimator $\|\widehat{s}_m\|^2 - \varepsilon^2 D_m$ leads to Mallows' C_p

$$- \|\widehat{s}_m\|^2 + 2\varepsilon^2 D_m.$$

The weakness of this analysis is that it relies on the computation of the expectation of $\|\widehat{s}_m\|^2$ for every given model but nothing warrants that $\|\widehat{s}_m\|^2$ will stay of the same order of magnitude as its expectation for all models simultaneously. This leads to consider some more general model selection criteria involving penalties which may differ from Mallows' penalty.

Statement of the Model Selection Theorem

The above heuristics can be justified (or corrected) if one can specify how close is $\|\widehat{s}_m\|^2$ from its expectation $\|s_m\|^2 + \varepsilon^2 D_m$, uniformly with respect to $m \in \mathcal{M}$. The Gaussian concentration inequality will precisely be the adequate tool to do that. The idea will be to consider more general penalized least squares criteria than Mallows' C_p. More precisely, we shall study criteria of the form $-\|\widehat{s}_m\|^2 + \text{pen}\,(m)$, where $\text{pen} : \mathcal{M} \to \mathbb{R}_+$ is an appropriate penalty function. Note that in the following theorem (see [23]), one simultaneously gets a precise form for the penalty and an "oracle" type inequality.

Theorem 4.2 *Let $\{x_m\}_{m\in\mathcal{M}}$ be some family of positive numbers such that*

$$\sum_{m\in\mathcal{M}} \exp\,(-x_m) = \Sigma < \infty. \tag{4.5}$$

Let $K > 1$ and assume that

$$\text{pen}\,(m) \geq K\varepsilon^2 \left(\sqrt{D_m} + \sqrt{2x_m} \right)^2. \tag{4.6}$$

Then, almost surely, there exists some minimizer $\widehat{m}$ of the penalized least-squares criterion

$$-\|\widehat{s}_m\|^2 + \text{pen}\,(m)$$

over $m \in \mathcal{M}$. Moreover the corresponding penalized least-squares estimator $\widehat{s}_{\widehat{m}}$ is unique and the following inequality is valid

$$\mathbb{E}_s \left[\|\widehat{s}_{\widehat{m}} - s\|^2 \right] \leq C\,(K) \left\{ \inf_{m\in\mathcal{M}} \left(\|s_m - s\|^2 + \text{pen}\,(m) \right) + (1 + \Sigma)\,\varepsilon^2 \right\}, \tag{4.7}$$

where $C\,(K)$ depends only on K.

We shall derive Theorem 4.2 as a consequence of a more general result (namely Theorem 4.18) to be proven below except for the uniqueness part of the statement which is in fact straightforward. Indeed if $S_m = S_{m'}$, then $\widehat{s}_m = \widehat{s}_{m'}$ so that either $\text{pen}\,(m) = \text{pen}\,(m')$ but then, obviously $-\|\widehat{s}_m\|^2 + \text{pen}\,(m) = -\|\widehat{s}_{m'}\|^2 + \text{pen}\,(m')$ or $\text{pen}\,(m) \neq \text{pen}\,(m')$ but then $-\|\widehat{s}_m\|^2 + \text{pen}\,(m) \neq -\|\widehat{s}_{m'}\|^2 + \text{pen}\,(m')$. Therefore, in order to prove uniqueness, it is enough to show that $-\|\widehat{s}_m\|^2 + \text{pen}\,(m) \neq -\|\widehat{s}_{m'}\|^2 + \text{pen}\,(m')$ as soon as $S_m \neq S_{m'}$. This is a consequence of the fact that, in this case, $\varepsilon^{-2} \left(\|\widehat{s}_m\|^2 - \|\widehat{s}_{m'}\|^2 \right)$ has a distribution which is absolutely continuous with respect to Lebesgue measure since it can be written as the difference between two independent noncentral chi-square random variables.

It is important to realize that Theorem 4.2 easily allows to compare the risk of the *penalized LSE* $\widehat{s}_{\widehat{m}}$ with the benchmark $\inf_{m\in\mathcal{M}} \mathbb{E}_s \|\widehat{s}_m - s\|^2$. To illustrate this idea, remembering that

$$\mathbb{E}_s \left[\|\hat{s}_m - s\|^2 \right] = \|s_m - s\|^2 + \varepsilon^2 D_m,$$

let us indeed consider the simple situation where one can take $\{x_m\}_{m \in \mathcal{M}}$ such that $x_m = L D_m$ for some positive constant L and $\sum_{m \in \mathcal{M}} \exp(-x_m) \leq 1$, say (1 is not a magic number here and we could use some other numerical constant as well). Then, taking

$$\text{pen}(m) = K \varepsilon^2 D_m \left(1 + \sqrt{2L} \right)^2,$$

the right-hand side in the risk bound is (up to constant) bounded by

$$\inf_{m \in \mathcal{M}} \mathbb{E}_s \|\hat{s}_m - s\|^2.$$

In such a case, we recover the desired benchmark, which means that the selected estimator performs (almost) as well as an "oracle".

It is also worth noticing that Theorem 4.2 provides a link with Approximation Theory. To see this let us assume that, for every integer D, the cardinality of the family of models with dimension D is finite. Then a typical choice of the weights is $x_m = x(D_m)$ with $x(D) = \alpha D + \ln |\{m \in \mathcal{M}; D_m = D\}|$ and $\alpha > 0$ so that those weights really represent the price to pay for redundancy (i.e., many models with the same dimension). The penalty can be taken as

$$\text{pen}(m) = \text{pen}(D_m) = K \varepsilon^2 \left(\sqrt{D_m} + \sqrt{2x(D_m)} \right)^2$$

and (4.7) becomes

$$\mathbb{E}_s \left[\|\hat{s}_{\hat{m}} - s\|^2 \right] \leq C' \inf_{D \geq 1} \left\{ b_D^2(s) + D \varepsilon^2 \left(1 + \sqrt{\frac{2x(D)}{D}} \right)^2 \right\},$$

where

$$b_D^2(s) = \inf_{m \in \mathcal{M}, D_m = D} \left(\|s_m - s\|^2 \right)$$

and the positive constant C' depends on K and α. This bound shows that the approximation properties of $\bigcup_{D_m = D} S_m$ are absolutely essential. One can hope substantial gains in the bias term when considering redundant models at some reasonable price since the dependency of $x(D)$ with respect to the number of models with the same dimension is logarithmic. This is typically what happens when one uses wavelet expansions to denoise some signal. More generally, most of the constructive methods of approximation of a function s that we know are based on infinite dimensional expansions with respect to some special bases (such as polynomials, piecewise polynomials, trigonometric polynomials, wavelets, splines, etc.) and therefore naturally lead to collections of finite dimensional linear models S_m for which the bias term $b_D^2(s)$ can be controlled in term of the various moduli of smoothness of s.

Many examples of applications of Theorem 4.2 are to be found in [23] and several of them will be detailed in Section 4.3. We first focus here on two cases example: variable selection and change points detection.

Variable Selection

Let $\{\varphi_j, j \in \Lambda\}$ be some collection of linearly independent functions. For every subset m of Λ we define S_m to be the linear span of $\{\varphi_j, j \in m\}$ and we consider some collection $\mathcal{M}$ of subsets of Λ. We first consider the *ordered* variable selection problem. In this case we take $\Lambda = \{1, \dots, N\}$ or $\Lambda = \mathbb{N}^*$ and define $\mathcal{M}$ as the collection of subsets of Λ of the form $\{1, \dots, D\}$. Then, one can take $\mathrm{pen}\,(m) = K' \,|m|\,/n$ with $K' > 1$. This leads to an oracle inequality of the form

$$\mathbb{E}_s\left[\left\|\widehat{s}_{\widehat{m}} - s\right\|^2\right] \leq C' \inf_{m \in \mathcal{M}} \mathbb{E}_s\left[\left\|\widehat{s}_m - s\right\|^2\right].$$

Hence the selected estimator behaves like an oracle. It can be shown that the restriction $K' > 1$ is sharp (see Section 4.2.2 below). Indeed, if $K' < 1$ the selection criterion typically explodes in the sense that it systematically selects models with large dimensions (or order N if $\Lambda = \{1, \dots, N\}$ or tending to infinity if $\Lambda = \mathbb{N}^*$) provided that ε is small enough.

In the *complete* variable selection context, $\mathcal{M}$ is the collection of all subsets of $\Lambda = \{1, \dots, N\}$. Taking $x_m = |m| \ln (N)$ leads to

$$\Sigma = \sum_{m \in \mathcal{M}} \exp\left(-x_m\right) = \sum_{D \leq N} \binom{N}{D} \exp\left(-D \ln (N)\right) \leq e$$

and

$$\mathrm{pen}\,(m) = K\varepsilon^2 \,|m|\left(1 + \sqrt{2\ln (N)}\right)^2$$

with $K > 1$. Then

$$\mathbb{E}_s\left[\left\|\widehat{s}_{\widehat{m}} - s\right\|^2\right] \leq C'\,(K) \inf_{D \geq 1} \left\{b_D^2\,(s) + D\,(1 + \ln (N))\,\varepsilon^2\right\}, \qquad (4.8)$$

where

$$b_D^2\,(s) = \inf_{m \in \mathcal{M}, |m| = D} \left(\|s_m - s\|^2\right)$$

and we see that the extra factor $\ln (N)$ is a rather modest price to pay as compared to the potential gain in the bias term provided by the redundancy of models with the same dimension. Interestingly, no orthogonality assumption is required on the system of functions $\{\varphi_j, j \leq N\}$ to derive this result. However whenever $\{\varphi_j, j \leq N\}$ is an orthonormal system, the penalized LSE can be explicitly computed and one recover the hard-thresholding estimator introduced by Donoho and Johnstone in the white noise framework (see [47]). Indeed it is easy to check that $\widehat{s}_{\widehat{m}}$ is simply equal to the thresholding estimator defined by

$$\widetilde{s}_T = \sum_{j=1}^{N} \widehat{\beta}_j \mathbb{1}_{\left|\widehat{\beta}_j\right| \geq T}\varphi_j \qquad (4.9)$$

where the $\widehat{\beta}_j$'s are the empirical coefficients (i.e., $\widehat{\beta}_j = Y_\varepsilon(\varphi_j)$) and $T = \sqrt{K}\varepsilon\left(1 + \sqrt{2\ln(N)}\right)$. Again the restriction $K > 1$ turns out to be sharp (see Section 4.2.2 below).

Note that the previous computations for the weights can be slightly refined. More precisely it is possible to replace the logarithmic factor $\ln(N)$ above by $\ln(N/|m|)$. Indeed, we first recall the classical upper bound for the binomial coefficient (which a fortiori derives from (2.9))

$$\ln\binom{N}{D} \le D\ln\left(\frac{eN}{D}\right). \tag{4.10}$$

So defining x_m as $x_m = |m|\,L(|m|)$ leads to

$$\Sigma = \sum_{D \le N}\binom{N}{D}\exp[-DL(D)] \le \sum_{D \le N}\left(\frac{eN}{D}\right)^D\exp[-DL(D)]$$

$$\le \sum_{D \le N}\exp\left[-D\left[L(D) - 1 - \ln\left(\frac{N}{D}\right)\right]\right].$$

Hence the choice $L(D) = 1 + \theta + \ln(N/D)$ with $\theta > 0$ leads to $\Sigma \le \sum_{D=0}^\infty e^{-D\theta} = \left[1 - e^{-\theta}\right]^{-1}$. Choosing $\theta = \ln 2$ for the sake of simplicity we may take

$$\mathrm{pen}\,(m) = K\varepsilon^2\,|m|\left(1 + \sqrt{2\left(1 + \ln\left(2N/|m|\right)\right)}\right)^2$$

with $K > 1$ and derive the following bound for the corresponding penalized LSE

$$\mathbb{E}_s\left[\|\widehat{s}_{\widehat{m}} - s\|^2\right] \le C''\inf_{1 \le D \le N}\left\{b_D^2(s) + D\left(1 + \ln(N/D)\right)\varepsilon^2\right\}, \tag{4.11}$$

where $b_D^2(s) = \inf_{m \in \mathcal{M},|m|=D}\left(\|s_m - s\|^2\right)$. This bound is slightly better than (4.8).

On the other hand, the penalized LSE is also rather easy to compute when the system $\{\varphi_j\}_{j \le N}$ is orthonormal. Indeed

$$\inf_{m \in \mathcal{M}}\left\{-\sum_{j \in m}\widehat{\beta}_j^2 + K\varepsilon^2|m|\left(1 + \sqrt{2L(|m|)}\right)^2\right\}$$

$$= \inf_{0 \le D \le N}\left\{-\sup_{\{m\,|\,|m|=D\}}\sum_{j \in m}\widehat{\beta}_j^2 + K\varepsilon^2|D|\left(1 + \sqrt{2L(|D|)}\right)^2\right\}$$

$$= \inf_{0 \le D \le N}\left\{-\sum_{j=1}^D\widehat{\beta}_{(j)}^2 + K\varepsilon^2|D|\left(1 + \sqrt{2L(|D|)}\right)^2\right\}$$

where $\widehat{\beta}_{(1)}^2 \geq \ldots \geq \widehat{\beta}_{(N)}^2$ are the squared estimated coefficients of s in decreasing order. We see that minimizing the penalized least squares criterion amounts to select a value $\widehat{D}$ of D which minimizes

$$-\sum_{j=1}^{D} \widehat{\beta}_{(j)}^2 + K\varepsilon^2 |D| \left(1 + \sqrt{2L(|D|)}\right)^2$$

and finally compute the penalized LSE as

$$\widehat{s_{\widehat{m}}} = \sum_{j=1}^{\widehat{D}} \widehat{\beta}_{(j)} \varphi_{(j)}. \tag{4.12}$$

The interesting point is that the risk bound 4.11 which holds true for this estimator cannot be further improved since it turns out to be optimal in a minimax sense on each set $\mathbb{S}_D^N = \bigcup_{|m|=D} S_m$, $D \leq N$ as we shall see in Section 4.3.1.

Change Points Detection

We consider the change points detection on the mean problem described above. Recall that one observes the noisy signal

$$\xi_j = s(j/n) + \varepsilon_j, \ 1 \leq j \leq n$$

where the errors are i.i.d. random standard normal variables. Defining S_m as the linear space of piecewise constant functions on the partition m, the change points detection problem amounts to select a model among the family $\{S_m\}_{m \in \mathcal{M}}$, where $\mathcal{M}$ denotes the collection of all possible partitions by intervals with end points on the grid $\{j/n, 0 \leq j \leq n\}$. Since the number of models with dimension D, i.e., the number of partitions with D pieces is equal to $\binom{n-1}{D-1}$, this collection of models has about the same combinatorial properties as the family of models corresponding to complete variable selection among $N = n - 1$ variables. Hence the same considerations concerning the penalty choice and the same resulting risk bounds as for complete variable selection hold true.

4.2.2 Lower Bounds for the Penalty Term

Following [23], our aim in this section is to show that a choice of $K < 1$ in (4.6) may lead to penalized LSE which behave in a quite unsatisfactory way. This means that the restriction $K > 1$ in Theorem 4.2 is, in some sense, necessary and that a choice of K smaller than one should be avoided. In the results presented below we prove that under-penalized least squares criteria explode when $s = 0$. Further results in the same direction are provided in [25] that include a study of the explosion phenomenon when s belongs to a low dimensional model for more general collections of model than below.

A Small Number of Models

We first assume that, for each D, the number of elements $m \in \mathcal{M}$ such that $D_m = D$ grows at most sub-exponentially with respect to D. In such a case, (4.5) holds with $x_m = L D_m$ for all $L > 0$ and one can apply Theorem 4.2 with a penalty of the form $\text{pen}\,(m) = K \varepsilon^2 \left(1 + \sqrt{2L}\right)^2 D_m$, where $K - 1$ and L are positive but arbitrarily close to 0. This means that, whatever $K' > 1$, the penalty $\text{pen}\,(m) = K' \varepsilon^2 D_m$ is allowed. Alternatively, the following result shows that if the penalty function satisfies $\text{pen}\,(\overline{m}) = K' \varepsilon^2 D_{\overline{m}}$ with $K' < 1$, even for *one* single model $S_{\overline{m}}$, provided that the dimension of this model is large enough (depending on K'), the resulting procedure behaves quite poorly if $s = 0$.

Proposition 4.3 *Let us assume that $s = 0$. Consider some collection of models $\{S_m\}_{m \in \mathcal{M}}$ such that*

$$\sum_{m \in \mathcal{M}} e^{-x D_m} < \infty, \quad \text{for any } x > 0. \tag{4.13}$$

Given $\text{pen} : \mathcal{M} \to \mathbb{R}_+$ *we set*

$$\text{crit}\,(m) = -\left\|\widehat{s}_m\right\|^2 + \text{pen}\,(m)$$

and either set $D_{\widehat{m}} = +\infty$ if $\inf_{m \in \mathcal{M}} \text{crit}\,(m) = -\infty$ or define $\widehat{m}$ such that $\text{crit}\,(\widehat{m}) = \inf_{m \in \mathcal{M}} \text{crit}\,(m)$ otherwise. Then, for any pair of real numbers $K, \delta \in (0,1)$, there exists some integer $\overline{N}$, depending only on K and δ, with the following property: if for some $\overline{m} \in \mathcal{M}$ with $D_{\overline{m}} \geq \overline{N}$

$$\text{pen}\,(\overline{m}) \leq K \varepsilon^2 D_{\overline{m}} \quad , \tag{4.14}$$

whatever the value of the penalty $\text{pen}\,(m)$ for $m \neq \overline{m}$ one has

$$\mathbb{P}_0 \left[D_{\widehat{m}} \geq \frac{(1-K)}{2} D_{\overline{m}} \right] \geq 1 - \delta \text{ and } \mathbb{E}_0 \left[\left\|\widehat{s}_{\widehat{m}}\right\|^2 \right] \geq \frac{(1-\delta)(1-K)}{4} D_{\overline{m}} \varepsilon^2.$$

Proof. Let us define, for any $m \in \mathcal{M}$, the nonnegative random variable χ_m by $\chi_m^2 = \varepsilon^{-2} \left\|\widehat{s}_m\right\|^2$. Then,

$$\text{crit}\,(m) - \text{crit}\,(\overline{m}) = \left\|\widehat{s}_{\overline{m}}\right\|^2 - \left\|\widehat{s}_m\right\|^2 + \text{pen}\,(m) - \text{pen}\,(\widehat{m}) \quad \text{for all } m \in \mathcal{M},$$

and therefore, by (4.14),

$$\varepsilon^{-2}[\text{crit}\,(m) - \text{crit}\,(\overline{m})] \geq \chi_{\overline{m}}^2 - \chi_m^2 - K D_{\overline{m}}. \tag{4.15}$$

The following proof relies on an argument about the concentration of the variables χ_m^2 around their expectations. Indeed choosing some orthonormal basis $\{\varphi_\lambda, \lambda \in \Lambda_m\}$ of S_m and recalling that $s = 0$, we have $\chi_m^2 = \sum_{\lambda \in \Lambda_m} W^2(\varphi_\lambda)$, which means that χ_m is the Euclidean norm of a standard Gaussian random vector. We may use the Gaussian concentration inequality. Indeed, setting $Z_m = \chi_m - \mathbb{E}_0[\chi_m]$ or $Z_m = -\chi_m + \mathbb{E}_0[\chi_m]$, on the one hand by Theorem 3.4 , we derive that

$$\mathbb{P}_0\left[Z_m \geq \sqrt{2x}\right] \leq e^{-x}$$

and on the other hand (3.5) implies that

$$0 \leq \mathbb{E}_0\left[\chi_m^2\right] - \left(\mathbb{E}_0[\chi_m]\right)^2 \leq 1.$$

Since $\mathbb{E}_0\left[\chi_m^2\right] = D_m$, combining these inequalities yields

$$\mathbb{P}_0\left[\chi_m \leq \sqrt{D_m - 1} - \sqrt{2x}\right] \leq e^{-x} \tag{4.16}$$

and

$$\mathbb{P}_0\left[\chi_m \geq \sqrt{D_m} + \sqrt{2x}\right] \leq e^{-x} \tag{4.17}$$

Let us now set

$$\eta = (1 - K)/4 < 1/4; \qquad D = 2D_{\overline{m}}\eta < D_{\overline{m}}/2; \qquad L = \eta^2/12 \tag{4.18}$$

and assume that $\overline{N}$ is large enough for the following inequalities to hold:

$$e^{-LD}\sum_{m \in \mathcal{M}} e^{-LD_m} \leq \delta; \qquad LD \geq 1/6. \tag{4.19}$$

Let us introduce the event

$$\overline{\Omega} = \left[\bigcap_{D_m < D}\left\{\chi_m \leq \sqrt{D_m} + \sqrt{2L(D_m + D)}\right\}\right]$$

$$\bigcap \left[\bigcap_{D_m \geq D}\left\{\chi_m \geq \sqrt{D_m - 1} - \sqrt{2L(D_m + D)}\right\}\right].$$

Using either (4.17) if $D_m < D$ or (4.16) if $D_m \geq D$, we get by (4.19)

$$\mathbb{P}_0\left[\overline{\Omega}^c\right] \leq \sum_{m \in \mathcal{M}} e^{-L(D_m + D)} \leq \delta.$$

Moreover, on $\overline{\Omega}$, $\chi_m^2 \leq \left(1 + 2\sqrt{L}\right)^2 D$, for all m such that $D_m < D$ and, by (4.18) and (4.19), $\chi_{\overline{m}} \geq \sqrt{D_{\overline{m}} - 1} - \sqrt{3LD_{\overline{m}}}$ and $LD_{\overline{m}} > 1/3$. Therefore $\chi_{\overline{m}}^2 \geq D_{\overline{m}}\left(1 - 2\sqrt{3L}\right)$. Hence, on $\overline{\Omega}$, (4.15) and (4.18) yield

$$\varepsilon^{-2}[\text{crit}(m) - \text{crit}(\overline{m})] \geq D_{\overline{m}}\left(1 - 2\sqrt{3L}\right) - \left(1 + 2\sqrt{L}\right)^2 D - KD_{\overline{m}}$$

$$> (1 - \eta)D_{\overline{m}} - 3\eta D_{\overline{m}} - (1 - 4\eta)D_{\overline{m}} = 0,$$

for all m such that $D_m < D$. This immediately implies that $D_{\widehat{m}}$ cannot be smaller than D on $\overline{\Omega}$ and therefore,

$$\mathbb{P}_0[D_{\widehat{m}} \geq D] \geq \mathbb{P}_0[\overline{\Omega}] \geq 1 - \delta. \tag{4.20}$$

Moreover, on the same set $\overline{\Omega}$, $\chi_m \geq \sqrt{D_m - 1} - \sqrt{2L(D_m + D)}$ if m is such that $D_m \geq D$. Noticing that $D > 32$ and recalling that $\eta \leq 1/4$, we derive that on the set $\overline{\Omega}$ if m is such that $D_m \geq D$

$$\chi_m \geq \sqrt{D}\left(\sqrt{1 - \frac{1}{32}} - \frac{1}{8}\right) > \sqrt{\frac{D}{2}} \geq \sqrt{\eta D_{\overline{m}}}.$$

Hence, on $\overline{\Omega}$, $D_{\widehat{m}} \geq D$ and $\chi_m \geq \sqrt{\eta D_{\overline{m}}}$ for all m such that $D_m \geq D$. Therefore $\chi_{\widehat{m}} \geq \sqrt{\eta D_{\overline{m}}}$. Finally,

$$\mathbb{E}_0\left[\|\widehat{s}_{\widehat{m}}\|^2\right] = \varepsilon^2 \mathbb{E}_0\left[\chi_{\widehat{m}}^2\right] \geq \varepsilon^2 \eta D_{\overline{m}} \mathbb{P}_0\left[\chi_{\widehat{m}} \geq \sqrt{\eta D_{\overline{m}}}\right] \geq \varepsilon^2 \eta D_{\overline{m}} \mathbb{P}_0[\overline{\Omega}],$$

which, together with (4.18) and (4.20) concludes the proof. ■

In order to illustrate the meaning of this proposition, let us assume that we are given some orthonormal basis $\{\varphi_j\}_{j \geq 1}$ of $\mathbb{H}$ and that S_m is the linear span of $\varphi_1, \ldots, \varphi_m$ for $m \in \mathbb{N}$. Assume that $s = 0$ and pen $(m) = K\varepsilon^2 m$ with $K < 1$. If $\mathcal{M} = \mathbb{N}$, then Proposition 4.3 applies with $D_{\overline{m}}$ arbitrarily large and letting $D_{\overline{m}}$ go to infinity and δ to zero, we conclude that $\inf_{m \in \mathcal{M}}$ crit $(m) = -\infty$ a.s. If we set $\mathcal{M} = \{0, 1, \ldots, N\}$, then $\widehat{m}$ is well defined but, setting $D_{\overline{m}} = N$, we see that $\mathbb{E}_0\left[\|\widehat{s}_{\widehat{m}}\|^2\right]$ is of the order of $N\varepsilon^2$ when N is large. If, on the contrary, we choose pen $(m) = Km\varepsilon^2$ with $K = 2$, for instance, as in Mallows' C_p, then

$$\mathbb{E}_0\left[\|\widehat{s}_{\widehat{m}}\|^2\right] \leq C\varepsilon^2$$

This means that choosing a penalty of the form pen $(m) = K\varepsilon^2 m$ with $K < 1$ is definitely not advisable..

A Large Number of Models

The previous result corresponds to a situation where the number of models having the same dimension D is moderate in which case we can choose the weights x_m as LD_m for an arbitrary small positive constant L. This means that the influence of the weights on the penalty is limited in the sense that they only play the role of a correction to the main term $K\varepsilon^2 D_m$. The situation becomes quite different when the number of models having the same dimension D grows much faster with D. More precisely, if we turn back to the case of complete variable selection as described above and take $\{\varphi_j, 1 \leq j \leq N\}$ to be some orthonormal system, $\mathcal{M}$ to be the collection of all subsets of $\{1,\dots,N\}$. If for every subset m of $\{1,\dots,N\}$ we define S_m to be the linear span of $\{\varphi_j, j \in m\}$, setting

$$\mathrm{pen}\,(m) = K\varepsilon^2 \,|m|\,\left(1 + \sqrt{2\ln(N)}\right)^2$$

with $K > 1$, then the penalized LSE is merely the thresholding estimator $\tilde{s}_T$ with $T = \varepsilon\sqrt{K}\left(1 + \sqrt{2\ln N}\right)$ as defined by (4.9). If $s = 0$, we can analyze the quadratic risk of the thresholding estimator. Indeed, if ξ is a standard normal random variable

$$\mathbb{E}_0\left[\|\tilde{s}_T\|^2\right] = = \sum_{\lambda=1}^{N} \mathbb{E}_0\left[\hat{\beta}_\lambda^2 \mathbb{1}_{\{|\hat{\beta}_\lambda|>T\}}\right] = N\varepsilon^2 \mathbb{E}\left[\xi^2 \mathbb{1}_{\{|\varepsilon\xi|>T\}}\right].$$

It suffices now to apply the next elementary lemma (the proof of which is left as an exercise).

Lemma 4.4 *If ξ is standard normal and $t \geq 0$, then*

$$\mathbb{E}\left[\xi^2 \mathbb{1}_{\{\xi>t\}}\right] \geq \left(\frac{t}{\sqrt{2\pi}} \vee \frac{1}{2}\right)\exp\left(-\frac{t^2}{2}\right).$$

Hence,

$$\mathbb{E}_0\left[\|\tilde{s}_T\|^2\right] \geq \frac{N}{2}\varepsilon^2\left(\frac{T\sqrt{2}}{\varepsilon\sqrt{\pi}} \vee 1\right)\exp\left(-\frac{T^2}{2\varepsilon^2}\right) \qquad (4.21)$$

so that if $T = \varepsilon\sqrt{K}\left(1 + \sqrt{2\ln N}\right)$

$$2\mathbb{E}_0\left[\|\tilde{s}_T\|^2\right] \geq \varepsilon^2 \exp\left[(1-K)\ln N - K\left(\sqrt{2\ln N} + 1/2\right)\right]. \qquad (4.22)$$

If $K < 1$, this grows like ε^2 times a power of N when N goes to infinity, as compared to the risk bound $C'(K)\varepsilon^2 \ln N$ which derives from (4.8) when $K > 1$. Clearly the choice $K < 1$ should be avoided. The situation for Mallows' C_p is even worse since if $\mathrm{pen}\,(m) = 2\varepsilon^2\,|m|$ then the penalized LSE is still a thresholding estimator $\tilde{s}_T$ but this time with $T = \varepsilon\sqrt{2}$ and therefore by (4.21)

$$2e\mathbb{E}_0\left[\|\tilde{s}_T\|^2\right] \geq N\varepsilon^2.$$

This means that Mallows' C_p is definitely not suitable for complete variable selection involving a large number of variables, although it is a rather common practice to use them in this situation, as more or less suggested for instance in [55] p. 299.

4.2.3 Mixing Several Strategies

Looking at model selection as a way of defining some adaptive estimators of s, the initial choice of a collection of models becomes a prior and heavily depends on the type of problem we consider or the type of result we are looking for.

Going back to one of our initial examples of a function s belonging to some unknown Sobolev ball $W^\alpha(R)$, a consequence of what we shall see in the next section is that a good strategy to estimate s is to consider the ordered variable selection strategy on the trigonometric basis $\{\varphi_i\}_{i\geq 1}$ as described above. The resulting estimator will be shown to be minimax, up to constants, over all Sobolev balls of radius $R \geq \varepsilon$. Unfortunately, such a strategy is good if s belongs to some Sobolev ball, but it may be definitely inadequate when s belongs to some particular Besov ball. In this case, one should use quite different strategies, for instance a thresholding method (which, as we have seen, is a specific strategy for complete variable selection) in connection with a wavelet basis, rather than the trigonometric one.

These examples are illustrations of a general recipe for designing simple strategies in view of solving the more elementary problems of adaptation: choose some orthonormal basis $\{\varphi_\lambda\}_{\lambda\in\Lambda}$ and a countable family $\mathcal{M}$ of finite subsets m of Λ, then define S_m to be the linear span of $\{\varphi_\lambda\}_{\lambda\in m}$ and find a family of weights $\{x_m\}_{m\in\mathcal{M}}$ satisfying (4.5). In this view, the choice of a proper value of m amounts to a problem of variable selection from an infinite set of variables which are the coordinates vectors in the Gaussian sequence framework associated with the basis $\{\varphi_\lambda\}_{\lambda\in\Lambda}$. Obviously, the choice of a basis influences the approximation properties of the induced families of models. For instance the Haar basis is not suitable for approximating functions s which are "too smooth" (such that $\int_0^1 [s''(x)]^2\, dx$ is not large, say). If we have at hand a collection of bases, the choice of a "best" basis given s, ε and a strategy for estimating within each of the bases would obviously increase our quality of estimation of s. Therefore one would like to be able to use all bases simultaneously rather than choosing one in advance. This is, in particular, a reason for preferring the Gaussian linear process approach to the Gaussian sequence framework.

The problem of the basis choice has been first considered and solved by Donoho and Johnstone (see [49]) for selecting among the different thresholding estimators built on the various bases. The following theorem provides a generic way of mixing several strategies in order to retain the "best one" which is not especially devoted to thresholding.

Theorem 4.5 *Let $\mathcal{J}$ be a finite or countable set and μ a probability distribution on $\mathcal{J}$. For each $j \in \mathcal{J}$ we are given a collection $\{S_m\}_{m\in\mathcal{M}_j}$ of finite dimensional linear models with respective dimensions D_m and a collection of weights $\{L_{m,j}\}_{m\in\mathcal{M}_j}$ and we assume that the distribution μ satisfies*

$$\sum_{j\in\mathcal{J}} \mu(\{j\}) \left(\sum_{\{m\in\mathcal{M}_j\,|\,D_m>0\}} \exp[-D_m L_{m,j}] \right) = \Sigma < +\infty.$$

Let us consider for each $j \in \mathcal{J}$ a penalty function $\mathrm{pen}_j(\cdot)$ on $\mathcal{M}_j$ such that

$$\mathrm{pen}_j(m) \geq K\varepsilon^2 D_m \left(1 + \sqrt{2L_{m,j}} \right)^2 \quad \text{with } K > 1,$$

and the corresponding penalized LSE $\tilde{s}_j = \widehat{s}_{\widehat{m}_j}$ where $\widehat{m}_j$ minimizes the penalized least squares criterion $-\|\widehat{s}_m\|^2 + \mathrm{pen}_j(m)$ over $\mathcal{M}_j$. Let $\widehat{j}$ be a minimizer with respect to $j \in \mathcal{J}$ of

$$-\|\tilde{s}_j\|^2 + \mathrm{pen}_j(\widehat{m}_j) + \frac{2xK}{1-x}\varepsilon^2 l_j \quad \text{with } K^{-1} < x < 1 \quad \text{and } l_j = -\ln[\mu(\{j\})].$$

The resulting estimator $\tilde{s} = \tilde{s}_{\widehat{j}}$ then satisfies for some constant $C(x, K)$

$$\mathbb{E}_s\left[\|\tilde{s} - s\|^2\right] \leq C(x, K)\left[\inf_{j \in \mathcal{J}}\left\{R_j + \frac{2xK}{1-x}\varepsilon^2 l_j\right\} + \varepsilon^2(1 + \Sigma)\right],$$

with

$$R_j = \inf_{m \in \mathcal{M}_j}\left\{d^2(s, S_m) + \mathrm{pen}_j(m)\right\}.$$

Proof. Let $\mathcal{M} = \bigoplus_{j \in \mathcal{J}} \mathcal{M}_j \times \{j\}$ and set for all $(m, j) \in \mathcal{M}$ such that $D_m > 0$, $L'_{(m,j)} = L_{m,j} + D_m^{-1}l_j$. Then

$$\sum_{\{(m,j) \in \mathcal{M} \mid D_m > 0\}} \exp[-D_m L'_{(m,j)}] = \Sigma.$$

Let $\mathrm{pen}(m, j) = \mathrm{pen}_j(m) + [(2xK)/(1-x)]\varepsilon^2 l_j$, for all $(m, j) \in \mathcal{M}$. Using $\sqrt{a+b} \leq \sqrt{a} + \sqrt{b}$, we derive that

$$\left(\sqrt{D_m} + \sqrt{2L_{m,j}D_m + 2l_j}\right)^2 \leq D_m\left(1 + \sqrt{2L_{m,j}}\right)^2 + 2l_j + 2\sqrt{2l_jD_m},$$

which implies since $2\sqrt{2l_jD_m} \leq 2l_jx/(1-x) + D_m(1-x)/x$ that

$$\left(\sqrt{D_m} + \sqrt{2L_{m,j}D_m + 2l_j}\right)^2 \leq x^{-1}D_m\left(1 + \sqrt{2L_{m,j}}\right)^2 + 2l_j/(1-x).$$

It then follows that

$$\mathrm{pen}(m, j) \geq xK\varepsilon^2\left[x^{-1}D_m\left(1 + \sqrt{2L_{m,j}}\right)^2 + 2l_j/(1-x)\right]$$

$$\geq xK\varepsilon^2\left(\sqrt{D_m} + \sqrt{2L_{m,j}D_m + 2l_j}\right)^2,$$

and therefore

$$\mathrm{pen}(m, j) \geq xK\varepsilon^2 D_m\left(1 + \sqrt{2L'_{(m,j)}}\right)^2. \tag{4.23}$$

We can now apply Theorem 4.2 to the strategy defined for all $(m, j) \in \mathcal{M}$ by the model S_m and the penalty $\mathrm{pen}(m, j)$. By definition, the resulting estimator is clearly $\tilde{s}$ and the risk bound follows from (4.7) with K replaced by $xK > 1$ because of (4.23). ∎

Comments.

- The definition of $\mathcal{M}$ that we used in the proof of the theorem may lead to situations where the same model S_m appears several times with possibly different weights. This is indeed not a problem since such a redundancy is perfectly allowed by Theorem 4.2.
- Note that the choice of a suitable value of x leads to the same difficulties as the choice of K and one should avoid to take xK close to 1.

The previous theorem gives indeed a solution to the problems we considered before. If one wants to mix a moderate number of strategies one can build a "superstrategy" as indicated in the theorem, with μ the uniform distribution on $\mathcal{J}$, and the price to pay in the risk is an extra term of order $\varepsilon^2 \ln(|\mathcal{J}|)$. In this case, the choice of $\widehat{j}$ is particularly simple since it should merely satisfy $\|\tilde{s}_{\widehat{j}}\|^2 - \operatorname{pen}_{\widehat{j}}(\widehat{m}_{\widehat{j}}) = \sup_{j \in \mathcal{J}} \left\{ \|\tilde{s}_j\|^2 - \operatorname{pen}_j(\widehat{m}_j) \right\}$. If $\mathcal{J}$ is too large, one should take a different "prior" than the uniform on the set of available strategies. One should put larger values of $\mu(\{j\})$ for the strategies corresponding to values of s we believe are more likely and smaller values for the other strategies. As for the choice of the weights the choice of μ may have some Bayesian flavor (see [23]).

4.3 Adaptive Estimation in the Minimax Sense

The main advantage of the oracle type inequality provided by Theorem 4.2 is that it holds for every given s. Its main drawback is that it allows a comparison of the risk of the penalized LSE with the risk of any estimator among the original collection $\{\widehat{s}_m\}_{m \in \mathcal{M}}$ but not with the risk of other possible estimators of s. Of course it is well known that there is no hope to make a pointwise risk comparison with an arbitrary estimator since a constant estimator equal to s_0 for instance is perfect at s_0 (but otherwise terrible). Therefore it is more reasonable to take into account the risk of estimators at different points simultaneously. One classical possibility is to consider the maximal risk over suitable subsets $\mathcal{T}$ of $\mathbb{H}$. This is the minimax point of view: an estimator is "good" if its maximal risk over $\mathcal{T}$ is close to the *minimax risk* given by

$$R_M(\mathcal{T}, \varepsilon) = \inf_{\widehat{s}} \sup_{s \in \mathcal{T}} \mathbb{E}_s \left[\|\widehat{s} - s\|^2 \right]$$

where the infimum is taken over all possible estimators of $\widehat{s}$, i.e., measurable functions of Y_ε which possibly also depend on $\mathcal{T}$. The performance of an estimator $\widehat{s}$ (generally depending on ε) can then be measured by the ratio

$$\sup_{s \in \mathcal{T}} \frac{\mathbb{E}_s \left[\|\widehat{s} - s\|^2 \right]}{R_M(\mathcal{T}, \varepsilon)},$$

and the closer this ratio to one, the better the estimator. In particular, if this ratio is bounded independently of ε, the estimator $\hat{s}$ will be called *approximately minimax* with respect to $\mathcal{T}$. Many approximately minimax estimators have been constructed for various sets $\mathcal{T}$. As for the case of Sobolev balls, they typically depend on $\mathcal{T}$ which is a serious drawback. One would like to design estimators which are approximately minimax for many subsets $\mathcal{T}$ simultaneously, for instance all Sobolev balls $W^{\alpha}(R)$, with $\alpha > 0$ and $R \geq \varepsilon$. The construction of such adaptive estimators has been the concern of many statisticians. We shall mention below some of Donoho, Johnstone, Kerkyacharian and Picard's works on hard thresholding of wavelet coefficients (see [53] for a review). This procedure can be interpreted as a model selection via penalization procedure but many other methods of adaptive estimation have been designed. All these methods rely on selection (or aggregation) procedures among a list of estimators. Of course the selection principles on which they are based (we typically have in mind Lepskii's method in [80] and [81]) may substantially differ from penalization. It is not our purpose here to open a general discussion on this topic and we refer to [12], Section 5 for a detailed discussion of adaptation with many bibliographic citations.

In order to see to what extent our method allows to build adaptive estimators in various situations, we shall consider below a number of examples and for any such example, use the same construction. Given a class of sets $\{\mathcal{T}_{\theta}\}_{\theta \in \Theta}$ we choose a family of models $\{S_m\}_{m \in \mathcal{M}}$ which adequately approximate those sets. This means that we choose the models in such a way that any s belonging to some $\mathcal{T}_{\theta}$ can be closely approximated by some model of the family. Then we choose a convenient family of weights $\{x_m\}_{m \in \mathcal{M}}$. Theses choices completely determine the construction of the penalized LSE $\tilde{s}$ (up to the choice of K which is irrelevant in term of rates since it only influences the constants). In order to analyze the performances of $\tilde{s}$, it is necessary to evaluate, for each $\theta \in \Theta$

$$\sup_{s \in \mathcal{T}_{\theta}} \inf_{m \in \mathcal{M}} \left(\|s_m - s\|^2 + \mathrm{pen}\,(m) \right)$$

Hence we first have to bound the bias term $\|s_m - s\|^2$ for each $m \in \mathcal{M}$, which derives from Approximation Theory, then proceed to the minimization with respect to $m \in \mathcal{M}$. In order to be able to compare the maximal risk of a penalized LSE on a set $\mathcal{T}_{\theta}$ with the minimax risk on $\mathcal{T}_{\theta}$ we need to establish lower bounds for the minimax risk for various classes of sets $\{\mathcal{T}_{\theta}\}_{\theta \in \Theta}$.

4.3.1 Minimax Lower Bounds

In order to use Birgé's lemma (more precisely Corollary 2.19), our first task is to compute the mutual Kullback–Leibler information numbers between Gaussian distributions given by (4.1). Note that in the white noise framework, identity (4.24) below is nothing else than the celebrated Girsanov formula.

Lemma 4.6 *Let us denote by* $\mathbb{P}_s$ *the distribution of* Y_ε *given by (4.1) on* $\mathbb{R}^{\mathbb{H}}$. *Then firstly* $\mathbb{P}_s$ *is absolutely continuous with respect to* $\mathbb{P}_0$ *with*

$$\frac{d\mathbb{P}_s}{d\mathbb{P}_0}(y) = \exp\left[\varepsilon^{-2}\left(y(s) - \frac{\|s\|^2}{2}\right)\right] \qquad (4.24)$$

and secondly, for every $t, s \in \mathbb{H}$

$$\mathbf{K}(\mathbb{P}_t, \mathbb{P}_s) = \frac{1}{2\varepsilon^2}\|s - t\|^2. \qquad (4.25)$$

Proof. Let W be an isonormal process on $\mathbb{H}$. To check that (4.24) holds true, setting

$$Z_\varepsilon(s) = \varepsilon^{-2}\left(\varepsilon W(s) - \frac{\|s\|^2}{2}\right)$$

it is enough to verify that for every finite subset m of $\mathbb{H}$ and any bounded and measurable function h on $\mathbb{R}^m$

$$\mathbb{E}\left[h\left((\varepsilon W(u))_{u \in m}\right)\exp\left(Z_\varepsilon(s)\right)\right] = \mathbb{E}\left[h\left((\langle s, u\rangle + \varepsilon W(u))_{u \in m}\right)\right]. \qquad (4.26)$$

Let Π_m denote the orthogonal projection operator onto the linear span $\langle m\rangle$ of m and $\{\varphi_j\}_{1 \leq j \leq k}$ be some orthonormal basis of $\langle m\rangle$. Since $W(s) - W(\Pi_m s)$ is independent from $\left((W(u))_{u \in m}, W(\Pi_m s)\right)$ the left-hand side of (4.26) equals

$$L = \mathbb{E}\left[h\left((\varepsilon W(u))_{u \in m}\right)\exp\left(\varepsilon^{-2}\left(\varepsilon W(\Pi_m s) - \frac{\|s\|^2}{2}\right)\right)\right]$$
$$\times \mathbb{E}\left[\exp\left(\frac{W(s) - W(\Pi_m s)}{\varepsilon}\right)\right].$$

But $W(s) - W(\Pi_m s)$ is a centered normal random variable with variance $\|s - \Pi_m s\|^2$, hence

$$\mathbb{E}\left[\exp\left(\frac{W(s) - W(\Pi_m s)}{\varepsilon}\right)\right] = \exp\left(\frac{\|s - \Pi_m s\|^2 \varepsilon^{-2}}{2}\right)$$

and therefore by Pythagore's identity

$$L = \mathbb{E}\left[h\left((\varepsilon W(u))_{u \in m}\right)e^{\varepsilon^{-2}\left(\varepsilon W(\Pi_m s) - (\|\Pi_m s\|^2/2)\right)}\right]. \qquad (4.27)$$

Now for every $t \in \langle m\rangle$, we can write

$$W(t) = \sum_{j=1}^{k}\langle t, \varphi_j\rangle W(\varphi_j)$$

and the $W(\varphi_j)$'s are i.i.d. standard normal random variables. So, setting $|x|_2^2 = \sum_{j=1}^k x_j^2$ for all $x \in \mathbb{R}^k$, the right-hand side of (4.26) can be written as

$$R = (2\pi)^{-k/2} \int_{\mathbb{R}^k} h\left(\left(\langle s, u \rangle + \varepsilon \sum_{j=1}^k \langle u, \varphi_j \rangle x_j\right)_{u \in m}\right) e^{-|x|_2^2/2} dx$$

which leads, setting $x_j = y_j - \langle s, \varphi_j \rangle \varepsilon^{-1}$ and, for every $t \in \langle m \rangle$, $w(t, y) = \sum_{j=1}^k \langle t, \varphi_j \rangle y_j$, to

$$R = (2\pi)^{-k/2} \int_{\mathbb{R}^k} h\left((\varepsilon w(u, y))_{u \in m}\right) e^{-\varepsilon^{-2}\left(\|\Pi_m s\|^2/2\right) + \varepsilon^{-1} w(\Pi_m s, y) - \left(|y|_2^2\right)/2} dy$$

and therefore $L = R$ via (4.27). Hence (4.26) and (4.24) hold true. We turn now to the computation of the Kullback–Leibler information number. By (4.24) we have

$$\mathbf{K}(\mathbb{P}_t, \mathbb{P}_s) = \frac{\varepsilon^{-2}}{2}\left(\|s\|^2 - \|t\|^2\right) + \varepsilon^{-1}\mathbb{E}\left[(W(t) - W(s)) e^{Z_\varepsilon(t)}\right]. \quad (4.28)$$

Introducing the orthogonal projection $\tilde{s} = \left(\langle s, t \rangle / \|t\|^2\right) t$ of s onto the linear span of t and using the fact that $W(\tilde{s}) - W(s)$ is centered at expectation and independent from $W(t)$ leads to

$$\mathbb{E}\left[(W(t) - W(s)) e^{\varepsilon^{-1} W(t)}\right] = \mathbb{E}\left[(W(t) - W(\tilde{s})) e^{\varepsilon^{-1} W(t)}\right]$$

$$= \frac{1}{\|t\|^2}\left(\|t\|^2 - \langle s, t \rangle\right) \mathbb{E}\left[W(t) e^{\varepsilon^{-1} W(t)}\right]$$

To finish the computation it remains to notice that $W(t) / \|t\|$ is a standard normal random variable and use the fact that for such a variable ξ one has for every real number λ

$$\mathbb{E}\left[\xi e^{\lambda \xi}\right] = \lambda e^{\lambda^2/2}.$$

Hence

$$\mathbb{E}\left[W(t) e^{\varepsilon^{-1} W(t)}\right] = \frac{\|t\|^2}{\varepsilon} e^{\varepsilon^{-2}\left(\|t\|^2/2\right)}$$

and therefore

$$\mathbb{E}\left[(W(t) - W(s)) e^{\varepsilon^{-1} W(t) - \varepsilon^{-2}\left(\|t\|^2/2\right)}\right] = \frac{\|t\|^2 - \langle s, t \rangle}{\varepsilon}.$$

Plugging this identity in (4.28) finally yields

$$\mathbf{K}(\mathbb{P}_t, \mathbb{P}_s) = \frac{1}{2\varepsilon^2}\left(\|s\|^2 - \|t\|^2\right) + \frac{1}{\varepsilon^2}\left(\|t\|^2 - \langle s, t \rangle\right)$$

and (4.25) follows. ∎

One readily derives from (4.24) (which is merely the Girsanov–Cameron–Martin formula in the white noise framework), that the least squares criterion is equivalent to the maximum likelihood criterion (exactly as in the finite dimensional case).

A Minimax Lower Bound on Hypercubes and ℓ_p-Bodies

Assume $\mathbb{H}$ to be infinite dimensional and consider some orthonormal basis $(\varphi_j)_{j\geq 1}$ of $\mathbb{H}$. Let $(c_j)_{j\geq 1}$ be a nonincreasing sequence converging to 0 as n goes to infinity. For $p \leq 2$, we consider the ℓ_p-body

$$\mathcal{E}_p(c) = \left\{ t \in \mathbb{H} : \sum_{j\geq 1} |\langle t, \varphi_j \rangle / c_j|^p \leq 1 \right\}.$$

Apart from Corollary 2.19 the main argument that we need is a combinatorial extraction Lemma known under the name of Varshamov–Gilbert's lemma and which derives from Chernoff's inequality for a symmetric binomial random variable.

Lemma 4.7 (*Varshamov–Gilbert's lemma*) Let $\{0,1\}^D$ be equipped with Hamming distance δ. Given $\alpha \in (0,1)$, there exists some subset Θ of $\{0,1\}^D$ with the following properties

$$\delta(\theta,\theta') > (1-\alpha)D/2 \text{ for every } (\theta,\theta') \in \Theta^2 \text{ with } \theta \neq \theta' \qquad (4.29)$$

$$\ln|\Theta| \geq \rho D/2, \qquad (4.30)$$

where $\rho = (1+\alpha)\ln(1+\alpha) + (1-\alpha)\ln(1-\alpha)$. In particular $\rho > 1/4$ when $\alpha = 1/2$.

Proof. Let Θ be a maximal family satisfying (4.29) and denote by $\mathcal{B}(\theta,r)$ the closed ball with radius r and center θ. Then, the maximality of Θ implies that

$$\bigcup_{\theta \in \Theta} \mathcal{B}(\theta,(1-\alpha)D/2) = \{0,1\}^D$$

and therefore

$$2^{-D} \sum_{\theta \in \Theta} |\mathcal{B}(\theta,(1-\alpha)D/2)| \geq 1.$$

Now, if S_D follows the binomial distribution $\text{Bin}(D,1/2)$ one clearly has for every $\theta \in \Theta$

$$2^{-D} |\mathcal{B}(\theta,(1-\alpha)D/2)| = P[S_D \leq (1-\alpha)D/2] = P[S_D \geq (1+\alpha)D/2],$$

which implies by Chernoff's inequality (2.2) and (2.37) that

$$2^{-D} |\mathcal{B}(\theta,(1-\alpha)D/2)| \leq e^{-Dh_{1/2}((1+\alpha)/2)} = e^{-\rho D/2}$$

and therefore

$$|\Theta| \geq e^{\rho D/2}.$$

Hence the result. ■

We can now prove a lower bound on a hypercube which will in turn lead to a minimax lower bound on some arbitrary ℓ_p-body.

Proposition 4.8 *Let D be some positive integer and r be some arbitrary positive number. Define the hypercube $\mathcal{C}_D(r) = \left\{ r \sum_{j=1}^{D} \theta_j \varphi_j, \theta \in [0,1]^D \right\}$, then there exists some absolute constant κ_0 such that for any estimator $\widehat{s}$ of s one has*

$$\sup_{s \in \mathcal{C}_D(r)} \mathbb{E}_s \left[\|\widehat{s} - s\|^2 \right] \geq \kappa_0 \left(r \wedge \varepsilon \right)^2 D. \tag{4.31}$$

Proof. Combining (4.25), Varshamov–Gilbert's lemma (Lemma 4.7) and Corollary 2.19, we derive that

$$4 \sup_{s \in \mathcal{C}_D(r)} \mathbb{E}_s \left[\|\widehat{s} - s\|^2 \right] \geq \frac{r^2 D}{4} (1 - \kappa)$$

provided that

$$\frac{\varepsilon^{-2} D r^2}{2} \leq \frac{\kappa D}{8}$$

i.e., $r \leq \varepsilon \sqrt{\kappa}/2$, which implies the desired result with $\kappa_0 = \kappa (1 - \kappa) / 64$. ∎

Choosing $r = \varepsilon$, a trivial consequence of Proposition 4.8 is that the LSE on a given D-dimensional linear subspace of $\mathbb{H}$ is approximately minimax. Of course a much more precise result can be proved. It is indeed well known that the LSE on a linear finite dimensional space is exactly minimax. In other words, the minimax risk is exactly equal to $D\varepsilon^2$ (this is a classical undergraduate exercise: one can compute exactly the Bayes quadratic risk when the prior is the D-dimensional centered Gaussian measure and covariance matrix $\sigma^2 I_D$ and see that it tends to $D\varepsilon^2$ when σ tends to infinity). It is now easy to derive from the previous lower bound on a hypercube, a minimax lower bound on some arbitrary ℓ_p-body, just by considering a sub-hypercube with maximal size of the ℓ_p-body.

Theorem 4.9 *Let κ_0 be the absolute constant of Proposition 4.8, then*

$$\inf_{\widehat{s}} \sup_{s \in \mathcal{E}_p(c)} \mathbb{E}_s \left[\|\widehat{s} - s\|^2 \right] \geq \kappa_0 \sup_{D \geq 1} \left(D^{1-2/p} c_D^2 \wedge D\varepsilon^2 \right) \tag{4.32}$$

where the infimum in the left-hand side is taken over the set of all estimators based on the observation of Y_ε given by (4.1).

Proof. Let D be some arbitrary positive integer and define $r = D^{-1/p} c_D$. Consider the hypercube $\mathcal{C}_D(r)$, then since by definition of r and monotonicity of $(c_j)_{j \geq 1}$ one has

$$r^p \sum_{j=1}^{D} c_j^{-p} \leq r^p D c_D^{-p} \leq 1,$$

the hypercube $\mathcal{C}_D(r)$ is included in the ℓ_p-body $\mathcal{E}_p(c)$. Hence for any estimator $\widehat{s}$, we derive from (4.31) that

$$\sup_{s \in \mathcal{E}_p(c)} \mathbb{E}_s\left[\|\widehat{s} - s\|^2\right] \geq \sup_{s \in \mathcal{C}_D(r)} \mathbb{E}_s\left[\|\widehat{s} - s\|^2\right] \geq \kappa_0 D\left(r^2 \wedge \varepsilon^2\right).$$

This leads to the desired lower bound. ∎

We shall see in the next section that this lower bound is indeed sharp (up to constant) for arbitrary ellipsoids and also for ℓ_p-bodies with $p < 2$ with some arithmetic decay of $(c_j)_{j \geq 1}$ (the Besov bodies). For arbitrary ℓ_p-bodies, the problem is much more delicate and this lower bound may miss some logarithmic factors (see the results on ℓ_p-balls in [23]). We turn now to a possible refinement of the previous technique which precisely allows to exhibit necessary logarithmic factors in the lower bounds. We shall not present here the application of this technique to ℓ_p-balls which would lead us too far from our main goal (we refer the interested reader to [23]). We content ourselves with the construction of a lower bound for complete variable selection.

A Minimax Lower Bound for Variable Selection

Our aim is here to analyze the minimax risk of estimation for the complete variable selection problem. Starting from a finite orthonormal system $\{\varphi_j, 1 \leq j \leq N\}$ of elements of $\mathbb{H}$, we want to show that the risk of estimation for a target of the form $s = \sum_{j=1}^{N} \beta_j \varphi_j$, knowing that at most D among the coefficients β's are nonzero is indeed influenced by the knowledge of the set where those coefficients are nonzero. In other words, there is a price to pay if you do not know in advance what is this set. More precisely, we consider for every subset m of $\{1, \ldots, N\}$ the linear span S_m of $\{\varphi_j, j \in m\}$ and the set $\mathbb{S}_D^N = \bigcup_{|m|=D} S_m$. Our purpose is to evaluate the minimax risk for the quadratic loss on the parameter space $\mathbb{S}_D^N = \bigcup_{|m|=D} S_m$. The approach is the same as in the previous section, except that we work with a more complicated object than a hypercube which forces us to use a somehow more subtle version of the Varshamov–Gilbert Lemma. Let us begin with the statement of this combinatorial Lemma due to [21]. The proof that we present here is borrowed from [102] and relies on exponential hypergeometric tail bounds rather than tail bounds for the symmetric binomial as in the previous section.

Lemma 4.10 *Let* $\{0,1\}^N$ *be equipped with Hamming distance* δ *and given* $1 \leq D < N$ *define* $\{0,1\}_D^N = \left\{x \in \{0,1\}^N : \delta(0,x) = D\right\}$. *For every* $\alpha \in (0,1)$ *and* $\beta \in (0,1)$ *such that* $D \leq \alpha\beta N$, *there exists some subset* Θ *of* $\{0,1\}_D^N$ *with the following properties*

$$\delta(\theta, \theta') > 2(1 - \alpha)D \text{ for every } (\theta, \theta') \in \Theta^2 \text{ with } \theta \neq \theta' \tag{4.33}$$

$$\ln|\Theta| \geq \rho D \ln\left(\frac{N}{D}\right), \tag{4.34}$$

where

$$\rho = \frac{\alpha}{-\ln(\alpha\beta)}\left(-\ln(\beta) + \beta - 1\right).$$

In particular, one has $\rho \geq 0.233$ for $\alpha = 3/4$ and $\beta = 1/3$.

Proof. Let Θ be a maximal subset of $\{0,1\}_D^N$ satisfying property (4.33), then the closed balls with radius $2(1-\alpha)D$ which centers belong to Θ are covering $\{0,1\}_D^N$. Hence

$$\binom{N}{D} \leq \sum_{x \in \Theta} |\mathcal{B}(x, 2(1-\alpha)D)|$$

and it remains to bound $|\mathcal{B}(x, 2(1-\alpha)D)|$. To do this we notice that for every $y \in \{0,1\}_D^N$ one has

$$\delta(x,y) = 2(D - |\{i : x_i = y_i = 1\}|)$$

so that

$$\mathcal{B}(x, 2(1-\alpha)D) = \left\{ y \in \{0,1\}_D^N : |\{i : y_i = x_i = 1\}| \geq \alpha D \right\}.$$

Now, on $\{0,1\}_D^N$ equipped with the uniform distribution, as a function of y, the number of indices i such that $x_i = y_i = 1$ appears as the number of success when sampling without replacement with D trials among a population of size N containing D favorable elements. Hence this variable follows an hypergeometric distribution $\mathcal{H}(N, D, D/N)$, so that if X is a random variable with distribution $\mathcal{H}(N, D, D/N)$ one has

$$1 \leq |\Theta| \, \mathbb{P}[X \geq \alpha D].$$

In order to get an exponential bound for $\mathbb{P}[X \geq \alpha D]$, we use Chernoff's inequality

$$\mathbb{P}[X \geq \alpha D] \leq \exp(-\psi_X^*(\alpha D))$$

and then an argument due to Aldous (see [3]), which allows to compare the Cramér transform of X with that of a binomial random variable. Indeed, one can define X and a variable Y with binomial distribution $\mathrm{Bin}(D, D/N)$ in such a way that $X = \mathbb{E}[Y \mid X]$. In particular Jensen's inequality implies that the moment generating function of X is not larger than the moment generating function of Y and therefore $\psi_X^*(\alpha D) \geq \psi_Y^*(\alpha D)$. But, according to (2.37) and (2.40), one has

$$\psi_Y^*(\alpha D) = D h_{D/N}(\alpha) \geq \frac{D^2}{N} h\left(\frac{\alpha - D/N}{D/N}\right),$$

where $h(u) = (1+u)\ln(1+u) - u$. Hence, collecting the above upper bounds leads to

$$D^{-1}\ln(\Theta) \geq \frac{D}{N} h\left(\frac{\alpha - D/N}{D/N}\right) = f_\alpha(D/N),$$

where, for every $u \leq \alpha$,

$$f_\alpha (u) = \alpha \ln \left(\frac{\alpha}{u} \right) - \alpha + u.$$

One can easily check that

$$u \rightarrow \frac{f_\alpha (u)}{\ln (1/u)}$$

is nonincreasing on $(0, \alpha)$ which implies that for every $u \leq \beta \alpha$

$$f_\alpha (u) \geq \frac{f_\alpha (\beta \alpha)}{- \ln (\alpha \beta)} \ln (1/u) = \rho \ln (1/u)$$

and therefore

$$D^{-1} \ln (\Theta) \geq f_\alpha (D/N) \geq \rho \ln (N/D) . \blacksquare$$

We can derive from this combinatorial lemma and Birgé's lemma the following analogue of Proposition 4.8.

Proposition 4.11 *Let D and N be integers such that $1 \leq D \leq N$ and r be some arbitrary positive number. Define*

$$\mathcal{C}_D^N (r) = \left\{ r \sum_{j=1}^{N} \theta_j \varphi_j, \theta \in [0, 1]^N \text{ with } \sum_{j=1}^{N} 1\!\!1_{\theta_j \neq 0} \leq D \right\},$$

then there exists some absolute positive constant κ_1 such that for any estimator $\widehat{s}$ of s one has

$$\sup_{s \in \mathcal{C}_D^N (r)} \mathbb{E}_s \left[\| \widehat{s} - s \|^2 \right] \geq \kappa_1 D \left(r^2 \wedge \varepsilon^2 \left(1 + \ln (N/D) \right) \right) . \tag{4.35}$$

Proof. Keeping the same notations as the statement of Lemma 4.10 we notice that

$$\{0,1\}_D^N \subseteq \left\{ \theta \in [0, 1]^N \text{ with } \sum_{j=1}^{N} 1\!\!1_{\theta_j \neq 0} \leq D \right\}.$$

Assuming first that $N \geq 4D$, we can therefore combine (4.25), Corollary 2.19 and Lemma 4.10 with $\alpha = 3/4$ and $\beta = 1/3$. Hence one has

$$4 \sup_{s \in \mathcal{C}_D^N (r)} \mathbb{E}_s \left[\| \widehat{s} - s \|^2 \right] \geq \frac{r^2 D}{2} (1 - \kappa)$$

provided that

$$\varepsilon^{-2} D r^2 \leq \kappa \rho D \ln (N/D)$$

i.e., $r^2 \leq \kappa \rho \varepsilon^2 \ln (N/D)$, where $\rho = 0.2$. Since $N \geq 4D$, we notice that

$$\frac{\ln (N/D)}{1 + \ln (N/D)} \geq \frac{2 \ln (2)}{1 + 2 \ln (2)},$$

and derive that (4.11) holds true provided that

$$\kappa_1 \leq \frac{\rho \kappa (1 - \kappa) \ln (2)}{4 (1 + 2 \ln (2))}.$$

If $N < 4D$, we may use this time that $\mathcal{C}_D (r) \subseteq \mathcal{C}_D^N (r)$ and derive from (4.31) that (4.35) is valid at least if

$$\kappa_1 \leq \frac{\kappa_0}{1 + 2 \ln (2)}.$$

Finally choosing

$$\kappa_1 = \frac{\kappa_0}{1 + 2 \ln (2)} \wedge \frac{\rho \kappa (1 - \kappa) \ln (2)}{4 (1 + 2 \ln (2))},$$

(4.35) holds true in any case. ■

Of course since any set $\mathcal{C}_D^N (r)$ is included in $\mathbb{S}_D^N$, choosing

$$r = \varepsilon (1 + \ln (N/D))^{1/2}$$

we derive an immediate corollary from the previous proposition.

Corollary 4.12 *Let D and N be integers such that $1 \leq D \leq N$, then for any estimator $\widehat{s}$ of s one has*

$$\sup_{s \in \mathbb{S}_D^N} \mathbb{E}_s \left[\| \widehat{s} - s \|^2 \right] \geq \kappa_1 D \varepsilon^2 (1 + \ln (N/D)). \tag{4.36}$$

This minimax lower bound obviously shows that since the penalized LSE defined by (4.12) satisfies to (4.11), it is simultaneously approximately minimax on each set $\mathbb{S}_D^N$ for every $D \leq N$.

A Lower Bound Under Metric Entropy Conditions

We turn now to a somehow more abstract version of what we have done before. Following [16], in order to build a lower bound based on metric properties of some totally bounded parameter space $S \subset \mathbb{H}$ with respect to the Hilbertian distance, the idea is to construct some δ-net (i.e., a maximal set of points which are δ-separated), such that the mutual distances between the elements of this net stay of order δ, less or equal to $2C\delta$ say for some constant $C > 1$. To do that, we may use an argument borrowed from [126] and goes as follows. Consider some δ-net $\mathcal{C}'$ and some $C\delta$-net $\mathcal{C}''$ of S. Any point of $\mathcal{C}'$ must belong to some ball with radius $C\delta$ centered at some point of $\mathcal{C}''$, hence if $\mathcal{C}$ denotes an intersection of $\mathcal{C}'$ with such a ball with maximal cardinality one has for every $t, t' \in \mathcal{C}$ with $t \neq t'$

$$\delta \leq \|t - t'\| \leq 2C\delta \tag{4.37}$$

and

$$\ln\left(|\mathcal{C}|\right) \geq H\left(\delta, S\right) - H\left(C\delta, S\right). \tag{4.38}$$

Combining (4.25) and Corollary 2.19 again, we derive from (4.37) and (4.38) that for any estimator $\widehat{s}$ one has

$$4 \sup_{s \in S} \mathbb{E}_s \left[\|s - \widehat{s}\|^2 \right] \geq \delta^2 \left(1 - \kappa\right),$$

provided that $2C^2\delta^2 \leq \kappa\varepsilon^2 \left(H\left(\delta, S\right) - H\left(C\delta, S\right)\right)$, where κ denotes the absolute constant of Lemma 2.18. The lower bound (4.38) is too crude to capture the metric structure of the unit ball of a Euclidean space for instance. A refined approach of this kind should involve the notion of metric dimension rather than metric entropy as explained in [16]. However if $H\left(\delta, S\right)$ behaves like a negative power of δ for instance, then it leads to some relevant minimax lower bound. Indeed, if we assume that for some $\delta \leq R$ one has

$$C_1 \left(\frac{R}{\delta}\right)^{1/\alpha} \leq H\left(\delta, S\right) \leq C_2 \left(\frac{R}{\delta}\right)^{1/\alpha}, \tag{4.39}$$

choosing $C = \left(2C_2/C_1\right)^{\alpha}$ warrants that

$$H\left(\delta, S\right) - H\left(C\delta, S\right) \geq \left(C_1/2\right) \left(\frac{R}{\delta}\right)^{1/\alpha}$$

for every $\delta \leq R/C$. It suffices now to take

$$\delta = \left(\frac{1}{C} \wedge \left(\frac{C_1\kappa}{4C^2}\right)^{\alpha/(2\alpha+1)}\right) \left(\frac{R}{\varepsilon}\right)^{1/(2\alpha+1)} \varepsilon$$

to obtain the following result.

Proposition 4.13 *Let S be some totally bounded subset of $\mathbb{H}$. Let $H\left(., S\right)$ denote the metric entropy of S and assume that (4.39) holds for every $\delta \leq R$ and some positive constants α, C_1 and C_2. Then, there exists some positive constant κ_1 (depending on α, C_1 and C_2) such that for every estimator $\widehat{s}$*

$$\sup_{s \in S} \mathbb{E}_s \left[\|s - \widehat{s}\|^2 \right] \geq \kappa_1 R^{2/(2\alpha+1)} \varepsilon^{4\alpha/(2\alpha+1)},$$

provided that $\varepsilon \leq R$.

It is a quite classical result of approximation theory that a Besov ellipsoid $\mathcal{B}_2 \left(\alpha, R\right)$ satisfies to (4.39). Indeed we can prove the following elementary bounds for the metric entropy of a Euclidean ball and then on a Besov ellipsoid. We recall that the notions of packing number N and metric entropy H that we use below are defined in Section 3.4.1.

Lemma 4.14 *Let $\mathbb{B}_D$ denote the unit ball of the Euclidean space $\mathbb{R}^D$. Then, for every positive numbers δ and r with $\delta \leq r$*

$$\left(\frac{r}{\delta}\right)^D \leq N\left(\delta, r\mathbb{B}_D\right) \leq \left(\frac{r}{\delta}\right)^D \left(2 + \frac{\delta}{r}\right)^D \tag{4.40}$$

Moreover for every positive α, R and δ with $\delta \leq R$, one has

$$\kappa_\alpha \left(\frac{R}{\delta}\right)^{1/\alpha} \leq H\left(\delta, \mathcal{B}_2\left(\alpha, R\right)\right) \leq \kappa'_\alpha \left(\frac{R}{\delta}\right)^{1/\alpha} \tag{4.41}$$

for some positive constants κ_α and κ'_α depending only on α.

Proof. By homogeneity we may assume that $r = 1$. Let S_δ be some δ-net of $\mathbb{B}_D$. Then by definition of S_δ the following properties hold true

$$\bigcup_{x \in S_\delta} (x + \delta\mathbb{B}_D) \supseteq \mathbb{B}_D$$

and conversely

$$\bigcup_{x \in S_\delta} \left(x + \frac{\delta}{2}\mathbb{B}_D\right) \subseteq \left(1 + \frac{\delta}{2}\right)\mathbb{B}_D.$$

Hence, since on the one hand each ball $x + \delta\mathbb{B}_D$ has volume $\delta^D \mathrm{Vol}\left(\mathbb{B}_D\right)$ and on the other hand the balls $x + (\delta/2)\mathbb{B}_D$ are disjoints with volume $(\delta/2)^D \mathrm{Vol}\left(\mathbb{B}_D\right)$, we derive that

$$\delta^D \left|S_\delta\right| \mathrm{Vol}\left(\mathbb{B}_D\right) \geq \mathrm{Vol}\left(\mathbb{B}_D\right)$$

and

$$(\delta/2)^D \left|S_\delta\right| \mathrm{Vol}\left(\mathbb{B}_D\right) \leq \left(1 + \frac{\delta}{2}\right)^D \mathrm{Vol}\left(\mathbb{B}_D\right)$$

which leads to (4.40). Turning now to the proof of (4.41) we first notice that since $\mathcal{B}_2\left(\alpha, R\right) = R\mathcal{B}_2\left(\alpha, 1\right)$, it is enough to prove (4.41) for $R = 1$. Introducing for every $j \geq 0$, the set of integers $I_j = \{2^j, 2^j + 1, \ldots, 2^{j+1} - 1\}$, we note that for every integer J,

$$\mathcal{B}_2\left(\alpha, 1\right) \supseteq \left\{\beta \in \ell_2 : \beta_k = 0 \text{ for every } k \notin I_J \text{ and } \sum_{k \in I_J} \beta_k^2 k^{2\alpha} \leq 1\right\}$$

$$\supseteq \left\{\beta \in \ell_2 : \beta_k = 0 \text{ for every } k \notin I_J \text{ and } \sum_{k \in I_J} \beta_k^2 \leq 2^{-2(J+1)\alpha}\right\}.$$

Since

$$\left\{\beta \in \ell_2 : \beta_k = 0 \text{ for every } k \notin I_J \text{ and } \sum_{k \in I_J} \beta_k^2 \leq 2^{-2(J+1)\alpha}\right\}$$

is isometric to the Euclidean ball $2^{-(J+1)\alpha}\mathbb{B}_D$ with $D = 2^J$, setting $\delta_J = 2^{-1-(J+1)\alpha}$, we derive from (4.40) that

$$H\left(\delta_J, \mathcal{B}_2\left(\alpha, 1\right)\right) \geq 2^J \ln\left(2\right). \tag{4.42}$$

Now either $\delta_0 \leq \delta$ and in this case we set $J = 0$ so that

$$H\left(\delta, \mathcal{B}_2\left(\alpha, 1\right)\right) \geq \ln\left(2\right) \geq \ln\left(2\right) \delta_0^{1/\alpha}\delta^{-1/\alpha},$$

or $\delta_0 > \delta$, in which case we take $J = \sup\left\{j \geq 0 : \delta_j > \delta\right\}$ and we deduce from (4.42) that

$$H\left(\delta, \mathcal{B}_2\left(\alpha, 1\right)\right) \geq 2^J \ln\left(2\right) = \delta_{J+1}^{-1/\alpha}2^{-1/\alpha-2} \ln\left(2\right) \geq \delta^{-1/\alpha}2^{-1/\alpha-2} \ln\left(2\right).$$

In any case we therefore have

$$H\left(\delta, \mathcal{B}_2\left(\alpha, 1\right)\right) \geq \delta^{-1/\alpha}2^{-1/\alpha-2} \ln\left(2\right).$$

Conversely, we notice that

$$\mathcal{B}_2\left(\alpha, 1\right) \subseteq \mathcal{E} = \left\{\beta \in \ell_2 : \sum_{j \geq 0} 2^{2j\alpha}\left(\sum_{k \in I_j} \beta_k^2\right) \leq 1\right\}.$$

Given $\delta \leq 1$, noticing that $\left\{j \geq 0 : 2^{-2j\alpha} > \delta^2/2\right\} \neq \emptyset$, let us define

$$J = \sup\left\{j \geq 0 : 2^{-2j\alpha} > \delta^2/2\right\}.$$

Then J satisfies to

$$2^{-2(J+1)\alpha} \leq \delta^2/2 \tag{4.43}$$

$$2^J \leq 2^{1/(2\alpha)}\delta^{-1/\alpha}. \tag{4.44}$$

Let us introduce the truncated ellipsoid

$$\mathcal{E}^J = \left\{\beta \in \ell_2 : \sum_{j=0}^{J} 2^{2j\alpha}\left(\sum_{k \in I_j} \beta_k^2\right) \leq 1 \text{ and } \beta_k = 0 \text{ whenever } k \notin \bigcup_{j=0}^{J} I_j\right\}.$$

Since for every $\beta \in \mathcal{E}$ one has

$$\sum_{j>J}^{\infty}\left(\sum_{k \in I_j} \beta_k^2\right) \leq 2^{-2(J+1)\alpha},$$

we derive from (4.43) that

$$N'\left(\delta, \mathcal{E}\right) \leq N'\left(\delta/\sqrt{2}, \mathcal{E}^J\right), \tag{4.45}$$

where we recall that $N'(.,S)$ denotes the covering number of S. Let us define for every $j \leq J$

$$\delta_j = \sqrt{3}2^{-J\alpha+j-J-1-\alpha},$$

then by (4.43)

$$\sum_{j=0}^{J} \delta_j^2 \leq 32^{-2J\alpha-2-2\alpha} \sum_{k=0}^{\infty} 2^{-2k} \leq 2^{-2(J+1)\alpha} \leq \delta^2/2.$$

Introducing for every integer $j \leq J$

$$B_j = \left\{ \beta \in \ell_2 : \beta_k = 0 \text{ if } k \notin I_j \text{ and } 2^{2j\alpha} \left(\sum_{k \in I_j} \beta_k^2 \right) \leq 1 \right\},$$

we notice that B_j is isometric to $2^{-j\alpha}\mathbb{B}_{2^j}$. Hence, in order to construct some $\delta/\sqrt{2}$ covering of $\mathcal{E}^J$, we simply use (4.40) which ensures, for every $j \leq J$, the existence of an δ_j-net $\{\beta(\lambda_j, j), \lambda_j \in \Lambda_j\}$ of B_j with

$$|\Lambda_j| \leq 3^{2^j} \left(\frac{2^{-j\alpha}}{\delta_j} \right)^{2^j}.$$

Then we define for every $\lambda \in \Lambda = \prod_{j=0}^{J} \Lambda_j$, the sequence $\beta(\lambda)$ belonging to $\mathcal{E}^J$ by

$$\beta_k(\lambda) = \beta_k(\lambda_j, j) \text{ whenever } k \in I_j \text{ for some } j \leq J \text{ and } \beta_k(\lambda) = 0 \text{ otherwise.}$$

If β is some arbitrary point in $\mathcal{E}^J$, for every $j \leq J$ there exists $\lambda_j \in \Lambda_j$ such that

$$\sum_{k \in I_j} (\beta_k - \beta_k(\lambda_j, j))^2 \leq \delta_j^2$$

and therefore

$$\|\beta - \beta(\lambda)\|^2 = \sum_{j=0}^{J} \sum_{k \in I_j} (\beta_k - \beta_k(\lambda_j, j))^2 \leq \sum_{j=0}^{J} \delta_j^2 \leq \delta^2/2.$$

This shows that the balls centered at some point of $\{\beta(\lambda), \lambda \in \Lambda\}$ with radius $\delta/\sqrt{2}$ are covering $\mathcal{E}^J$. Hence

$$\ln N'\left(\delta/\sqrt{2}, \mathcal{E}^J\right) \leq \sum_{j=0}^{J} \ln|\Lambda_j| \leq \sum_{j=0}^{J} 2^j \ln\left(32^{-j\alpha}/\delta_j\right)$$

and therefore, since $\sum_{j=0}^{J} 2^j = 2^{J+1} = 2^J \sum_{k=0}^{\infty} 2^{-k}k$,

$$\ln N' \left(\delta/\sqrt{2}, \mathcal{E}^J\right) \leq \ln \left(\sqrt{3}2^{1+\alpha}\right) \left(\sum_{j=0}^{J} 2^j\right) + 2^J \sum_{k=0}^{\infty} 2^{-k} \ln \left(2^{k(1+\alpha)}\right)$$

$$\leq 2^{J+1} \ln \left(\sqrt{3}4^{1+\alpha}\right).$$

Combining this inequality with (4.44) and (4.45) yields

$$\ln N' (\delta, \mathcal{E}) \leq 2^{1+(1/2\alpha)} \ln \left(\sqrt{3}4^{1+\alpha}\right) \delta^{-1/\alpha}$$

and the conclusion follows since $H(\delta, \mathcal{B}_2(\alpha, 1)) \leq \ln N' (\delta/2, \mathcal{E})$. ∎

As a consequence we can use Proposition 4.13 to re-derive (4.32) in the special case of Besov ellipsoids.

4.3.2 Adaptive Properties of Penalized Estimators for Gaussian Sequences

The ideas presented in Section 4.1.1 are now part of the statistical folklore. To summarize: for a suitable choice of an orthonormal basis $\{\varphi_j\}_{j\geq 1}$ of some Hilbert space $\mathbb{H}$ of functions, smoothness properties of the elements of $\mathbb{H}$ can be translated into properties of their coefficients in the space ℓ_2. Sobolev or Besov balls in the function spaces correspond to ellipsoids or more generally ℓ_p-bodies when the basis is well chosen. Moreover, once we have chosen a suitable basis, an isonormal process can be turned to an associated Gaussian sequence of the form

$$\widehat{\beta}_j = \beta_j + \varepsilon \xi_j, \quad j \geq 1, \tag{4.46}$$

for some sequence of i.i.d. standard normal variables ξ_js. Since Theorem 4.5 allows to mix model selection procedures by using possibly different basis, we can now concentrate on what can be done with a given basis i.e., on the Gaussian sequence framework. More precisely we shall focus on the search for good strategies for estimating $(\beta_j)_{j\geq 1} \in \ell_2$ from the sequence $(\widehat{\beta}_j)_{j\geq 1}$ under the assumption that it belongs to various types of ℓ_p-bodies. These strategies will all consist in selecting (from the data) some adequate *finite* subset $\widehat{m}$ of $\mathbb{N}^*$ and consider $\widetilde{\beta} = (\widehat{\beta}_j)_{j\in\widehat{m}}$ as an estimator of $\beta = (\beta_j)_{j\geq 1}$. In other words, we consider our original nonparametric problem as an *infinite dimensional* variable selection problem: we are looking at some finite family of "most significant" coordinates among a countable collection.

In our treatment of the Gaussian sequence framework (4.46), we shall stick to the following notations: the family $\{\Lambda_m\}_{m\in\mathcal{M}}$ is a countable family of finite subsets of $\mathbb{N}^*$ and for each $m \in \mathcal{M}$, S_m is the finite dimensional linear space of sequences $(\beta_j)_{j\geq 1}$ such that $\beta_j = 0$ whenever $j \notin \Lambda_m$. Selecting a value of m amounts to select a set Λ_m orequivalently some finite subset of the coordinates.

Our purpose will be to define proper collections $\{\Lambda_m\}_{m\in\mathcal{M}}$ of subsets of $\mathbb{N}^*$ together with weights $\{x_m\}_{m\in\mathcal{M}}$ of the form $x_m = D_m L_m$ such that

$$\sum_{m\in\mathcal{M}} \exp\left(-D_m L_m\right) = \Sigma < \infty. \tag{4.47}$$

and consider the penalized LSE corresponding to the penalty

$$\mathrm{pen}\,(m) = K\varepsilon^2 D_m \left(1 + \sqrt{2L_m}\right)^2, \text{ for every } m \in \mathcal{M} \tag{4.48}$$

with $K > 1$. In this context Theorem 4.2 takes the following form.

Corollary 4.15 *Let $\{L_m\}_{m\in\mathcal{M}}$ be some family of positive numbers such that (4.47) holds. Let $K > 1$ and take the penalty function* pen *satisfying (4.48). Then, almost surely, there exists some minimizer $\widehat{m}$ of the penalized least-squares criterion*

$$-\sum_{j\in\Lambda_m} \widehat{\beta}_j^2 + \mathrm{pen}\,(m)\,,$$

over $\mathcal{M}$ and the corresponding penalized LSE $\widetilde{\beta} = (\widehat{\beta}_j)_{j\in\widehat{m}}$ is unique. Moreover and upper bound for the quadratic risk $\mathbb{E}_\beta\left[\left\|\widetilde{\beta} - \beta\right\|^2\right]$ is given by

$$C\,(K)\left\{\inf_{m\in\mathcal{M}}\left(\left(\sum_{j\notin\Lambda_m} \beta_j^2\right) + \varepsilon^2 D_m\left(1 + \sqrt{2L_m}\right)^2\right) + \Sigma\varepsilon^2\right\}, \tag{4.49}$$

where $C\,(K)$ depends only on K.

Finally we shall optimize the oracle type inequality (4.49) using the knowledge that $(\beta_j)_{j\geq 1}$ belongs to some typical ℓ_p-bodies. Such computations will involve the approximation properties of the models S_m in the collection with respect to the considered ℓ_p bodies since our work will consist in bounding the bias term $\sum_{j\notin\Lambda_m} \beta_j^2$.

4.3.3 Adaptation with Respect to Ellipsoids

We first consider the strategy of ordered variable selection which is suitable when β belongs to some *unknown* ellipsoid. This strategy is given by $\mathcal{M} = \mathbb{N}$, $\Lambda_0 = \emptyset$, $\Lambda_m = \{1, 2, \ldots, m\}$ for $m > 0$ and $L_m \equiv L$ where L is some arbitrary positive constant. Hence we may take

$$\mathrm{pen}\,(m) = K'm\varepsilon^2$$

where $K' > 1$ and (4.49) becomes

$$\mathbb{E}_\beta\left[\left\|\widetilde{\beta} - \beta\right\|^2\right] \leq C'\,(K')\left\{\inf_{m\geq 1}\left(\sum_{j>m} \beta_j^2 + \varepsilon^2 m\right)\right\}.$$

Since $\sum_{j>m}\beta_j^2$ converges to zero when m goes to infinity, it follows that $\mathbb{E}_\beta\left[\left\|\tilde{\beta}-\beta\right\|^2\right]$ goes to zero with ε and our strategy leads to consistent estimators for all $\beta\in\ell_2$.

Let us now assume that β belongs to some ellipsoid $\mathcal{E}_2(c)$. We deduce from the monotonicity of the sequence c that

$$\sum_{j>m}\beta_j^2 = \sum_{j>m}c_j^2\left(\beta_j^2/c_j^2\right) \le c_{m+1}^2\sum_{j>m}\left(\beta_j^2/c_j^2\right) \le c_{m+1}^2$$

and therefore

$$\sup_{\beta\in\mathcal{E}_2(c)}\mathbb{E}_\beta\left[\left\|\tilde{\beta}-\beta\right\|^2\right] \le C'(K')\left\{\inf_{m\ge 1}\left(c_{m+1}^2+\varepsilon^2 m\right)\right\}. \tag{4.50}$$

Note that if one assumes that $c_1\ge\varepsilon$, then by monotonicity of c

$$\sup_{D\ge 1}\left(c_D^2\wedge D\varepsilon^2\right) = \max\left(c_{m_0+1}^2, m_0\varepsilon^2\right) \ge \frac{1}{2}\left(c_{m_0+1}^2+m_0\varepsilon^2\right),$$

where $m_0 = \sup\left\{D\ge 1 : D\varepsilon^2\le c_D^2\right\}$ and therefore

$$\sup_{D\ge 1}\left(c_D^2\wedge D\varepsilon^2\right) \ge \frac{1}{2}\inf_{m\ge 1}\left(c_{m+1}^2+\varepsilon^2 m\right).$$

Combining this inequality with (4.50) and the minimax lower bound (4.32) we derive that

$$\sup_{\beta\in\mathcal{E}_2(c)}\mathbb{E}_\beta\left[\left\|\tilde{\beta}-\beta\right\|^2\right] \le \frac{2C'(K')}{\kappa_0}\inf_{\hat{\beta}}\sup_{\beta\in\mathcal{E}_2(c)}\mathbb{E}_s\left[\left\|\hat{\beta}-\beta\right\|^2\right],$$

which means that $\tilde{\beta}$ is simultaneously minimax (up to some constant) on all ellipsoids $\mathcal{E}_2(c)$ which are nondegenerate, i.e., such that $c_1\ge\varepsilon$.

4.3.4 Adaptation with Respect to Arbitrary ℓ_p-Bodies

We choose for $\mathcal{M}$ the collection of all finite subsets m of $\mathbb{N}^*$ and set $\Lambda_m = m$ and $N_m = \sup m$; then, if $m\ne\emptyset$, $1\le D_m = |m|\le N_m$. Finally, in order to define the weights, fix some $\theta>0$ and set for all $m\ne\emptyset$, $L_m = L(D_m, N_m)$ with

$$L(D,N) = \ln\left(\frac{N}{D}\right) + (1+\theta)\left(1+\frac{\ln N}{D}\right). \tag{4.51}$$

Let us now check that (4.47) is satisfied with Σ bounded by some Σ_θ depending only on θ. We first observe that $\mathcal{M}\setminus\emptyset$ is the disjoint union of all the sets $\mathcal{M}(D,N)$, $1\le D\le N$, where

$$\mathcal{M}(D,N) = \{m\in\mathcal{M}\,|\,D_m = D \text{ and } N_m = N\}, \tag{4.52}$$

and that by (2.9)

$$|\mathcal{M}(D, N)| = \binom{N-1}{D-1} \leq \binom{N}{D} \leq \left(\frac{eN}{D}\right)^D,$$

from which we derive that

$$\Sigma \leq \sum_{N \geq 1} \sum_{D=1}^{N} |\mathcal{M}(D, N)| \exp[-D \ln(N/D) - (1+\theta)(D + \ln N)]$$

$$\leq \sum_{N \geq 1} \sum_{D \geq 1} \exp[-\theta D] N^{-\theta-1} \leq \frac{e^{-\theta}}{1 - e^{-\theta}} \int_{1/2}^{+\infty} x^{-\theta-1} \, dx$$

$$\leq \frac{e^{-\theta}}{1 - e^{-\theta}} \frac{2^\theta}{\theta} = \Sigma_\theta. \tag{4.53}$$

Computation of the Estimator

The penalty is a function of D_m and N_m and can therefore be written as $\text{pen}'(D_m, N_m)$. In order to compute the penalized LSE one has to find the minimizer $\widehat{m}$ of

$$\text{crit}(m) = -\sum_{j \in m} \widehat{\beta}_j^2 + \text{pen}'(D_m, N_m).$$

Given N and D, the minimization of crit over the set $\mathcal{M}(D, N)$ amounts to the maximization of $\sum_{j \in m} \widehat{\beta}_j^2$ over this set. Since by definition all such m's contain N and $D - 1$ elements of the set $\{1, 2, \ldots, N-1\}$, it follows that the minimizer $m(D, N)$ of crit over $\mathcal{M}(D, N)$ is the set containing N and the indices of the $D - 1$ largest elements $\widehat{\beta}_j^2$ for $1 \leq j \leq N - 1$ denoted by $\overline{\{1, 2, \ldots, N-1\}}[D-1]$. So

$$\inf_{m \in \mathcal{M}(D,N)} \text{crit}(m) = -\sum_{j \in m(D,N)} \widehat{\beta}_j^2 + \text{pen}'(D, N),$$

with $m(D, N) = \{N\} \cup \overline{\{1, 2, \ldots, N-1\}}[D-1]$. The computation of $\widehat{m}$ then results from an optimization with respect to N and D. In order to perform this optimization, let us observe that if $J = \max \overline{\{1, 2, \ldots, N\}}[D] < N$, then $\sum_{j \in m(D,N)} \widehat{\beta}_j^2 \leq \sum_{j \in m(D,J)} \widehat{\beta}_j^2$. On the other hand, it follows from the definition of $L(D, \cdot)$ that $L(D, J) < L(D, N)$ and therefore $\text{crit}(m(D, N)) > \text{crit}(m(D, J))$. This implies that, given D, the optimization with respect to N should be restricted to those N's such that $\max \overline{\{1, 2, \ldots, N\}}[D] = N$. It can easily be deduced from an iterative computation of the sets $\{\widehat{\beta}_\lambda^2\}_{\lambda \in \overline{\{1, 2, \ldots, N\}}[D]}$ starting with $N = D$. It then remains to optimize our criterion with respect to D.

Performance of the Estimator

We observe that $(\ln N)/D < \ln(N/D) + 0.37$ for any pair of positive integers $D \leq N$, which implies that

$$L(D, N) \leq (2 + \theta) \ln \left(\frac{N}{D}\right) + 1.37(1 + \theta). \tag{4.54}$$

We derive from (4.54), (4.53) and (4.49) that for some positive constant $C(K, \theta)$

$$\mathbb{E}_\beta \left\| \widetilde{\beta} - \beta \right\|^2 \leq C(K, \theta) \inf_{\{(D,N) \mid 1 \leq D \leq N\}} \left\{ b_D^2 + \varepsilon^2 D \left(\ln \left(\frac{N}{D}\right) + 1 \right) \right\},$$

where $b_D^2 = \left(\inf_{m \in \mathcal{M}(D,N)} \sum_{j \notin m} \beta_j^2 \right)$. The basic remark concerning the performance of $\widetilde{\beta}$ is therefore that it is simultaneously approximately minimax on the collection of all sets $\mathbb{S}_D^N$ when D and N vary with $D \leq N$. On the other hand, if we restrict to those ms such that $N_m = D_m$, then $L_m \leq 1.37(1 + \theta)$. Moreover, if $\beta \in \mathcal{E}_p(c)$ with $0 < p \leq 2$

$$\sum_{j>N} \beta_j^2 \leq \left(\sum_{j>N} |\beta_j|^p \right)^{2/p} \leq c_{N+1}^2, \tag{4.55}$$

which leads to the following analogue of (4.50)

$$\sup_{\beta \in \mathcal{E}_p(c)} \mathbb{E}_\beta \left[\left\| \widetilde{\beta} - \beta \right\|^2 \right] \leq C(K, \theta) \inf_{N \geq 1} \left\{ c_{N+1}^2 + \varepsilon^2 N \right\}. \tag{4.56}$$

This means that at least this estimator performs as well as the previous one and is therefore approximately minimax on all nondegenerate ellipsoids simultaneously.

Let us now turn to an improved bound for ℓ_p-bodies with $p < 2$. This bound is based on the following:

Lemma 4.16 *Given N nonnegative numbers $\{a_i\}_{i \in I}$, we consider a permutation $\{a_{(j)}\}_{1 \leq j \leq N}$ of the set $\{a_i\}_{i \in I}$ such that $a_{(1)} \geq \ldots \geq a_{(N)}$. Then for every real number $0 < p \leq 2$ and any integer n satisfying $0 \leq n \leq N - 1$,*

$$\sum_{j=n+1}^{N} a_{(j)}^2 \leq \left(\sum_{i \in I} a_i^p \right)^{2/p} (n+1)^{1-2/p}.$$

Proof. The result being clearly true when $n = 0$, we can assume that $n \geq 1$. Let $a = a_{(n+1)}$. Then $a \leq a_{(j)}$ whatever $j \leq n$ and therefore $(1 + n)a^p \leq \sum_{i \in I} a_i^p$. We then conclude from

$$\sum_{j=n+1}^{N} a_{(j)}^2 \leq a^{2-p} \sum_{j=n+1}^{N} a_{(j)}^p \leq \left(\frac{\sum_{i \in I} a_i^p}{1+n} \right)^{2/p-1} \sum_{i \in I} a_i^p. \quad \blacksquare$$

We are now in position to control the bias term in (4.49). More precisely we intend to prove the following result.

Proposition 4.17 *Given $\theta > 0$, let $\widetilde{\beta}$ be the penalized LSE with penalty* $\text{pen}(m) = K\varepsilon^2 D_m \left(1 + \sqrt{2L_m}\right)^2$ *on every finite subset of $\mathbb{N}^*$, where $L_m = L(|m|, \sup(m))$, the function L (depending on θ) being defined by (4.51), then there exists some positive constant $C'(K, \theta)$ such that for every $\beta \in \mathcal{E}_p(c)$*

$$\mathbb{E}_\beta\left[\left\|\widetilde{\beta} - \beta\right\|^2\right] \leq C'(K, \theta) \inf_{1 \leq D \leq N} \left\{c_{N+1}^2 + c_D^2 D^{1-2/p} + \varepsilon^2 D \ln\left(\frac{eN}{D}\right)\right\}. \tag{4.57}$$

Proof. Setting $\overline{\mathcal{M}}(J, M) = \cup_{J \leq N \leq M} \mathcal{M}(J, N)$ where $\mathcal{M}(J, N)$ is defined by (4.52), we derive from (4.54), (4.53) and (4.49) that for some positive constant $C(K, \theta)$

$$C^{-1}(K, \theta)\,\mathbb{E}_\beta\left[\left\|\widetilde{\beta} - \beta\right\|^2\right]$$

$$\leq \inf_{\{(J,N)\,|\,1 \leq J \leq N\}} \left\{\left(\inf_{m \in \mathcal{M}(J,N)} \sum_{j \notin m} \beta_j^2\right) + \varepsilon^2 J\left(L(J, N) + 1\right)\right\}$$

$$\leq \inf_{\{(J,M)\,|\,1 \leq J < M\}} \left\{F(J, M)\right\}, \tag{4.58}$$

where

$$F(J, M) = \inf_{m \in \overline{\mathcal{M}}(J,M)} \sum_{j \notin m} \beta_j^2 + \varepsilon^2 J\left[\ln\left(\frac{M}{J}\right) + 1\right].$$

Let us fix some pair (J, M) and some $m \in \overline{\mathcal{M}}(J, M)$. It follows from (4.55) that

$$\sum_{j \notin m} \beta_j^2 = \sum_{j > M} \beta_j^2 + \sum_{1 \leq j \leq M, j \notin m} \beta_j^2 \leq c_{M+1}^2 + \sum_{1 \leq j \leq M, j \notin m} \beta_j^2,$$

so that

$$F(J, M) \leq c_{M+1}^2 + \sum_{j=J+1}^{M} \beta_{(j)}^2 + \varepsilon^2 J\left[\ln\left(\frac{M}{J}\right) + 1\right],$$

where $\left(\beta_{(j)}^2\right)_{1 \leq j \leq M}$ are the ordered squared coefficients $\left(\beta_j^2\right)_{1 \leq j \leq M}$ such that $\beta_{(1)}^2 \geq \ldots \geq \beta_{(M)}^2$. It then follows from Lemma 4.16 with $N = M - D + 1$ and $n = J - D + 1$ that

$$\sum_{j=J+1}^{M} \beta_{(j)}^2 \leq \left(\sum_{j=D}^{M} \beta_{(j)}^p\right)^{2/p} (J - D + 2)^{1-2/p}.$$

Observing that if $1 \leq D \leq J + 1 \leq M$,

$$\sum_{j=D}^{M} \beta_{(j)}^{p} \leq \sum_{j=D}^{M} |\beta_j|^p \leq c_D^p,$$

we derive from the previous inequality that

$$\sum_{j=J+1}^{M} \beta_{(j)}^2 \leq c_D^2 (J - D + 2)^{1-2/p}.$$

Let us now define

$$\lceil x \rceil = \inf\{n \in \mathbb{N} \mid n \geq x\} \tag{4.59}$$

and fix $D = \lceil (J + 1)/2 \rceil$ and $N = \lceil (M - 1)/2 \rceil$. Then $J - D + 2 \geq D$ and $J/2 < D \leq J$, which implies that

$$F(J, M) \leq c_{N+1}^2 + c_D^2 D^{1-2/p} + 2\varepsilon^2 D \left(\ln \left(\frac{2N+1}{D} \right) + 1 \right).$$

Finally, since $N \geq D$ implies that $M > J$, (4.57) easily follows from (4.58). ∎

The Case of Besov Bodies

In the case of general ℓ_p-bodies, we cannot, unfortunately, handle the minimization of the right-hand side of (4.57) as we did for the ellipsoids since it involves c_D and c_{N+1} simultaneously. We now need to be able to compare $c_D^2 D^{1-2/p}$ with c_{N+1}^2 which requires a rather precise knowledge about the rate of decay of c_j as a function of j. This is why we shall restrict ourselves to some particular ℓ_p-bodies. Following [52], we define the Besov body $\mathcal{B}_p(\alpha, R)$ where $\alpha > 1/p - 1/2$ as an ℓ_p-body $\mathcal{E}_p(c)$ with $c_j = Rj^{-(\alpha+1/2-1/p)}$ for every integer $j \geq 1$ (see also [23] for the definition of extended Besov bodies which allow to study sharp effects in the borderline case where $\alpha = 1/p - 1/2$). As briefly recalled in the appendix these geometrical objects correspond to Besov balls in function spaces for a convenient choice of the basis.

In order to keep the discussion as simple as possible we shall assume that $R/\varepsilon \geq e$. From (4.57) we derive the following upper bound for every $\beta \in \mathcal{E}_p(c)$

$$\mathbb{E}_\beta \left[\left\| \tilde{\beta} - \beta \right\|^2 \right]$$
$$\leq C' \inf_{1 \leq D \leq N} \left\{ N^{-2(\alpha+1/2-1/p)} R^2 + D^{-2\alpha} R^2 + \varepsilon^2 D \ln \left(\frac{eN}{D} \right) \right\}.$$

Setting

$$\Delta = (R/\varepsilon)^{\frac{2}{2\alpha+1}} (\ln(R/\varepsilon))^{-1/(2\alpha+1)},$$

we choose $D = \lceil \Delta \rceil$ and $N = \left\lceil \Delta^{\frac{\alpha}{\alpha+1/2-1/p}} \right\rceil$ and some elementary computations lead to

$$\sup_{\beta \in \mathcal{E}_p(c)} \mathbb{E}_\beta \left[\left\| \tilde{\beta} - \beta \right\|^2 \right] \leq C'' R^{2/(2\alpha+1)} \varepsilon^{4\alpha/(2\alpha+1)} \left(\ln(R/\varepsilon) \right)^{2\alpha/(2\alpha+1)}, \quad (4.60)$$

where C'' depends on K, θ, p and α.

As for the minimax lower bound, we simply use (4.32) which warrants that the minimax risk is bounded from below by

$$\kappa_0 \sup_{D \geq 1} \left(R^2 D^{-2\alpha} \wedge D\varepsilon^2 \right) \geq \kappa_\alpha R^{2/(2\alpha+1)} \varepsilon^{4\alpha/(2\alpha+1)}. \quad (4.61)$$

We can conclude that the upper bound (4.60) matches this lower bound up to a power of $\ln(R/\varepsilon)$. As we shall see below, the lower bound is actually sharp and a refined strategy, especially designed for estimation in Besov bodies, can improve the upper bound.

4.3.5 A Special Strategy for Besov Bodies

Let us recall from the previous section that we have at hand a strategy for model selection in the Gaussian sequence model which is, up to constants, minimax over all sets $\mathbb{S}_D^N$ and all ellipsoids, but fails to be minimax for Besov bodies since its risk contains some extra power of $\ln(R/\varepsilon)$ as a nuisance factor. We want here to describe a strategy, especially directed towards estimation in Besov bodies, which will turn to be minimax for all Besov bodies $\mathcal{B}_p(\alpha, R)$ when $\alpha > 1/p - 1/2$.

The Strategy

The construction of the models is based on a decomposition of $\Lambda = \mathbb{N}^\star$ into a partition $\Lambda = \cup_{j \geq 0} \Lambda(j)$ with $\mu_0 = 1$ and

$$\Lambda(j) = \{\mu_j, \ldots, \mu_{j+1} - 1\} \text{ with } 2^j \leq \mu_{j+1} - \mu_j \leq M2^j \text{ for } j \geq 0, \quad (4.62)$$

where M denotes some given constant that we shall assume to be equal to 1 in the sequel for the sake of simplicity. Typically, this kind of dyadic decomposition fits with the natural structure of a conveniently ordered wavelet basis as explained in the Appendix but piecewise polynomials could be considered as well (see for instance [22]). We also have to choose a real parameter $\theta > 2$ (the choice $\theta = 3$ being quite reasonable) and set for $J, j \in \mathbb{N}$,

$$A(J, j) = \lfloor 2^{-j}(j+1)^{-\theta} |\Lambda(J+j)| \rfloor \text{ with } \lfloor x \rfloor = \sup\{n \in \mathbb{N} \mid n \leq x\}.$$

It follows that

$$\lfloor 2^J(j+1)^{-\theta} \rfloor \geq A(J, j) > 2^J(j+1)^{-\theta} - 1, \quad (4.63)$$

which in particular implies that $A(J, j) = 0$ for j large enough (depending on J). Let us now set for $J \in \mathbb{N}$

$$\mathcal{M}_J = \left\{ m \subset \Lambda \ \middle| \ m = \left[\bigcup_{j \geq 0} m(J + j) \right] \cup \left[\bigcup_{j=0}^{J-1} \Lambda(j) \right] \right\},$$

with

$$m(J + j) \subset \Lambda(J + j) \text{ and } |m(J + j)| = A(J, j).$$

Clearly, each $m \in \mathcal{M}_J$ is finite with cardinality $D_m = M(J)$ satisfying

$$M(J) = \sum_{j=0}^{J-1} |\Lambda(j)| + \sum_{j \geq 0} A(J, j)$$

and therefore by (4.63)

$$2^J \leq M(J) \leq 2^J \left[1 + \sum_{n \geq 1} n^{-\theta} \right]. \tag{4.64}$$

We turn now to the fundamental combinatorial property of the collection $\mathcal{M}_J$. Since $x \to x \ln (eN/x)$ is increasing on $[0, N]$, (4.10) via (4.63) yields

$$\ln |\mathcal{M}_J| \leq \sum_{j=0}^n \ln \binom{|\Lambda (J + j)|}{A (J, j)} \leq \sum_{j=0}^n A (J, j) \ln \left(\frac{e \, |\Lambda (J + j)|}{A (J, j)} \right)$$

$$\leq 2^J \sum_{j=0}^\infty (j + 1)^{-\theta} \ln \left(e 2^j (j + 1)^\theta \right)$$

so that

$$\ln |\mathcal{M}_J| \leq c_\theta 2^J, \tag{4.65}$$

with some constant c_θ depending only on θ. Let us now set $\mathcal{M} = \cup_{J \geq 0} \mathcal{M}_J$ and $L_m = c_\theta + L$ with $L > 0$ for all m. Then by (4.64) and (4.65)

$$\sum_{m \in \mathcal{M}} e^{-L_m D_m} \leq \sum_{J \geq 0} |\mathcal{M}_J| \exp \left[-c_\theta 2^J - L 2^J \right] \leq \sum_{J \geq 0} \exp \left[-L 2^J \right] = \Sigma_L,$$

and it follows that (4.47) is satisfied with $\Sigma \leq \Sigma_L$.

The Construction of the Estimator

One has to compute the minimizer $\hat{m}$ of $\text{crit} (m) = \text{pen} (m) - \sum_{\lambda \in m} \hat{\beta}_\lambda^2$. The penalty function, as defined by (4.48), only depends on J when $m \in \mathcal{M}_J$ since L_m is constant and $D_m = M(J)$. Setting $\text{pen} (m) = \text{pen}'(J)$ when $m \in \mathcal{M}_J$, we see that $\widehat{m}$ is the minimizer with respect to J of $\mathrm{pen}'(J) - \sum_{\lambda \in \widehat{m}_J} \widehat{\beta}_\lambda^2$ where $\widehat{m}_J \in \mathcal{M}_J$ maximizes

$$\sum_{k \geq 0} \sum_{\lambda \in m(J+k)} \widehat{\beta}_\lambda^2 + \sum_{j=0}^{J-1} \sum_{\lambda \in \Lambda(j)} \widehat{\beta}_\lambda^2$$

or equivalently $\sum_{k \geq 0} \sum_{\lambda \in m(J+k)} \widehat{\beta}_\lambda^2$ with respect to $m \in \mathcal{M}_J$. Since the cardinality $A(J,k)$ of $m(J+k)$ only depends of J and k, one should choose for the $m(J+k)$ corresponding to $\widehat{m}_J$ the subset of $\Lambda(J+k)$ of those $A(J,k)$ indices corresponding to the $A(J,k)$ largest values of $\widehat{\beta}_\lambda^2$ for $\lambda \in \Lambda(J+k)$. In practice of course, the number of coefficients $\widehat{\beta}_\lambda$ at hand, and therefore the maximal value of J is bounded. A practical implementation of this estimator is therefore feasible and has actually been completed in [96].

The Performances of the Estimator

We derive from (4.49) that

$$\mathbb{E}_\beta\left[\left\|\widetilde{\beta} - \beta\right\|^2\right] \leq C\left(K, \theta, L\right) \left\{ \inf_{J \geq 0} \left(\inf_{m \in \mathcal{M}_J} \left(\sum_{\lambda \notin m} \beta_\lambda^2 \right) + \varepsilon^2 2^J \right) \right\} \quad (4.66)$$

so that the main issue is to bound the bias term $\inf_{m \in \mathcal{M}_J} \left(\sum_{\lambda \notin m} \beta_\lambda^2 \right)$ knowing that $\beta \in \mathcal{B}_p\left(\alpha, R\right)$. For each $j \geq 0$ we set $B_j = \left[\sum_{\lambda \in \Lambda(J+j)} |\beta_\lambda|^p \right]^{1/p}$ and denote the coefficients $|\beta_\lambda|$ in decreasing order, for $\lambda \in \Lambda(J+j)$, $k \geq 0$ by $|\beta_{(k),j}|$. The arguments we just used to define $\widehat{m}_J$ immediately show that

$$\inf_{m \in \mathcal{M}_J} \left(\sum_{\lambda \notin m} \beta_\lambda^2 \right) = \sum_{j \geq 0} \sum_{k=A(J,j)+1}^{\mu_{J+j+1} - \mu_{J+j}} \beta_{(k),j}^2,$$

and it follows from Lemma 4.16 and (4.63) that

$$\sum_{k=A(J,j)+1}^{\mu_{J+j+1} - \mu_{J+j}} \beta_{(k),j}^2 \leq B_j^2 \left(A\left(J,j\right) + 1 \right)^{1-2/p}$$

$$\leq B_j^2 2^{J(1-2/p)} (j+1)^{\theta(2/p-1)},$$

from which we get

$$\inf_{m \in \mathcal{M}_J} \left(\sum_{\lambda \notin m} \beta_\lambda^2 \right) \leq \sum_{j \geq 0} B_j^2 2^{J(1-2/p)} (j+1)^{\theta(2/p-1)}. \quad (4.67)$$

We recall that $\beta \in \mathcal{B}_p\left(\alpha, R\right)$ means that $\beta \in \mathcal{E}_p\left(c\right)$ with $c_\lambda = R\lambda^{-(\alpha+1/2-1/p)}$. Observe now that for every j

$$c_{\mu_{J+j}} \leq R2^{-(J+j)(\alpha+1/2-1/p)}, \tag{4.68}$$

since by (4.62) $\mu_{J+j} \geq 2^{J+j}$. So $\beta \in \mathcal{B}_p(\alpha, R)$ also implies that $B_j \leq \sup_{\lambda \in \Lambda(J+j)} c_\lambda = c_{\mu_{J+j}}$ and it then follows from (4.67) that

$$\inf_{m \in \mathcal{M}_J} \left(\sum_{\lambda \notin m} \beta_\lambda^2 \right) \leq R^2 2^{-2J\alpha} \sum_{j \geq 0} 2^{-2j(\alpha+1/2-1/p)} (j+1)^{\theta(2/p-1)}. \tag{4.69}$$

Now, the series in (4.69) converges with a sum bounded by $C = C(\alpha, p, \theta)$ and we finally derive from (4.69) and (4.66) that

$$\sup_{\beta \in \mathcal{B}_p(\alpha, R)} \mathbb{E}_\beta \left[\left\| \tilde{\beta} - \beta \right\|^2 \right] \leq C(\alpha, p, K, L, \theta) \inf_{J \geq 0} \{R^2 2^{-2J\alpha} + 2^J \varepsilon^2\}. \tag{4.70}$$

To optimize (4.70) it suffices to take $J = \inf\left\{ j \geq 0 \,\middle|\, 2^j \geq (R/\varepsilon)^{2/(2\alpha+1)} \right\}$. Then $2^J = \rho\Delta(R/\varepsilon)$ with $1 \leq \rho < 2$ provided that $R/\varepsilon \geq 1$ and we finally get

$$\sup_{\beta \in \mathcal{B}_p(\alpha, R)} \mathbb{E}_\beta \left[\left\| \tilde{\beta} - \beta \right\|^2 \right] \leq C'(\alpha, p, K, L, \theta) R^{2/(2\alpha+1)} \varepsilon^{4\alpha/(2\alpha+1)}.$$

If we compare this upper bound with the lower bound (4.61) we see that they match up to constant. In other words the penalized LSE is simultaneously approximately minimax over all Besov bodies $\mathcal{B}_p(\alpha, R)$ with $\alpha > 1/p - 1/2$ which are nondegenerate, i.e., with $R \geq \varepsilon$.

Using the model selection theorem again, it would be possible to mix the two above strategies in order to build a new one with the advantages of both. Pushing further this idea one can also (in the spirit of [49]) use several bases at the same time and not a single one as we did in the Gaussian sequence framework.

4.4 A General Model Selection Theorem

4.4.1 Statement

Our purpose is to propose a model selection procedure among a collection of possibly nonlinear models. This procedure is based on a penalized least squares criterion which involves a penalty depending on some extended notion of dimension allowing to deal with nonlinear models.

Theorem 4.18 *Let* $\{S_m\}_{m \in \mathcal{M}}$ *be some finite or countable collection of subsets of* $\mathbb{H}$. *We assume that for any* $m \in \mathcal{M}$, *there exists some a.s. continuous version* W *of the isonormal process on* S_m. *Assume furthermore the existence of some positive and nondecreasing continuous function* ϕ_m *defined on* $(0, +\infty)$ *such that* $\phi_m(x)/x$ *is nonincreasing and*

$$2\mathbb{E} \left[\sup_{t \in S_m} \left(\frac{W(t) - W(u)}{\|t - u\|^2 + x^2} \right) \right] \leq x^{-2} \phi_m(x) \tag{4.71}$$

for any positive x and any point u in S_m. Let us define $\tau_m = 1$ if S_m is closed and convex and $\tau_m = 2$ otherwise. Let us define $D_m > 0$ such that

$$\phi_m \left(\tau_m \varepsilon \sqrt{D_m} \right) = \varepsilon D_m \tag{4.72}$$

and consider some family of weights $\{x_m\}_{m \in \mathcal{M}}$ such that

$$\sum_{m \in \mathcal{M}} e^{-x_m} = \Sigma < \infty.$$

Let K be some constant with $K > 1$ and take

$$\mathrm{pen}\,(m) \geq K\varepsilon^2 \left(\sqrt{D_m} + \sqrt{2x_m} \right)^2. \tag{4.73}$$

We set for all $t \in \mathbb{H}$, $\gamma_\varepsilon(t) = \|t\|^2 - 2Y_\varepsilon(t)$ and consider some collection of ρ-LSEs $(\widehat{s}_m)_{m \in \mathcal{M}}$ i.e., for any $m \in \mathcal{M}$,

$$\gamma_\varepsilon(\widehat{s}_m) \leq \gamma_\varepsilon(t) + \rho, \text{ for all } t \in S_m.$$

Then, almost surely, there exists some minimizer $\widehat{m}$ of $\gamma_\varepsilon(\widehat{s}_m) + \mathrm{pen}\,(m)$ over $\mathcal{M}$. Defining a penalized ρ-LSE as $\widetilde{s} = \widehat{s}_{\widehat{m}}$, the following risk bound holds for all $s \in \mathbb{H}$

$$\mathbb{E}_s \left[\|\widetilde{s} - s\|^2 \right] \leq C(K) \left[\inf_{m \in \mathcal{M}} \left(d^2(s, S_m) + \mathrm{pen}\,(m) \right) + \varepsilon^2 (\Sigma + 1) + \rho \right] \tag{4.74}$$

Proof. We first notice that by definition of τ_m in any case for every positive number η and any point $s \in \mathbb{H}$, there exists some point $s_m \in S_m$ satisfying to the following conditions

$$\|s - s_m\| \leq (1 + \eta) d(s, S_m) \tag{4.75}$$
$$\|s_m - t\| \leq \tau_m (1 + \eta) \|s - t\|, \text{ for all } t \in S_m. \tag{4.76}$$

Indeed, whenever S_m is closed and convex, one can take s_m as the projection of s onto S_m. Then $\|s - s_m\| = d(s, S_m)$ and since the projection is known to be a contraction one has for every $s, t \in \mathbb{H}$

$$\|s_m - t_m\| \leq \|s - t\|$$

which a fortiori implies (4.76) with $\eta = 0$. In the general case, one can always take s_m such that (4.75) holds. Then for every $t \in S_m$

$$\|s_m - t\| \leq \|s_m - s\| + \|s - t\| \leq 2(1 + \eta) \|s - t\|$$

so that (4.76) holds.

Let us first assume for the sake of simplicity that $\rho = 0$ and that conditions (4.75) and (4.76) hold with $\eta = 0$. We now fix some $m \in \mathcal{M}$ and define $\mathcal{M}' = \{m' \in \mathcal{M}, \gamma_\varepsilon(\widehat{s}_{m'}) + \text{pen}(m') \leq \gamma_\varepsilon(\widehat{s}_m) + \text{pen}(m)\}$. By definition, for every $m' \in \mathcal{M}'$

$$\gamma_\varepsilon(\widehat{s}_{m'}) + \text{pen}(m') \leq \gamma_\varepsilon(\widehat{s}_m) + \text{pen}(m) \leq \gamma_\varepsilon(s_m) + \text{pen}(m)$$

which implies, since $\|s\|^2 + \gamma_\varepsilon(t) = \|t - s\|^2 - 2\varepsilon W(t)$, that

$$\|\widehat{s}_{m'} - s\|^2 \leq \|s - s_m\|^2 + 2\varepsilon\left[W(\widehat{s}_{m'}) - W(s_m)\right] - \text{pen}(m') + \text{pen}(m). \quad (4.77)$$

For any $m' \in \mathcal{M}$, we consider some positive number $y_{m'}$ to be chosen later, define for any $t \in S_{m'}$

$$2w_{m'}(t) = [\|s - s_m\| + \|s - t\|]^2 + y_{m'}^2$$

and finally set

$$V_{m'} = \sup_{t \in S_{m'}} \left[\frac{W(t) - W(s_m)}{w_{m'}(t)}\right].$$

Taking these definitions into account, we get from (4.77)

$$\|\widehat{s}_{m'} - s\|^2 \leq \|s - s_m\|^2 + 2\varepsilon w_{m'}(\widehat{s}_{m'}) V_{m'} - \text{pen}(m') + \text{pen}(m). \quad (4.78)$$

It now remains to control the variables $V_{m'}$ for all possible values of m' in $\mathcal{M}$. To do this we use the concentration inequality for the suprema of Gaussian processes (i.e., Proposition 3.19) which ensures that, given $z > 0$, for any $m' \in \mathcal{M}$,

$$\mathbb{P}\left[V_{m'} \geq \mathbb{E}[V_{m'}] + \sqrt{2v_{m'}(x_{m'} + z)}\right] \leq e^{-x_{m'}}e^{-z} \quad (4.79)$$

where

$$v_{m'} = \sup_{t \in S_{m'}} \text{Var}\left[W(t) - W(s_m)/w_{m'}(t)\right] = \sup_{t \in S_{m'}}\left[\|t - s_m\|^2/w_{m'}^2(t)\right].$$

Since $w_{m'}(t) \geq \|t - s_m\| y_{m'}$, then $v_{m'} \leq y_{m'}^{-2}$ and therefore, summing up inequalities (4.79) over $m' \in \mathcal{M}$ we derive that, on some event Ω_z with probability larger than $1 - \Sigma e^{-z}$, for all $m' \in \mathcal{M}$

$$V_{m'} \leq \mathbb{E}[V_{m'}] + y_{m'}^{-1}\sqrt{2(x_{m'} + z)}. \quad (4.80)$$

We now use assumption (4.71) to bound $\mathbb{E}[V_{m'}]$. Indeed

$$\mathbb{E}[V_{m'}] \leq \mathbb{E}\left[\sup_{t \in S_{m'}}\left(\frac{W(t) - W(s_{m'})}{w_{m'}(t)}\right)\right]$$
$$+ \mathbb{E}\left[\frac{(W(s_{m'}) - W(s_m))_+}{\inf_{t \in S_{m'}}[w_{m'}(t)]}\right] \quad (4.81)$$

and since by (4.76) (with $\eta = 0$), $2w_{m'}(t) \geq \tau_{m'}^{-2} \|t - s_{m'}\|^2 + y_{m'}^2$ for all $t \in S_{m'}$ we derive from (4.71) with $u = s_{m'}$ and the monotonicity assumption on $\phi_{m'}$ that

$$\mathbb{E}\left[\sup_{t \in S_{m'}} \left(\frac{W(t) - W(s_{m'})}{w_{m'}(t)}\right)\right] \leq y_{m'}^{-2}\phi_{m'}(\tau_{m'}y_{m'})$$

$$\leq y_{m'}^{-1}\phi_{m'}\left(\tau_{m'}\varepsilon\sqrt{D_{m'}}\right)\varepsilon^{-1}D_{m'}^{-1/2}$$

whenever $y_{m'} \geq \varepsilon\sqrt{D_{m'}}$. Hence, by (4.72),

$$\mathbb{E}\left[\sup_{t \in S_{m'}} \left(\frac{W(t) - W(s_{m'})}{w_{m'}(t)}\right)\right] \leq y_{m'}^{-1}\sqrt{D_{m'}}$$

which achieves the control of the first term in the right-hand side of (4.81). For the second term in (4.81), we use (4.75) (with $\eta = 0$) to get

$$\inf_{t \in S_{m'}}[w_{m'}(t)] \geq 2^{-1}\left[\|s_m - s_{m'}\|^2 + y_{m'}^2\right] \geq y_{m'}\|s_m - s_{m'}\|,$$

hence

$$\mathbb{E}\left[\frac{(W(s_{m'}) - W(s_m))_+}{\inf_{t \in S_{m'}}[w_{m'}(t)]}\right] \leq y_{m'}^{-1}\mathbb{E}\left[\frac{W(s_{m'}) - W(s_m)}{\|s_m - s_{m'}\|}\right]_+$$

and since $[W(s_{m'}) - W(s_m)]/\|s_m - s_{m'}\|$ is a standard normal variable

$$\mathbb{E}\left[\frac{(W(s_{m'}) - W(s_m))_+}{\inf_{t \in S_{m'}}[w_{m'}(t)]}\right] \leq y_{m'}^{-1}(2\pi)^{-1/2}.$$

Collecting these inequalities we get from (4.81)

$$\mathbb{E}[V_{m'}] \leq y_{m'}^{-1}\left[\sqrt{D_{m'}} + (2\pi)^{-1/2}\right], \text{ for all } m' \in \mathcal{M}.$$

Hence, (4.80) implies that on the event Ω_z

$$V_{m'} \leq y_{m'}^{-1}\left[\sqrt{D_{m'}} + \sqrt{2x_{m'}} + (2\pi)^{-1/2} + \sqrt{2z}\right], \text{ for all } m' \in \mathcal{M}.$$

Given $K' \in \left(1, \sqrt{K}\right]$ to be chosen later, if we define

$$y_{m'} = K'\varepsilon\left[\sqrt{D_{m'}} + \sqrt{2x_{m'}} + (2\pi)^{-1/2} + \sqrt{2z}\right],$$

we know that, on the event Ω_z, $\varepsilon V_{m'} \leq K'^{-1}$ for all $m' \in \mathcal{M}$, which in particular implies via (4.78) for every $m' \in \mathcal{M}'$

$$\|\hat{s}_{m'} - s\|^2 \leq \|s - s_m\|^2 + 2K'^{-1}w_{m'}(\hat{s}_{m'}) - \text{pen}(m') + \text{pen}(m)$$

and therefore

$$\|\widehat{s}_{m'} - s\|^2 \le \|s - s_m\|^2 + K'^{-1} \left[[\|s - s_m\| + \|s - \widehat{s}_{m'}\|]^2 + y_{m'}^2 \right]$$
$$- \operatorname{pen}(m') + \operatorname{pen}(m).$$

Using repeatedly the elementary inequality

$$(a + b)^2 \le (1 + \theta) a^2 + (1 + \theta^{-1}) b^2$$

for various values of $\theta > 0$, we derive that for every $m' \in \mathcal{M}'$, on the one hand,

$$y_{m'}^2 \le K'^2 \varepsilon^2 \left[K' \left(\sqrt{D_{m'}} + \sqrt{2x_{m'}} \right)^2 + \frac{2K'}{K' - 1} \left(\frac{1}{2\pi} + 2z \right) \right]$$

and on the other hand

$$[\|s - s_m\| + \|s - \widehat{s}_{m'}\|]^2 \le \sqrt{K'} \left(\|s - \widehat{s}_{m'}\|^2 + \frac{\|s - s_m\|^2}{\sqrt{K'} - 1} \right).$$

Hence, setting $A' = \left(1 + K'^{-1/2} \left(\sqrt{K'} - 1 \right)^{-1} \right)$, on the event Ω_z, the following inequality is valid for every $m' \in \mathcal{M}'$

$$\|\widehat{s}_{m'} - s\|^2 \le A' \|s - s_m\|^2 + K'^{-1/2} \|s - \widehat{s}_{m'}\|^2$$
$$+ K'^2 \varepsilon^2 \left(\sqrt{D_{m'}} + \sqrt{2x_{m'}} \right)^2 - \operatorname{pen}(m')$$
$$+ \operatorname{pen}(m) + \frac{2K'^2 \varepsilon^2}{K' - 1} \left(\frac{1}{2\pi} + 2z \right),$$

which in turn implies because of condition (4.73)

$$\left(\frac{\sqrt{K'} - 1}{\sqrt{K'}} \right) \|\widehat{s}_{m'} - s\|^2 + \left(1 - \frac{K'^2}{K} \right) \operatorname{pen}(m')$$
$$\le A' \|s - s_m\|^2 + \operatorname{pen}(m) + \frac{2K'^2 \varepsilon^2}{K' - 1} \left(\frac{1}{2\pi} + 2z \right). \tag{4.82}$$

We may now use (4.82) in two different ways. Choosing first

$$K' = \sqrt{\frac{1 + K}{2}},$$

we derive from (4.82) that, on the event Ω_z, the following inequality holds for every $m' \in \mathcal{M}'$

$$\left(\frac{K - 1}{2K} \right) \operatorname{pen}(m') \le A' \|s - s_m\|^2 + \operatorname{pen}(m) + \frac{2K'^2 \varepsilon^2}{K' - 1} \left(\frac{1}{2\pi} + 2z \right).$$

Hence, the random variable $M = \sup_{m' \in \mathcal{M}'} \mathrm{pen}\,(m')$ is finite a.s. Now observe that by (4.73), $2K x_{m'} \varepsilon^2 \leq M$ for every $m' \in \mathcal{M}'$, hence

$$\Sigma \geq \sum_{m' \in \mathcal{M}'} \exp\left(-x_{m'}\right) \geq |\mathcal{M}'| \exp\left(-\frac{M}{2K\varepsilon^2}\right)$$

and therefore $\mathcal{M}'$ is almost surely a finite set. This proves of course that some minimizer $\widehat{m}$ of $\gamma_\varepsilon\,(\widehat{s}_{m'}) + \mathrm{pen}\,(m')$ over $\mathcal{M}'$ and hence over $\mathcal{M}$ does exist. Choosing this time $K' = \sqrt{K}$, we derive from (4.82) that on the event Ω_z

$$\left(\frac{K^{1/4} - 1}{K^{1/4}}\right) \|\widehat{s}_{\widehat{m}} - s\|^2 \leq \|s - s_m\|^2 \left(1 + K^{-1/4}\left(K^{1/4} - 1\right)^{-1}\right)$$

$$+ \mathrm{pen}\,(m) + \frac{2K\varepsilon^2}{\sqrt{K} - 1}\left(\frac{1}{2\pi} + 2z\right).$$

Integrating this inequality with respect to z straightforwardly leads to the required risk bound (4.74) at least whenever $\rho = 0$ and $\eta = 0$. It is easy to check that our proof remains valid (up to straightforward modifications of the constants involved) under the more general assumptions given in the statement of the theorem. ■

In the statement of Theorem 4.18, the function ϕ_m plays a crucial role in order to define D_m and therefore the penalty $\mathrm{pen}\,(m)$. Let us study the simplest example where S_m is linear with dimension D'_m in which case $\tau_m = 1$. Then, noticing that $\|t - u\|^2 + x^2 \geq 2\|t - u\|\,x$ we get via Jensen's inequality

$$2\mathbb{E}\left[\sup_{t \in S_m}\left(\frac{W\,(t) - W\,(u)}{\|t - u\|^2 + x^2}\right)\right] \leq x^{-1}\mathbb{E}\left[\sup_{t \in S_m}\left(\frac{W\,(t) - W\,(u)}{\|t - u\|}\right)\right]$$

$$\leq x^{-1}\left[\mathbb{E}\left[\sup_{t \in S_m}\left(\frac{W\,(t) - W\,(u)}{\|t - u\|}\right)^2\right]\right]^{1/2}.$$

Now taking some orthonormal basis $\{\varphi_j, j = 1, \ldots, D'_m\}$ of S_m we have by linearity of W and Cauchy–Schwarz inequality

$$\sup_{t \in S_m}\left(\frac{W\,(t) - W\,(u)}{\|t - u\|}\right)^2 = \sum_{j=1}^{D'_m} W^2\,(\varphi_j).$$

Since $W\,(\varphi_j)$ is a standard normal random variable for any j, we derive that

$$2\mathbb{E}\left[\sup_{t \in S_m}\left(\frac{W\,(t) - W\,(u)}{\|t - u\|^2 + x^2}\right)\right] \leq x^{-1}\left[\mathbb{E}\left[\sum_{j=1}^{D'_m} W^2\,(\varphi_j)\right]\right]^{1/2} \leq x^{-1}\sqrt{D'_m}.$$

Therefore assumption (4.71) is fulfilled with $\phi_m\,(x) = x\sqrt{D'_m}$, which implies that the solution of equation (4.72) is exactly $D_m = D'_m$. This shows that our Theorem 4.18 indeed implies Theorem 4.2.

The main interest of the present statement is that it allows to deal with nonlinear models. For a linear model S_m we have seen that the quantity D_m can be taken as the dimension of S_m. Let us see what this gives now for ellipsoids.

4.4.2 Selecting Ellipsoids: A Link with Regularization

Considering some Hilbert–Schmidt ellipsoid $\mathcal{E}_2(c)$ our purpose is to evaluate

$$2x^2 \mathbb{E}\left[\sup_{t \in \mathcal{E}_2(c)} \left(\frac{W(t) - W(u)}{\|t - u\|^2 + x^2}\right)\right]$$

where W is an a.s. continuous version of the isonormal process on $\mathbb{H}$, in order to be able to apply the model selection theorem for possibly nonlinear models. We prove the following

Lemma 4.19 *Let* $\{\varphi_j\}_{j \geq 1}$ *be some orthonormal basis of* $\mathbb{H}$ *and* $(c_j)_{j \geq 1}$ *be some nonincreasing sequence belonging to* ℓ_2. *Consider the Hilbert–Schmidt ellipsoid*

$$\mathcal{E}_2(c) = \left\{\sum_{j \geq 1} \beta_j \varphi_j : \sum_{j \geq 1} \frac{\beta_j^2}{c_j^2} \leq 1\right\}.$$

Then for every positive real number x, *one has*

$$2x^2 \mathbb{E}\left[\sup_{t \in \mathcal{E}_2(c)} \left(\frac{W(t)}{\|t\|^2 + x^2}\right)\right] \leq \sqrt{5 \sum_{j \geq 1} (c_j^2 \wedge x^2)}. \tag{4.83}$$

Proof. Let $t = \sum_{j \geq 1} \beta_j \varphi_j$ be some element of $\mathcal{E}_2(c)$ and consider

$$t_x = \frac{2x^2 t}{\|t\|^2 + x^2}.$$

Defining $\gamma_j = \theta(c_j \wedge x)$, where θ is some positive constant to be chosen later, we notice that

$$\langle t_x, \varphi_j \rangle^2 \leq (4\beta_j^2) \wedge \left(\frac{x^2 \beta_j^2}{\|t\|^2}\right).$$

Hence

$$\sum_{j \geq 1} \frac{\langle t_x, \varphi_j \rangle^2}{\gamma_j^2} \leq \frac{1}{\theta^2}\left(\sum_{j \geq 1} \frac{\langle t_x, \varphi_j \rangle^2}{c_j^2} + \sum_{j \geq 1} \frac{\langle t_x, \varphi_j \rangle^2}{x^2}\right)$$

$$\leq \frac{1}{\theta^2}\left(\sum_{j \geq 1} \frac{4\beta_j^2}{c_j^2} + \sum_{j \geq 1} \frac{\beta_j^2}{\|t\|^2}\right) \leq \frac{5}{\theta^2},$$

which means that $t_x \in \mathcal{E}_2(\gamma)$ if we choose $\theta = \sqrt{5}$. Therefore (4.83) follows from (3.42). ∎

Since $t - u$ belongs to $\mathcal{E}_2(2c)$ when t and u belong to $\mathcal{E}_2(c)$, Lemma 4.19 also implies that

$$2x^2 \mathbb{E}\left[\sup_{t \in \mathcal{E}_2(c)} \left[\frac{W(t) - W(u)}{\|t - u\|^2 + x^2} \right] \right] \leq \sqrt{20 \sum_{j \geq 1} \left(c_j^2 \wedge x^2 \right)} \leq 5 \sqrt{\sum_{j \geq 1} \left(c_j^2 \wedge x^2 \right)}.$$

We have now in view to apply Theorem 4.18 in order to select among some collection of ellipsoids, using penalties which will depend on the pseudodimensions of the ellipsoids as defined by (4.72). Defining such a pseudodimension for a given ellipsoid $\mathcal{E}_2(c)$ requires to use the preceding calculations. Since the function $\phi : x \to \sqrt{\sum_{j \geq 1} \left(c_j^2 \wedge x^2 \right)}$ is nondecreasing and $x \to \phi(x)/x$ is nonincreasing, (4.83) implies that we may define the pseudodimension $D(\varepsilon)$ of the ellipsoid $\mathcal{E}_2(c)$ by solving equation (4.72), i.e.,

$$5\phi\left(\varepsilon \sqrt{D(\varepsilon)} \right) = \varepsilon D(\varepsilon). \tag{4.84}$$

Adaptation over Besov Ellipsoids

Let us consider the Gaussian sequence framework (4.46) and specialize to Besov ellipsoids $\mathcal{B}_2(r, R)$, i.e., those for which $c_j = Rj^{-r}$ for every $j \geq 1$, where r is some real number larger than $1/2$. It is then easy to compute $\phi(x)$, at least if $R \geq x$. Indeed, introducing $N = \sup\{j : x \leq Rj^{-r}\}$ we observe that

$$\frac{1}{2}\left(\frac{R}{x} \right)^{1/r} \leq N \leq \left(\frac{R}{x} \right)^{1/r}. \tag{4.85}$$

Moreover

$$\phi^2(x) = x^2 N + \sum_{j \geq N+1} R^2 j^{-2r} \leq x^2 N + \frac{1}{2r - 1} N^{-2r+1} R^2$$

and we get from (4.85)

$$\phi^2(x) \leq R^{1/r} x^{2-1/r} + \frac{2^{2r-1}}{2r - 1} \left(\frac{R}{x} \right)^{(-2r+1)/r} R^2.$$

Hence

$$\phi(x) \leq \kappa_r x^{1-1/(2r)} R^{1/(2r)}$$

where

$$\kappa_r = \sqrt{1 + \frac{2^{2r-1}}{2r - 1}}$$

and, provided that $R \geq \varepsilon$, we derive from (4.84) that

$$\varepsilon D\left(\varepsilon\right) \leq 5\kappa_r \varepsilon^{1-1/(2r)} R^{1/(2r)} \left(D\left(\varepsilon\right)\right)^{(1/2)-1/(4r)}.$$

Finally

$$D\left(\varepsilon\right) \leq C_r \left(\frac{R}{\varepsilon}\right)^{2/(2r+1)}, \tag{4.86}$$

with $C_r = (5\kappa_r)^{4r/(2r+1)}$.

Given $r > 1/2$, let us now consider as a collection of models the family $\{S_m\}_{m \geq 1}$ of Besov ellipsoids $S_m = \mathcal{B}_2\left(r, m\varepsilon\right)$. If we apply Theorem 4.18, we derive from (4.86) that, setting $D_m = C_r m^{2/(2r+1)}$ and $x_m = m^{2/(2r+1)}$, we can take a penalty pen (m) which satisfies

$$\operatorname{pen}\left(m\right) \geq K \left(\sqrt{C_r} + \sqrt{2}\right)^2 m^{2/(2r+1)} \varepsilon^2, \tag{4.87}$$

where $K > 1$ is a given constant. This leads to a risk bound for the corresponding penalized LSE $\widetilde{\beta}$ of β which has the following form:

$$\mathbb{E}_\beta \left[\left\|\widetilde{\beta} - \beta\right\|^2\right] \leq C\left(K\right) \left[\inf_{m \geq 1} \left(d^2\left(\beta, S_m\right) + \operatorname{pen}\left(m\right)\right) + \varepsilon^2\left(\Sigma_r + 1\right),\right]$$

where $\Sigma_r = \Sigma_{m \geq 1} \exp\left(-m^{2/(2r+1)}\right)$ and therefore, since pen $(m) \geq \varepsilon^2$,

$$\mathbb{E}_\beta \left[\left\|\widetilde{\beta} - \beta\right\|^2\right] \leq C\left(K, r\right) \left[\inf_{m \geq 1} \left(d^2\left(\beta, \mathcal{B}_2\left(r, m\varepsilon\right)\right) + \operatorname{pen}\left(m\right)\right).\right] \tag{4.88}$$

A First Penalization Strategy

In view of the constraint (4.87) on the penalty, it is tempting to choose the penalty as

$$\operatorname{pen}\left(m\right) = K_r m^{2/(2r+1)} \varepsilon^2, \tag{4.89}$$

where K_r is an adequate constant depending only on r. In this case we may rewrite (4.88) as

$$\mathbb{E}_\beta \left[\left\|\widetilde{\beta} - \beta\right\|^2\right] \leq C\left(r\right) \inf_{m \geq 1} \left(d^2\left(\beta, \mathcal{B}_2\left(r, m\varepsilon\right)\right) + m^{2/(2r+1)} \varepsilon^2\right). \tag{4.90}$$

Of course whenever $\beta \in \mathcal{B}_2\left(r, R\right)$ with $R \geq \varepsilon$, we may consider the smallest integer m larger than R/ε, then $\beta \in S_m$ and the oracle type inequality (4.90) implies that

$$\mathbb{E}_\beta \left[\left\|\widetilde{\beta} - \beta\right\|^2\right] \leq C\left(r\right) m^{2/(2r+1)} \varepsilon^2.$$

But $m \leq 2R/\varepsilon$, so that

$$\mathbb{E}_\beta \left[\left\| \tilde{\beta} - \beta \right\|^2 \right] \leq 2C\left(r\right) R^{2/(2r+1)} \varepsilon^{4r/(2r+1)},$$

which, via the minimax lower bound (4.61), means that $\tilde{\beta}$ is minimax over all Besov ellipsoids $\mathcal{B}_2\left(r, R\right)$ when R varies with $R \geq \varepsilon$. This is of course an expected result. The penalized LSE being built on the collection of (discretized) homothetics of the given Besov ellipsoid $\mathcal{B}_2\left(r, 1\right)$, it is indeed expected that the resulting selected estimator should be minimax over all these homothetics simultaneously. This simply means that the penalty is well designed and that the discretization does not cause any nuisance. The next result is maybe more surprising. Indeed Proposition 4.20 below together with the minimax lower bound (4.61) ensure that $\tilde{\beta}$ is also minimax on the whole collection of Besov ellipsoids $\mathcal{B}_2\left(\alpha, R\right)$ when R varies with $R \geq \varepsilon$, for *all* regularity indices $\alpha \in (0, r]$.

Proposition 4.20 *Let $r > 1/2$ be given. Let $\tilde{\beta}$ be the penalized LSE defined from the collection of Besov ellipsoids $\left(\mathcal{B}_2\left(r, m\varepsilon\right)\right)_{m \geq 1}$ with a penalty given by (4.89), then there exists some constant $C'\left(r\right)$ such that for every $\alpha \in (0, r]$ and every real number $R \geq \varepsilon$ one has*

$$\sup_{\beta \in \mathcal{B}_2(\alpha, R)} \mathbb{E}_\beta \left[\left\| \tilde{\beta} - \beta \right\|^2 \right] \leq C'\left(r\right) R^{2/(2\alpha+1)} \varepsilon^{4\alpha/(2\alpha+1)}. \tag{4.91}$$

Proof. The main issue is to bound the bias term $d^2\left(\beta, \mathcal{B}_2\left(r, m\varepsilon\right)\right)$ appearing in (4.90) knowing that $\beta \in \mathcal{B}_2\left(\alpha, R\right)$. Let us consider

$$D = \left\lfloor \left(\frac{R}{\varepsilon}\right)^{2/(2\alpha+1)} \right\rfloor.$$

Since $R \geq \varepsilon$ we have

$$\left(\frac{R}{\varepsilon}\right)^{2/(2\alpha+1)} - 1 < D \leq \left(\frac{R}{\varepsilon}\right)^{2/(2\alpha+1)}. \tag{4.92}$$

Let us now take as an approximation of β the D-dimensional approximating sequence $\beta\left(D\right)$ defined by $\beta_j\left(D\right) = \beta_j$ whenever $j \leq D$ and $\beta_j\left(D\right) = 0$ otherwise. Then since $\beta \in \mathcal{B}_2\left(\alpha, R\right)$, we derive from (4.92) that

$$d^2\left(\beta, \beta\left(D\right)\right) = \sum_{j>D} \beta_j^2 \leq (D+1)^{-2\alpha} R^2 \leq R^{2/(2\alpha+1)} \varepsilon^{4\alpha/(2\alpha+1)}. \tag{4.93}$$

Moreover using again that $\beta \in \mathcal{B}_2\left(\alpha, R\right)$ and the fact that $r \geq \alpha$ we also get via (4.92)

$$\sum_{j=1}^D \beta_j^2 j^{2r} \leq D^{2(r-\alpha)} R^2 \leq \left(\frac{R}{\varepsilon}\right)^{4(r-\alpha)/(2\alpha+1)} R^2.$$

Hence, defining

$$m = \left\lceil \left(\frac{R}{\varepsilon}\right)^{(2r+1)/(2\alpha+1)} \right\rceil$$

we derive that

$$m \geq \left(\frac{R}{\varepsilon}\right)^{(2r+1)/(2\alpha+1)}$$

and therefore that

$$\sum_{j=1}^{D} \beta_j^2 j^{2r} \leq \left(\frac{R}{\varepsilon}\right)^{4(r-\alpha)/(2\alpha+1)} R^2 \leq m^2 \varepsilon^2.$$

The later inequality expresses that $\beta(D) \in \mathcal{B}_2(r, m\varepsilon)$ so that we get from (4.90)

$$\mathbb{E}_\beta\left[\left\|\tilde{\beta} - \beta\right\|^2\right] \leq C(r)\left(d^2(\beta, \beta(D)) + m^{2/(2r+1)}\varepsilon^2\right),$$

and finally from (4.93) and the definition of m

$$\mathbb{E}_\beta\left[\left\|\tilde{\beta} - \beta\right\|^2\right] \leq C(r)\left(1 + 2^{2/(2r+1)}\right) R^{2/(2\alpha+1)}\varepsilon^{4\alpha/(2\alpha+1)}. \quad \blacksquare$$

We now wish to turn to another penalization strategy for selecting ellipsoids which is directly connected to what is usually called *regularization*.

Regularization

Here is some alternative view on ellipsoids selection. Given some nonincreasing sequence $c = (c_j)_{j \geq 1}$ in ℓ_2, one can define the space $\mathbb{H}_c$ of elements θ of ℓ_2 such that $\sum_{j \geq 1}(\theta_j^2/c_j^2) < \infty$ and consider on $\mathbb{H}_c$ the new Hilbertian norm

$$\|\theta\|_c = \left(\sum_{j \geq 1}(\theta_j^2/c_j^2)\right)^{1/2}.$$

Considering the least squares criterion

$$\gamma_\varepsilon(\theta) = -2\sum_{j \geq 1}\hat{\beta}_j\theta_j + \sum_{j \geq 1}\theta_j^2,$$

if pen is some nondecreasing function on $\mathbb{R}_+$, one has

$$\inf_{\theta \in \mathbb{H}_c}\{\gamma_\varepsilon(\theta) + \text{pen}(\|\theta\|_c)\} = \inf_{R > 0}\left\{\inf_{\|\theta\|_c \leq R}\{\gamma_\varepsilon(\theta) + \text{pen}(R)\}\right\}. \quad (4.94)$$

This means that, provided that all these quantities do exist, if $\widehat{\beta}(R)$ denotes some minimizer of $\gamma_\varepsilon(\theta)$ on the ellipsoid $\{\theta \in \ell_2 : \|\theta\|_c \leq R\}$ and $\widehat{R}$ denotes a minimizer of the penalized least squares criterion

$$\gamma_\varepsilon\left(\widehat{\beta}(R)\right) + \text{pen}(R),$$

then the corresponding penalized LSE $\widetilde{\beta} = \widehat{\beta}(\widehat{R})$ can also be interpreted as a minimizer of $\gamma_\varepsilon(\theta) + \text{pen}(\|\theta\|_c)$ over $\mathbb{H}_c$. The case where $\text{pen}(R) = \lambda_\varepsilon R^2$ corresponds to what is usually called *regularization*. The special role of the $R \to R^2$ function comes from the fact that in this case one can explicitly compute the minimizer $\widetilde{\beta}$ of

$$\gamma_\varepsilon(\theta) + \lambda_\varepsilon \|\theta\|_c^2$$

on $\mathbb{H}_c$. Indeed, it is not hard to check that it is merely a linear estimator given by

$$\widetilde{\beta}_j = \frac{\widehat{\beta}_j}{1 + \lambda_\varepsilon c_j^{-2}}. \tag{4.95}$$

One can find in the literature many papers devoted to the regularization methods. The special instance that we are looking at here is especially easy to develop and to explain due to the simplicity of the Gaussian sequence framework. Nevertheless it remains true even in more sophisticated frameworks that the main reasons for choosing a quadratic shape for the penalty function is computational. For regularization in a fixed design regression setting with a Sobolev smoothing norm for instance, the regularized LSE can be nicely computed as a spline with knots on the design. A very interesting account of the theory of regularization is given in [124].

Turning now to the case where $c_j = j^{-r}$, in view of the results of the previous section one would be tempted to take in (4.94) $\text{pen}(R) = \lambda_\varepsilon R^{2/(2r+1)}$ rather than $\text{pen}(R) = \lambda_\varepsilon R^2$. Of course Theorem 4.18 allows to overpenalize (which then be the case for values of the radius R larger than 1). But if we do so we should be ready to pay the price in the risk bound (4.74). In particular we can expect that if the true β really belongs to a given ellipsoid $\mathcal{E}_2(Rc)$ we should get a risk of order $\lambda_\varepsilon R^2$. Since we do not know in advance the value of R, it is reasonable to take $\lambda_\varepsilon = K\varepsilon^{4r/(2r+1)}$ in order to preserve our chance to match the minimax rate. But then the factor R^2 is suboptimal as compared to $R^{2/(2r+1)}$, at least for reasonably large values of R. This is indeed what we intend to prove now. More than that, we shall show that unlike the penalized LSE of the previous section, the regularized estimator misses the right minimax rate on each Besov ellipsoid $\mathcal{B}_2(\alpha, 1)$ where $0 < \alpha < r$.

Our result is based on the following elementary lower bound.

Lemma 4.21 *Assume that $\widetilde{\beta}$ is the estimator given by (4.95) with $\lambda_\varepsilon \leq c_1^2$ and define $J = \sup\{j : \lambda_\varepsilon c_j^{-2} \leq 1\}$. Let $R > 0$ be given and define the sequence β as*

$$\beta_j = \frac{Rc_J}{\sqrt{J}} \text{ if } j \leq J \text{ and } \beta_j = 0 \text{ if } j > J, \tag{4.96}$$

then

$$4\mathbb{E}_\beta\left[\left\|\widetilde{\beta} - \beta\right\|^2\right] \geq \left(\frac{R^2 c_J^2}{J}\right)\lambda_\varepsilon^2\left(\sum_{j=1}^{J} c_j^{-4}\right) + J\varepsilon^2 \tag{4.97}$$

which in particular implies that

$$\mathbb{E}_\beta\left[\left\|\widetilde{\beta} - \beta\right\|^2\right] \geq \left(\frac{2^{-2r-2}}{4r+1}\right)\left\{\lambda_\varepsilon R^2 + \varepsilon^2\lambda_\varepsilon^{-1/(2r)}\right\}, \tag{4.98}$$

whenever $c_j = j^{-r}$ for every $j \geq 1$.

Proof. It follows from (4.95) that $\widetilde{\beta}_j = \theta_j\widehat{\beta}_j$ with $\theta_j = \left(1 + \lambda_\varepsilon c_j^{-2}\right)^{-1}$, hence

$$\mathbb{E}_\beta\left[\left\|\widetilde{\beta} - \beta\right\|^2\right] = \sum_{j\geq 1}(\theta_j - 1)^2\beta_j^2 + \varepsilon^2\sum_{j\geq 1}\theta_j^2$$

$$= \lambda_\varepsilon^2\left(\sum_{j\geq 1}c_j^{-4}\theta_j^2\beta_j^2\right) + \varepsilon^2\sum_{j\geq 1}\theta_j^2.$$

Hence, noticing that J has been defined such that $\theta_j \geq 1/2$ whenever $j \leq J$, we derive from the previous identity that

$$4\mathbb{E}_\beta\left[\left\|\widetilde{\beta} - \beta\right\|^2\right] \geq \lambda_\varepsilon^2\left(\sum_{j=1}^{J}c_j^{-4}\beta_j^2\right) + J\varepsilon^2$$

and (4.97) follows from the definition of β. Assuming now that $c_j = j^{-r}$, we have on the one hand

$$\sum_{j=1}^{J}c_j^{-4} \geq \int_0^J x^{4r}dx = \frac{J^{4r+1}}{(4r+1)}$$

and on the other hand, by the definition of J,

$$2J \geq J + 1 \geq \lambda_\varepsilon^{-1/(2r)}.$$

Using these two lower bounds we easily derive (4.98) from (4.97). ∎

Note that the sequence β defined by (4.96) belongs to the ellipsoid $\mathcal{E}_2(Rc)$ or equivalently to $\mathcal{B}_2(r, R)$ whenever $c_j = j^{-r}$. We learn from the first term in the right-hand side of (4.98) that (up to some constant) the quantity $\lambda_\varepsilon R^2$ that we used as a penalty does appear in the risk. As for the second term, it tells us that too small values of λ_ε should be avoided. We can say even more about the choice of λ_ε. More precisely, taking $R = 1$, we seefrom (4.98) that if one wants to recover the minimax rate of convergence $\varepsilon^{4r/(2r+1)}$ on the Besov ellipsoid $\mathcal{B}_2(r,1)$ for the maximal risk of $\widetilde{\beta}$, i.e., if we assume that

$$\sup_{\beta \in \mathcal{B}_2(r,1)} \mathbb{E}_\beta \left[\left\| \widetilde{\beta} - \beta \right\|^2 \right] \leq C\varepsilon^{4r/(2r+1)},$$

then on the one hand one should have

$$\left(\frac{2^{-2r-2}}{4r+1} \right) \lambda_\varepsilon \leq C\varepsilon^{4r/(2r+1)}$$

and on the other hand

$$\left(\frac{2^{-2r-2}}{4r+1} \right) \varepsilon^2 \lambda_\varepsilon^{-1/(2r)} \leq C\varepsilon^{4r/(2r+1)}.$$

This means that for adequate positive constants $C_1(r)$ and $C_2(r)$ one should have

$$C_1(r)\,\varepsilon^{4r/(2r+1)} \leq \lambda_\varepsilon \leq C_2(r)\,\varepsilon^{4r/(2r+1)}.$$

In other words λ_ε must be taken of the order of $\varepsilon^{4r/(2r+1)}$. So, let us take $\lambda_\varepsilon = K\varepsilon^{4r/(2r+1)}$ to be simple. Then, we derive from Lemma 4.21 the result that we were expecting i.e.,

$$\sup_{\beta \in \mathcal{B}_2(r,R)} \mathbb{E}_\beta \left[\left\| \widetilde{\beta} - \beta \right\|^2 \right] \geq K \left(\frac{2^{-2r-2}}{4r+1} \right) R^2 \varepsilon^{4r/(2r+1)}.$$

This tells us that, at least when ε is small and R is large, the estimator $\widetilde{\beta}$ is not approximately minimax on $\mathcal{B}_2(r,R)$ since we do not recover the $R^{2/(2r+1)}$ factor. Since the rate $\varepsilon^{4r/(2r+1)}$ is correct, one could think that we only loose in the constants. This is indeed not the case.

Corollary 4.22 *Let $r > 1/2$, $\lambda_\varepsilon = K\varepsilon^{4r/(2r+1)}$ and $\widetilde{\beta}$ be the estimator defined by*

$$\widetilde{\beta}_j = \frac{\widehat{\beta}_j}{1 + \lambda_\varepsilon j^{2r}}.$$

Then, there exists some positive constant κ_r such that for every $\alpha \in (0,r]$, the following lower bound holds

$$\sup_{\beta \in \mathcal{B}_2(\alpha,1)} \mathbb{E}_\beta \left[\left\| \widetilde{\beta} - \beta \right\|^2 \right] \geq \kappa_r (K \wedge 1) \varepsilon^{4\alpha/(2r+1)}. \tag{4.99}$$

Proof. The proof is quite trivial, it is enough to notice that the sequence β defined by (4.96) satisfies

$$\sum_{j \geq 1} \beta_j^2 j^{2\alpha} = R^2 J^{-2r-1} \sum_{j=1}^{J} j^{2\alpha} \leq R^2 J^{-2r-1} \frac{(J+1)^{2\alpha+1}}{(2\alpha+1)}$$

$$\leq R^2 J^{2(\alpha-r)} \frac{2^{2\alpha+1}}{(2\alpha+1)} \leq R^2 J^{2(\alpha-r)} 2^{2r}.$$

Hence, choosing $R = J^{r-\alpha} 2^{-r}$ warrants that $\beta \in \mathcal{B}_2(\alpha, 1)$ and (4.98) leads to

$$\mathbb{E}_\beta\left[\left\|\widetilde{\beta} - \beta\right\|^2\right] \geq \left(\frac{2^{-4r-2}}{4r+1}\right) \lambda_\varepsilon J^{2(r-\alpha)}.$$

To finish the proof, we have to remember that by the definition of J, $2J \geq \lambda_\varepsilon^{-1/(2r)}$ and therefore

$$\mathbb{E}_\beta\left[\left\|\widetilde{\beta} - \beta\right\|^2\right] \geq \left(\frac{2^{-6r-2}}{4r+1}\right) \lambda_\varepsilon^{\alpha/r}$$

which yields (4.99). ∎

Since when $\alpha < r$, $4\alpha/(2r+1) < 4\alpha/(2\alpha+1)$, Corollary 4.22 proves that the regularized estimator $\widetilde{\beta}$ is suboptimal in terms of minimax rates of convergence on the Besov ellipsoids $\mathcal{B}_2(\alpha, 1)$ with $\alpha < r$.

4.4.3 Selecting Nets Toward Adaptive Estimation for Arbitrary Compact Sets

We intend to close this chapter with less constructive but somehow conceptually more simple and more general adaptive estimation methods based on metric properties of statistical models. Our purpose is first to show that the quantity D_m defined in Theorem 4.18 is generally speaking measuring the massiveness of S_m and therefore can be viewed as a pseudodimension (which depends on the scale parameter ε).

A Maximal Inequality for Weighted Processes

We more precisely intend to relate D_m with the metric entropy numbers. In order to do so we need a tool which allows, for a given process, to derive global maximal inequalities for a conveniently weighted version of the initial process from local maximal inequalities. This is the purpose of the following Lemma which is more or less classical and well known. We present some (short) proof of it for the sake of completeness. This proof is based on the *pealing device* introduced by Alexander in [4]. Note that in the statement and the proof of Lemma 4.23 below we use the convention that $\sup_{t \in A} g(t) = 0$ whenever A is the empty set.

Lemma 4.23 (*Pealing lemma*) *Let S be some countable set, $u \in S$ and $a : S \to \mathbb{R}_+$ such that $a(u) = \inf_{t \in S} a(t)$. Let Z be some process indexed by S and assume that the nonnegative random variable $\sup_{t \in \mathcal{B}(\sigma)}[Z(t) - Z(u)]$ has finite expectation for any positive number σ, where*

$$\mathcal{B}(\sigma) = \{t \in S, \, a(t) \leq \sigma\}.$$

Then, for any function ψ on $\mathbb{R}_+$ such that $\psi(x)/x$ is nonincreasing on $\mathbb{R}_+$ and satisfies to

$$\mathbb{E}\left[\sup_{t\in\mathcal{B}(\sigma)} Z(t) - Z(u)\right] \leq \psi(\sigma), \quad \text{for any } \sigma \geq \sigma_* \geq 0$$

one has for any positive number $x \geq \sigma_*$

$$\mathbb{E}\left[\sup_{t\in S}\left[\frac{Z(t) - Z(u)}{a^2(t) + x^2}\right]\right] \leq 4x^{-2}\psi(x).$$

Proof. Let us introduce for any integer j

$$\mathcal{C}_j = \left\{t \in S, \, r^j x < a(t) \leq r^{j+1} x\right\},$$

with $r > 1$ to be chosen later. Then $\left\{\mathcal{B}(x), \{\mathcal{C}_j\}_{j\geq 0}\right\}$ is a partition of S and therefore

$$\sup_{t\in S}\left[\frac{Z(t) - Z(u)}{a^2(t) + x^2}\right] \leq \sup_{t\in\mathcal{B}(x)}\left[\frac{(Z(t) - Z(u))_+}{a^2(t) + x^2}\right]$$
$$+ \sum_{j\geq 0}\sup_{t\in\mathcal{C}_j}\left[\frac{(Z(t) - Z(u))_+}{a^2(t) + x^2}\right],$$

which in turn implies that

$$x^2\sup_{t\in S}\left[\frac{Z(t) - Z(u)}{a^2(t) + x^2}\right] \leq \sup_{t\in\mathcal{B}(x)}(Z(t) - Z(u))_+$$
$$+ \sum_{j\geq 0}\left(1 + r^{2j}\right)^{-1}\sup_{t\in\mathcal{B}(r^{j+1}x)}(Z(t) - Z(u))_+.$$

$$(4.100)$$

Since $a(u) = \inf_{t\in S} a(t)$, $u \in \mathcal{B}(r^k x)$ for every integer k for which $\mathcal{B}(r^k x)$ is nonempty and therefore

$$\sup_{t\in\mathcal{B}(r^k x)}(Z(t) - Z(u))_+ = \sup_{t\in\mathcal{B}(r^k x)}(Z(t) - Z(u)).$$

Hence, taking expectation in (4.100) yields

$$x^2\mathbb{E}\left[\sup_{t\in S}\left[\frac{Z(t) - Z(u)}{a^2(t) + x^2}\right]\right] \leq \psi(x) + \sum_{j\geq 0}\left(1 + r^{2j}\right)^{-1}\psi\left(r^{j+1}x\right).$$

Now by our monotonicity assumption, $\psi\left(r^{j+1}x\right) \leq r^{j+1}\psi(x)$, thus

$$x^2 \mathbb{E} \left[\sup_{t \in S} \left[\frac{Z(t) - Z(u)}{a^2(t) + x^2} \right] \right] \leq \psi(x) \left[1 + r \sum_{j \geq 0} r^j \left(1 + r^{2j} \right)^{-1} \right]$$

$$\leq \psi(x) \left[1 + r \left(\frac{1}{2} + \sum_{j \geq 1} r^{-j} \right) \right]$$

$$\leq \psi(x) \left[1 + r \left(\frac{1}{2} + \frac{1}{r-1} \right) \right]$$

and the result follows by choosing $r = 1 + \sqrt{2}$. ∎

This Lemma warrants that in order to check assumption (4.71) it is enough to consider some nondecreasing function ϕ_m such that $\phi_m(x)/x$ is nonincreasing and satisfies

$$\mathbb{E} \left[\sup_{t \in S_m, \|u-t\| \leq \sigma} [W(t) - W(u)] \right] \leq \phi_m(\sigma)/8 \qquad (4.101)$$

for every $u \in S_m$ and any positive σ. Now we know from Theorem 3.18 that one can always take

$$\phi_m(\sigma) = \kappa \int_0^\sigma \sqrt{H(x, S_m)} dx,$$

where $H(., S_m)$ is the metric entropy of S_m and κ is some absolute constant. Then (4.101) and therefore (4.71) are satisfied. This shows that the pseudo-dimension D_m defined by (4.72) is directly connected to the metric entropy and is therefore really measuring the massiveness of model S_m. In particular if S_m is a finite set with cardinality $\exp(\Delta_m)$, we can take $\phi_m(\sigma) = \kappa \sigma \sqrt{\Delta_m}$ (we could even specify that in this case $\kappa = 8\sqrt{2}$ works because of (2.3)) and the solution D_m of (4.72) with $\tau_m = 2$ is given by $D_m = 4\kappa^2 \Delta_m$. This leads to the following completely discrete version of Theorem 4.18 (in the spirit of [10]).

Corollary 4.24 *Let $\{S_m\}_{m \in \mathcal{M}}$ be some at most countable collection of finite subsets of $\mathbb{H}$. Consider some family of positive real numbers $\{\Delta_m\}_{m \in \mathcal{M}}$ and $\{x_m\}_{m \in \mathcal{M}}$ such that*

$$\text{for every } m \in \mathcal{M}, \ \ln|S_m| \leq \Delta_m \ \text{ and } \ \sum_{m \in \mathcal{M}} e^{-x_m} = \Sigma < \infty.$$

Take

$$\text{pen}(m) = \kappa' \varepsilon^2 (\Delta_m + x_m), \qquad (4.102)$$

where κ' is a suitable numerical constant. If, for every $m \in \mathcal{M}$ we denote by $\hat{s}_m$ the LSE over S_m, we select $\hat{m}$ minimizing $\gamma_\varepsilon(\hat{s}_m) + \text{pen}(m)$ over $\mathcal{M}$ and *define the penalized LSE $\tilde{s} = \hat{s}_{\hat{m}}$. Then, the following risk bound holds for all $s \in \mathbb{H}$*

$$\mathbb{E}_s \left[\|\tilde{s} - s\|^2 \right] \leq C' \left[\inf_{m \in \mathcal{M}} \left(d^2 \left(s, S_m \right) + \mathrm{pen} \left(m \right) \right) + \varepsilon^2 \left(\Sigma + 1 \right) \right]. \qquad (4.103)$$

The general idea for using such a result in order to build some adaptive estimator on a collection of compact sets of $\mathbb{H}$ can be roughly described as follows. Given some countable collection of totally bounded subsets $(\mathcal{T}_m)_{m \in \mathcal{M}}$ of $\mathbb{H}$, one can consider, for each $m \in \mathcal{M}$, some δ_m-net S_m of $\mathcal{T}_m$, where δ_m will be chosen later. We derive from Corollary 4.24 that, setting $\Delta_m = \left(\varepsilon^{-2} \delta_m^2 \right) \vee H \left(\delta_m, \mathcal{T}_m \right)$ and $x_m = \Delta_m$, provided that

$$\Sigma = \sum_{m \in \mathcal{M}} e^{-\Delta_m} < \infty,$$

the penalized LSE with penalty given by (4.102) satisfies for all $m \in \mathcal{M}$

$$\mathbb{E}_s \left[\|\hat{s}_{\hat{m}} - s\|^2 \right] \leq C' \varepsilon^2 \left(2\kappa' \Delta_m + (1 + \Sigma) \right), \text{ for every } s \in \mathcal{T}_m. \qquad (4.104)$$

Now the point is that choosing δ_m^2 in such a way that $\varepsilon^2 H \left(\delta_m, \mathcal{T}_m \right)$ is of order δ_m leads to a risk bound of order δ_m^2 (at least if Σ remains under control). In view of Proposition 4.13, we know that, under some proper conditions, δ_m^2 can be taken of the order of the minimax risk on $\mathcal{T}_m$, so that if Σ is kept under control, (4.104) implies that $\hat{s}_{\hat{m}}$ is approximately minimax on each parameter set $\mathcal{T}_m$. One can even hope that the δ_m-nets which are tuned for the countable collection of parameter spaces $(\mathcal{T}_m)_{m \in \mathcal{M}}$ can also provide adequate approximations on a wider class of parameter spaces.

As an exercise illustrating these general ideas, let us apply Corollary 4.24 to build an adaptive estimator in the minimax sense over some collection of compact sets of $\mathbb{H}$

$$\left\{ \mathcal{S}_{\alpha,R} = R \mathcal{S}_{\alpha,1}, \alpha \in \mathbb{N}^*, R > 0 \right\},$$

where $\mathcal{S}_{\alpha,1}$ is star shaped at 0 (i.e., $\theta t \in \mathcal{S}_{\alpha,1}$ for all $t \in \mathcal{S}_{\alpha,1}$ and $\theta \in [0,1]$) and satisfies for some positive constants $C_1 \left(\alpha \right)$ and $C_2 \left(\alpha \right)$,

$$C_1 \left(\alpha \right) \delta^{-1/\alpha} \leq H \left(\delta, \mathcal{S}_{\alpha,1} \right) \leq C_2 \left(\alpha \right) \delta^{-1/\alpha}$$

for every $\delta \leq 1$. This implies that $\mathcal{S}_{\alpha,R}$ fulfills (4.39). Hence, it comes from Proposition 4.13 that for some positive constant $\kappa \left(\alpha \right)$ depending only on α

$$\sup_{s \in \mathcal{S}_{\alpha,R}} \mathbb{E}_s \left[\|s - \hat{s}\|^2 \right] \geq \kappa \left(\alpha \right) R^{2/(2\alpha+1)} \varepsilon^{2\alpha/(2\alpha+1)}.$$

In order to build an estimator which achieves (up to constants) the minimax risk over all the compact sets $\mathcal{S}_{r,R}$, we simply consider for every positive integers α and k a $k^{1/(2\alpha+1)}\varepsilon$-net $S_{\alpha,k}$ of $\mathcal{S}_{\alpha,k\varepsilon}$ and apply Corollary 4.24 to the collection $(S_{\alpha,k})_{\alpha\geq 1, k\geq 1}$. Since

$$H(\delta, \mathcal{S}_{\alpha,k\varepsilon}) \leq C_2(\alpha) \left(\frac{k\varepsilon}{\delta}\right)^{1/\alpha},$$

we may take $\Delta_{\alpha,k} = C_2(\alpha) k^{2/(2\alpha+1)}$. Defining $x_{\alpha,k} = 4\alpha k^{2/(2\alpha+1)}$ leads to the penalty

$$\text{pen}(\alpha, k) = (C_2(\alpha) + 4\alpha) k^{2/(2\alpha+1)} \varepsilon^2.$$

Noticing that $x_{\alpha,k} \geq \alpha + 2\ln(k)$ one has

$$\Sigma = \sum_{\alpha,k} e^{-x_{\alpha,k}} \leq \left(\sum_{\alpha\geq 1} e^{-\alpha}\right)\left(\sum_{k\geq 1} k^{-2}\right) < 1$$

and it follows from (4.103) that if $\widetilde{s}$ denotes the penalized LSE one has

$$\mathbb{E}_s\left[\|s - \widetilde{s}\|^2\right] \leq C(\alpha) \inf_{\alpha,k} \left(d^2(s, \mathcal{S}_{\alpha,k}) + k^{2/(2\alpha+1)}\varepsilon^2\right).$$

Because of the star shape property, given α, the family $\{\mathcal{S}_{\alpha,R}; R > 0\}$ is nested. In particular if $s \in \mathcal{S}_{\alpha,R}$ for some integer α and some real number $R \geq \varepsilon$, setting $k = \lceil R/\varepsilon \rceil$ we have $s \in \mathcal{S}_{\alpha,k\varepsilon}$ and since $S_{\alpha,k}$ is a $k^{1/(2\alpha+1)}\varepsilon$-net of $\mathcal{S}_{\alpha,k\varepsilon}$, the previous inequality implies that

$$\sup_{s\in\mathcal{S}_{\alpha,R}} \mathbb{E}_s\left[\|s - \widetilde{s}\|^2\right] \leq 2C(\alpha) k^{2/(2\alpha+1)}\varepsilon^2$$

$$\leq 2C(\alpha) 2^{2/(2\alpha+1)} \left(\frac{R}{\varepsilon}\right)^{2/(2\alpha+1)} \varepsilon^2.$$

Hence $\widetilde{s}$ is minimax (up to constants) on each compact set $\mathcal{S}_{\alpha,R}$ with $\alpha \in \mathbb{N}^*$ and $R \geq \varepsilon$. Note that Lemma 4.14 implies that this approach can be applied to the case where $\mathcal{S}_{\alpha,R}$ is the Besov ellipsoid $\mathcal{B}_2(\alpha, R)$. Constructing minimax estimators via optimally calibrated nets in the parameter space is quite an old idea which goes back to Le Cam (see [75]) for parameter spaces with finite metric dimension. Indeed MLEs on nets are not always the right procedures to consider in full generality (from this perspective the "ideal" white noise framework does not reflect the difficulties that may be encountered with MLEs in other functional estimation contexts). Hence Le Cam's estimators differ from discretized MLEs in a rather subtle way. They are based on families of tests between balls rather than between their centers (which is in fact what discretized MLEs do when they are viewed as testing procedures) in order to warrant their robustness . The same kind of ideas have been developed at length by Birgé in [16] for arbitrary parameter spaces and in various functional estimation frameworks. Recently Birgé in [18] has shown that the same estimation procedure based on discretization of parameter spaces and robust tests becomes adaptive if one considers families of nets instead of a given net. In some sense, the results that we have presented above in the especially simple Gaussian framework where MLEs on nets are making a good job, are illustrative of the *metric point of view* for adaptive estimation promoted by Birgé in [18].

4.5 Appendix: From Function Spaces to Sequence Spaces

Our purpose here is to briefly recall, following more or less [52], why it is natural to search for adaptive procedures over various types of ℓ_p-bodies and particularly Besov bodies.

We recall that, given three positive numbers $p, q \in (0, +\infty]$ and $\alpha > 1/p - 1/2$ one defines the Besov seminorm $|t|_{B_q^\alpha(L_p)}$ of any function $t \in \mathbb{L}_2([0,1])$ by

$$|t|_{B_q^\alpha(L_p)} = \begin{cases} \left(\sum_{j=0}^\infty \left[2^{j\alpha} w_r(t, 2^{-j}, [0,1])_p \right]^q \right)^{1/q} & \text{when } q < +\infty, \\ \sup_{j \geq 0} 2^{j\alpha} w_r(t, 2^{-j}, [0,1])_p & \text{when } q = +\infty, \end{cases} \quad (4.105)$$

where $w_r(t, x, [0,1])_p$ denotes the modulus of smoothness of t, as defined by DeVore and Lorentz (see [44] p. 44) and $r = \lfloor \alpha \rfloor + 1$. When $p \geq 2$, since

$$w_r(t, 2^{-j}, [0,1])_p \geq w_r(t, 2^{-j}, [0,1])_2,$$

then $\left\{ t \mid |t|_{B_q^\alpha(L_p)} \leq R \right\} \subset \left\{ t \mid |t|_{B_q^\alpha(L_2)} \leq R \right\}$. Keeping in mind that we are interested in adaptation and therefore comparing the risk of our estimators with the minimax risk over such Besov balls, we can restrict our study to the case $p \leq 2$. Indeed, our nonasymptotic computations can only be done up to constants and it is known that the influence of p on the minimax risk is limited to those constants. It is therefore natural to ignore the smaller balls corresponding to $p > 2$.

If one chooses of a convenient wavelet basis, the Besov balls

$$\left\{ t \mid |t|_{B_q^\alpha(L_p)} \leq R \right\}$$

can be identified with subsets of ℓ_2 that have some nice geometrical properties. Given a pair (father and mother) of compactly supported orthonormal wavelets $(\bar\psi, \psi)$, any $t \in \mathbb{L}_2([0,1])$ can be written on $[0,1]$ as

$$t = \sum_{k \in \Lambda(-1)} \alpha_k \bar\psi_k + \sum_{j=0}^\infty \sum_{k \in \Lambda(j)} \beta_{j,k} \psi_{j,k}, \quad (4.106)$$

with

$$|\Lambda(-1)| = M' < +\infty \quad \text{and} \quad 2^j \leq |\Lambda(j)| \leq M 2^j \quad \text{for all } j \geq 0. \quad (4.107)$$

For a suitable choice of the wavelet basis and provided that the integer r satisfies $1 \le r \le \bar{r}$ with $\bar{r}$ depending on the basis,

$$2^{j(1/2-1/p)} \left(\sum_{k \in \Lambda(j)} |\beta_{j,k}|^p \right)^{1/p} \le C \omega_r(t, 2^{-j}, [0,1])_p, \qquad (4.108)$$

for all $j \ge 0$, $p \ge 1$, with a constant $C > 0$ depending only on the basis (see [38] and Theorem 2 in [52]). This result remains true if one replaces the wavelet basis by a piecewise polynomial basis generating dyadic piecewise polynomial expansions as shown in [22] Section 4.1.1. With some suitable restrictions on ω_r, this inequality still holds for $0 < p < 1$ and C depending on p (see [43] or [22]). In particular, if we fix $p \in [1,2]$, $q \in (0, +\infty]$, $\alpha > 1/p - 1/2$ and $R' > 0$ and consider those ts satisfying $|t|_{B_q^\alpha(L_p)} \le R'$, one derives from (4.108) that the coefficients $\beta_{j,k}$ of t in the expansion (4.106) satisfy

$$\sum_{j=0}^{\infty} \left[(R'C)^{-1} 2^{j(\alpha+1/2-1/p)} \left(\sum_{k \in \Lambda(j)} |\beta_{j,k}|^p \right)^{1/p} \right]^q \le 1 \quad \text{when } q < +\infty,$$

$$(4.109)$$

$$\sup_{j \ge 0} (R'C)^{-1} 2^{j(\alpha+1/2-1/p)} \left(\sum_{k \in \Lambda(j)} |\beta_{j,k}|^p \right)^{1/p} \le 1 \quad \text{when } q = +\infty, \qquad (4.110)$$

and one can show that such inequalities still hold for $p < 1$ (with C depending on p). Clearly, if (4.109) is satisfied for some q, it is also satisfied for all $q' > q$. The choice $q = +\infty$ dominates all other choices but does not allow us to deal with the limiting case $\alpha = 1/p - 1/2$ (when $p < 2$) since, with such a choice of α, (4.110) does not warrant that the coefficients $\beta_{j,k}$ belong to $\ell_2(\Lambda)$. It is therefore necessary, in this case, to restrict to $q = p$. For this reason, only two values of q are of interest for us: $q = p$ and $q = +\infty$, results for other values deriving from the results concerning those two ones. For the sake of simplicity, we shall actually focus on the case $q = p$, only minor modifications being needed to extend the results, when $\alpha > 1/p - 1/2$, to the case $q = +\infty$.

If $q = p \le 2$, (4.109) becomes

$$\sum_{j=0}^{\infty} \left[2^{jp(1/2-1/p)} \left[\omega(2^{-j}) \right]^{-p} \sum_{k \in \Lambda(j)} |\beta_{j,k}|^p \right] \le 1, \qquad (4.111)$$

with $\omega(x) = Rx^\alpha$ and $R = R'C$. Apart from the fact that it corresponds to some smoothness of order α in the usual sense, there is no special reason to restrict to functions ω of this particular form . If for instance,

$$\sum_{j=0}^{\infty} \left[\frac{\omega_r(t, 2^{-j}, [0,1])_p}{\omega(2^{-j})} \right]^p \le C^{-p}$$

for some nonnegative continuous function ω such that $x^{1/2-1/p}\omega(x)$ is bounded on $[0, 1]$, it follows from (4.108) that (4.111) still holds and the set of βs satisfying (4.111) is a subset of $\ell_2(\Lambda)$, where $\Lambda = \bigcup_{j \geq 0} \{(j, k), k \in \Lambda(j)\}$. Now one can order Λ according to the lexicographical order and this correspondence gives

$$\text{if } j \geq 0, \ k \in \Lambda(j), \quad (j, k) \longleftrightarrow \lambda \quad \text{with } M2^j \leq \lambda \leq M\left(2^{j+1} - 1\right). \quad (4.112)$$

Identifying Λ and $\mathbb{N}^*$ through this correspondence, if the function $x \mapsto x^{1/2-1/p}\omega(x)$ is nondecreasing and tends to zero when $x \to 0$, the above set is indeed an ℓ_p-body. These considerations are in fact the main motivation for the introduction of the notion of ℓ_p-body. Besov bodies are suitable for analyzing sets of functions of the form $\{t \mid |t|_{B_p^\alpha(L_p)} \leq R\}$. Indeed assuming that (4.108) holds and $\alpha' = \alpha + 1/2 - 1/p > 0$, we derive that $\{t \mid |t|_{B_p^\alpha(L_p)} \leq R\}$ is included in the set of t's with coefficients satisfying

$$\sum_{j=0}^{\infty} \left[2^{jp\alpha'}(CR)^{-p} \sum_{k \in \Lambda(j)} |\beta_{j,k}|^p\right] = \sum_{j=0}^{\infty} \sum_{k \in \Lambda(j)} \left|\frac{\beta_{j,k}}{2^{-j\alpha'}RC}\right|^p \leq 1,$$

and it follows from (4.112) that the coefficients βs of t belong to the ℓ_p-body $\mathcal{E}_p(c)$ with a sequence $(c_\lambda)_{\lambda \geq 1}$ defined by

$$c_\lambda = R'\lambda^{-\alpha'}$$

with $R' = RC(2M)^{\alpha+1/2-1/p}$, which means that it indeed belongs to the Besov body $\mathcal{B}_p(\alpha, R')$.

5

Concentration Inequalities

5.1 Introduction

The purpose of this chapter is to present concentration inequalities for real valued random variables Z of the form $Z = \zeta(X_1, \ldots, X_n)$ where $(X_1, \ldots, X_n)$ are independent random variables under some assumptions on ζ. Typically we have already seen that if ζ is Lipschitz on $\mathbb{R}^n$ with Lipschitz constant 1 and if $X_1, \ldots, X_n$ are i.i.d. standard normal random variables then, for every $x \geq 0$,

$$\mathbb{P}\left[Z - M \geq x\right] \leq \exp\left(-\frac{x^2}{2}\right), \qquad (5.1)$$

where M denotes either the mean or the median of Z. Extending such results to more general product measures is not easy. Talagrand's approach to this problem relies on isoperimetric ideas in the sense that concentration inequalities for functionals around their median are derived from probability inequalities for enlargements of sets with respect to various distances. A typical result which can be obtained by his methods is as follows (see Corollary 2.2.3. in [112]). Let Ω^n be equipped with the Hamming distance and ζ be some Lipschitz function on Ω^n with Lipschitz constant 1. Let P be some product probability measure on Ω^n and M be some median of ζ (with respect to the probability P), then, for every $x \geq 0$

$$P\left[\zeta - M \geq \left(x + \sqrt{\frac{\ln(2)}{2}}\right)\sqrt{n}\right] \leq \exp\left(-2x^2\right). \qquad (5.2)$$

Moreover, the same inequality holds for $-\zeta$ instead of ζ. For this problem, the isoperimetric approach developed by Talagrand consists in proving that for any measurable set A of Ω^n,

$$P\left[d(.,A) \geq \left(x + \sqrt{\frac{\ln(1/P(A))}{2}}\right)\sqrt{n}\right] \leq \exp\left(-2x^2\right),$$

where $d(.,A)$ denotes the Euclidean distance function to A. The latter inequality can be proved by at least two methods: the original proof by Talagrand in [112] relies on a control of the moment generating function of $d(.,A)$ which is proved by induction on the number of coordinates while Marton'sor Dembo's proofs (see [87] and [42] respectively) are based on some transportation cost inequalities. For the applications that we have in view, deviation inequalities of a functional from its mean (rather than from its median) are more suitable and therefore we shall focus on proofs which directly lead to such kind of results. We begin with Hoeffding type inequalities which have been originally obtained by using martingale arguments (see [93]) and are closely connected to Talagrand's results (at least those mentioned above which are in some sense the basic ones) as we shall see below. The proof that we give here is completely straightforward when starting from Marton's transportation cost inequality presented in Chapter 2. We shall sometimes need some sharper bounds (namely Bernstein type inequalities) which cannot be obtained through by this way. This will be the main motivation for introducing the "entropy method" in Section 5.3 which will lead to refined inequalities, especially for empirical processes.

5.2 The Bounded Difference Inequality via Marton's Coupling

As mentioned above Marton's transportation cost inequality leads to a simple and elegant proof of the bounded difference inequality (also called Mc Diarmid's inequality). Note that the usual proof of this result relies on a martingale argument.

Theorem 5.1 *Let $\mathcal{X}^n = \mathcal{X}_1 \times \cdots \times \mathcal{X}_n$ be some product measurable space and $\zeta : \mathcal{X}^n \to \mathbb{R}$ be some measurable functional satisfying for some positive constants $(c_i)_{1 \leq i \leq n}$, the bounded difference condition*

$$|\zeta(x_1,\ldots,x_i,\ldots,x_n) - \zeta(x_1,\ldots,y_i,\ldots,x_n)| \leq c_i$$

for all $x \in \mathcal{X}^n$, $y \in \mathcal{X}^n$ and all integer $i \in [1,n]$. Then the random variable $Z = \zeta(X_1,\ldots,X_n)$ satisfies to

$$\psi_{Z-\mathbb{E}[Z]}(\lambda) \leq \frac{\lambda^2 \left(\sum_{i=1}^n c_i^2\right)}{8}, \text{ for all } \lambda \in \mathbb{R}.$$

Hence, for any positive x

$$\mathbb{P}[Z - \mathbb{E}[Z] \geq x] \leq \exp\left(-\frac{2x^2}{\sum_{i=1}^n c_i^2}\right)$$

and similarly

$$\mathbb{P}[\mathbb{E}[Z] - Z \geq x] \leq \exp\left(-\frac{2x^2}{\sum_{i=1}^n c_i^2}\right).$$

Proof. From the bounded difference condition we derive that for every $x \in \mathcal{X}^n$ and $y \in \mathcal{X}^n$

$$
\begin{aligned}
|\zeta(x) - \zeta(y)| &\leq |\zeta(x_1, \ldots, x_n) - \zeta(y_1, x_2, \ldots, x_n)| \\
&\quad + |\zeta(y_1, x_2, \ldots, x_n) - \zeta(y_1, y_2, x_3, \ldots, x_n)| \\
&\quad + \cdots \\
&\quad + |\zeta(y_1, \ldots, y_{n-1}, x_n) - \zeta(y_1, y_2, \ldots, y_n)| \\
&\leq \sum_{i=1}^{n} c_i \mathbb{1}_{x_i \neq y_i}.
\end{aligned}
$$

This means that the bounded difference condition is equivalent to the following Lipschitz type condition:

$$
|\zeta(x) - \zeta(y)| \leq \sum_{i=1}^{n} c_i \mathbb{1}_{x_i \neq y_i} \text{ for all } x \in \mathcal{X}^n \text{ and } y \in \mathcal{X}^n.
$$

Denoting by P the (product) probability distribution of $(X_1, \ldots, X_n)$ on $\mathcal{X}^n$, let Q be some probability distribution which is absolutely continuous with respect to P and $\mathbb{Q} \in \mathcal{P}(P, Q)$. Then

$$
\begin{aligned}
E_Q[\zeta] - E_P[\zeta] &= \int_{\mathcal{X}^n \times \mathcal{X}^n} [\zeta(y) - \zeta(x)] \, d\mathbb{Q}(x, y) \\
&\leq \sum_{i=1}^{n} c_i \left(\int_{\mathcal{X}^n \times \mathcal{X}^n} \mathbb{1}_{x_i \neq y_i} \, d\mathbb{Q}(x, y) \right)
\end{aligned}
$$

and therefore by Cauchy–Schwarz inequality

$$
E_Q[\zeta] - E_P[\zeta] \leq \left[\sum_{i=1}^{n} c_i^2 \right]^{1/2} \left[\sum_{i=1}^{n} \mathbb{Q}^2 \{(x, y) \in \mathcal{X}^n \times \mathcal{X}^n; x_i \neq y_i\} \right]^{1/2}.
$$

So, it comes from Lemma 2.21 that some clever choice of $\mathbb{Q}$ leads to

$$
E_Q[\zeta] - E_P[\zeta] \leq \sqrt{2v\mathbf{K}(Q, P)},
$$

where $v = \left[\sum_{i=1}^{n} c_i^2 \right] / 4$ and we derive from Lemma 2.13 that for any positive λ

$$
E_P[\exp[\lambda(\zeta - E_P[\zeta])]] \leq e^{\lambda^2 v / 2}.
$$

Since we can change ζ into $-\zeta$ the same inequality remains valid for negative values of λ. The conclusion follows via Chernoff's inequality. ∎

Comment. Note that the transportation method which is used above to derive the bounded difference inequality from Marton's transportation cost inequality can also be used to derive the Gaussian concentration inequality from Talagrand's transportation cost inequality for the Gaussian measure.

It is indeed proved in [114] that if P denotes the standard Gaussian measure on $\mathbb{R}^N$, Q is absolutely continuous with respect to P, and d denotes the Euclidean distance then

$$\min_{Q \in \mathcal{P}(P,Q)} \mathbb{E}_{\mathbb{Q}} \left[d\left(X, Y \right) \right] \leq \sqrt{2\mathbf{K}\left(Q, P \right)}.$$

Hence if $\zeta : \mathbb{R}^N \to \mathbb{R}$ is some 1-Lipschitz function with respect to d, choosing $\mathbb{Q}$ as an optimal coupling (i.e., achieving the minimum in the left-hand side of the above transportation cost inequality)

$$E_Q\left[\zeta\right] - E_P\left[\zeta\right] = \mathbb{E}_{\mathbb{Q}}\left[\zeta\left(Y\right) - \zeta\left(X\right)\right]$$
$$\leq \mathbb{E}_{\mathbb{Q}}\left[d\left(X, Y\right)\right] \leq \sqrt{2\mathbf{K}\left(Q, P\right)}$$

and therefore by Lemma 2.13

$$E_P\left[\exp\left[\lambda\left(\zeta - E_P\left[\zeta\right]\right)\right]\right] \leq e^{\lambda^2 v/2},$$

for any positive λ.

The connection between Talagrand's isoperimetric approach and the bounded difference inequality can be made through the following straightforward consequence of Theorem 5.1.

Corollary 5.2 *Let $\Omega^n = \prod_{i=1}^n \Omega_i$, where, for all $i \leq n$, (Ω_i, d_i) is a metric space with diameter c_i. Let P be some product probability measure on Ω^n and ζ be some 1-Lipschitz function $\zeta : \Omega^n \to \mathbb{R}$, in the sense that*

$$\left| \zeta\left(x\right) - \zeta\left(y\right) \right| \leq \sum_{i=1}^n d_i\left(x_i, y_i\right).$$

Then, for any positive x

$$P\left[\zeta - E_P\left[\zeta\right] \geq x\right] \leq \exp\left(-\frac{2x^2}{\sum_{i=1}^n c_i^2}\right). \tag{5.3}$$

Moreover, if $M_P\left[\zeta\right]$ is a median of ζ under P, then for any positive x

$$P\left[\zeta - M_P\left[\zeta\right] \geq x + \sqrt{\left[\sum_{i=1}^n c_i^2\right]\frac{\ln\left(2\right)}{2}}\right] \leq \exp\left(-\frac{2x^2}{\sum_{i=1}^n c_i^2}\right). \tag{5.4}$$

Proof. Obviously ζ fulfills the bounded difference condition and therefore Theorem 5.1 implies that (5.3) holds. Since (5.3) also holds for $-\zeta$ instead of ζ, we get

$$\left| E_P\left[\zeta\right] - M_P\left[\zeta\right] \right| \leq \sqrt{\left[\sum_{i=1}^n \Delta_i^2\right]\frac{\ln\left(2\right)}{2}}$$

and therefore (5.3) yields (5.4). ∎

Comments.

- Note that Corollary 5.2 can be applied to the Hamming distance. Indeed, if one considers on some arbitrary set Ω the trivial distance d defined by

$$d(s,t) = 0 \text{ if } s = t \text{ and } d(s,t) = 1 \text{ if } s \neq t$$

then, the Hamming distance on Ω^n is defined by

$$(x,y) \rightarrow \sum_{i=1}^{n} d(x_i, y_i)$$

and since Ω has diameter equal to 1, we derive from Corollary 5.2 that for any functional ζ which is $1-$Lipschitz on Ω^n with respect to Hamming distance and any product probability measure P on Ω^n, one has for any positive x

$$P\left[\zeta - E_P[\zeta] \geq x\sqrt{n}\right] \leq \exp\left(-2x^2\right)$$

and

$$P\left[\zeta - M_P[\zeta] \geq \left(x + \sqrt{\frac{\ln(2)}{2}}\right)\sqrt{n}\right] \leq \exp\left(-2x^2\right).$$

We exactly recover here the concentration inequality for the Hamming distance due to Talagrand for the median (easy consequence of Corollary 2.2.3. in [112]).

- We can also derive from Corollary 5.2 an isoperimetric type inequality. Under the assumptions of Corollary 5.2, let us consider the distance δ : $(x,y) \rightarrow \sum_{i=1}^{n} d_i(x_i, y_i)$ on Ω^n. Then for any measurable subset A of Ω^n, $\delta(.,A) : x \rightarrow \delta(x,A)$ is 1-Lipschitz. Assume that $P(A) > 0$. Since $\delta(.,A) = 0$ on the set A, $P[\delta(.,A) \leq t] < P(A)$ implies that $t < 0$ and therefore we derive from (5.3) with $\zeta = -\delta(.,A)$ that

$$E_P[\delta(.,A)] \leq \sqrt{\left[\sum_{i=1}^{n} \Delta_i^2\right]\frac{\ln(1/P(A))}{2}}$$

which finally yields via (5.3) with $\zeta = \delta(.,A)$

$$P\left[\delta(.,A) \geq x + \sqrt{\left[\sum_{i=1}^{n} \Delta_i^2\right]\frac{\ln(1/P(A))}{2}}\right] \leq \exp\left(-\frac{2x^2}{\sum_{i=1}^{n} \Delta_i^2}\right). \quad (5.5)$$

Inequality (5.5) generalizes on Talagrand's isoperimetric inequality for the Hamming distance (see Corollary 2.2.3 in [112]). We turn now to the application of Corollary 5.2 to sums of independent infinite dimensional and bounded random vectors (or empirical processes).

Theorem 5.3 *Let $X_1, \ldots, X_n$ be independent random variables with values in $\mathbb{R}^T$, where T is a finite set. We assume that for some real numbers $a_{i,t}$ and $b_{i,t}$*

$$a_{i,t} \leq X_{i,t} \leq b_{i,t}, \text{ for all } i \leq n \text{ and all } t \in T,$$

and set $L^2 = \sum_{i=1}^{n} \sup_{t \in T} (b_{i,t} - a_{i,t})^2$. Setting

$$Z = \sup_{t \in T} \sum_{i=1}^{n} X_{i,t} \quad or \quad Z = \sup_{t \in T} \left| \sum_{i=1}^{n} X_{i,t} \right|,$$

one has for every $x \geq 0$

$$\mathbb{P}\left[Z - \mathbb{E}\left[Z\right] \geq x\right] \leq \exp\left(-\frac{2x^2}{L^2}\right). \tag{5.6}$$

Moreover, the same inequality holds for $-Z$ instead of Z.

Proof.

The proof is immediate from Theorem 5.2. We define for every $i \leq n$

$$\Omega_i = \left\{ u \in \mathbb{R}^T : a_{i,t} \leq u_t \leq b_{i,t} \right\}$$

and $\zeta : \prod_{i=1}^{n} \Omega_i \to \mathbb{R}$ as

$$\zeta(x_1, \ldots, x_n) = \sup_{t \in T} \sum_{i=1}^{n} x_{i,t} \quad or \quad \zeta(x_1, \ldots, x_n) = \sup_{t \in T} \left| \sum_{i=1}^{n} x_{i,t} \right|.$$

Then

$$\left| \zeta(x_1, \ldots, x_n) - \zeta(y_1, \ldots, y_n) \right| \leq \sum_{i=1}^{n} \sup_{t \in T} \left| x_{i,t} - y_{i,t} \right|$$

which shows that ζ is 1-Lipschitz in the sense of Theorem 5.2 when setting for all $i \leq n$, $d_i(u, v) = \sup_{t \in T} |u_t - v_t|$ for all $u, v \in \Omega_i$. Since the diameter of Ω_i is equal to $\sup_{t \in T} (b_{i,t} - a_{i,t})$, applying Theorem 5.2 leads to the conclusion. ∎

Comments.

- It should be noticed that a similar concentration inequality holds for the median instead of the mean. Indeed, since (5.6) holds for $-Z$ instead of Z, denoting by M a median of Z, one has

$$|M - \mathbb{E}\left[Z\right]| \leq L\sqrt{\frac{\ln(2)}{2}}$$

and therefore (5.6) implies that for every $x \geq 0$

$$\mathbb{P}\left[Z - M \geq x + L\sqrt{\frac{\ln(2)}{2}}\right] \leq \exp\left(-\frac{2x^2}{L^2}\right)$$

- The constant 2 involved in the exponential probability bound of Theorem 5.3 is of course optimal since actually, it cannot be improved in the one dimensional case where $|T| = 1$.
- We would prefer to get $\sup_{t \in T} \sum_{i=1}^{n} (b_{i,t} - a_{i,t})^2$ as a variance factor rather than $\sum_{i=1}^{n} \sup_{t \in T} (b_{i,t} - a_{i,t})^2$. Hence the present statement is more interesting for situations where

$$\sum_{i=1}^{n} \sup_{t \in T} (b_{i,t} - a_{i,t})^2 = \sup_{t \in T} \sum_{i=1}^{n} (b_{i,t} - a_{i,t})^2.$$

This is typically the case when one wants to study independent random variables $\xi_1, \ldots, \xi_n$ with values in a separable Banach space, which are strongly bounded in the sense that $\|\xi_i\| \leq b_i$ for all $i \leq n$. Then one can apply Theorem 5.3 with T being arbitrary finite subset of a countable and dense subset of the unit ball of the dual of the Banach space, $X_{i,t} = \langle t, \xi_i \rangle - \mathbb{E}\left[\langle t, \xi_i \rangle\right]$ and $-a_{i,t} - \mathbb{E}\left[\langle t, \xi_i \rangle\right] = b_{i,t} + \mathbb{E}\left[\langle t, \xi_i \rangle\right] = b_i$ for all $i \leq n$ and $t \in T$. Setting $S_n = \xi_1 + \cdots + \xi_n$, this leads by monotone convergence to the following concentration inequality for $Z = \|S_n - \mathbb{E}\left[S_n\right]\|$ around its expectation

$$\mathbb{P}\left[|Z - \mathbb{E}\left[Z\right]| \geq x\right] \leq 2\exp\left(-\frac{x^2}{2\sum_{i=1}^{n} b_i^2}\right).$$

A useful consequence of Theorem 5.3 concerns empirical processes. Indeed, if $\xi_1, \ldots, \xi_n$ are independent random variables and $\mathcal{F}$ is a finite or countable class of functions such that, for some real numbers a and b, one has $a \leq f \leq b$ for every $f \in \mathcal{F}$, then setting $Z = \sup_{f \in \mathcal{F}} \sum_{i=1}^{n} f(\xi_i) - \mathbb{E}\left[f(\xi_i)\right]$, we get by monotone convergence

$$\mathbb{P}\left[Z - \mathbb{E}\left[Z\right] \geq x\right] \leq \exp\left(-\frac{2x^2}{n(b-a)^2}\right), \tag{5.7}$$

(the same inequality remaining true if one changes Z into $-Z$).

It should be noticed that (5.7) does not generally provide a sub-Gaussian inequality. The reason is that the maximal variance

$$\sigma^2 = \sup_{f \in \mathcal{F}} \text{Var}\left[\sum_{i=1}^{n} f(\xi_i)\right]$$

can be substantially smaller than $n(b-a)^2/4$ and therefore (5.7) can be much worse than its "sub-Gaussian" version which should make σ^2 appear instead of $n(b-a)^2/4$. It is precisely our purpose now to provide sharper bounds than Hoeffding type inequalities. The method that we shall use to derive such bounds has been initiated by Michel Ledoux (see [77]) and further developed in [90], [31], [32], [103], [30] or [33].

5.3 Concentration Inequalities via the Entropy Method

At the root of this method is the tensorization inequality for ϕ-entropy. We recall from Chapter 2 that this inequality holds under the condition that ϕ belongs to the Latala and Oleskiewicz class of functions $\mathcal{LO}$. The case $\phi : x \to x \ln(x)$ leads to the classical definition of entropy while another case of interest is when ϕ is a power function $x \to x^p$ with exponent $p \in [1, 2]$. All along this section, for every integer $i \in [1, n]$, we shall denote by $X^{(i)}$ the random vector $(X_1, \ldots, X_{i-1}, X_{i+1}, \ldots, X_n)$.

The Class $\mathcal{LO}$

In the sequel, we shall need the following properties of the elements of $\mathcal{LO}$.

Proposition 5.4 *Let* $\phi \in \mathcal{LO}$*, then both* ϕ' *and* $x \to (\phi(x) - \phi(0))/x$ *are concave functions on* $\mathbb{R}_+^*$*.*

Proof. Without loss of generality we may assume that $\phi(0) = 0$. The concavity of $1/\phi''$ implies a fortiori that for every $\lambda \in (0, 1)$ and every positive x and u

$$\lambda \phi'' ((1 - \lambda) u + \lambda x) \leq \phi''(x),$$

which implies that for every positive t

$$\lambda \phi'' (t + \lambda x) \leq \phi''(x).$$

Letting λ tend to 1, we derive from the above inequality that ϕ'' is nonincreasing i.e., ϕ' is concave. Setting $\psi(x) = \phi(x)/x$, one has

$$x^3 \psi''(x) = x^2 \phi''(x) - 2x\phi'(x) + 2\phi(x) = f(x).$$

The convexity of ϕ and its continuity at point 0 imply that $x\phi'(x)$ tends to 0 as x goes to 0. Also, the concavity of ϕ' implies that

$$x^2 \phi''(x) \leq 2x (\phi'(x) - \phi'(x/2))$$

hence $x^2 \phi''(x)$ tends to 0 as x goes to 0 and therefore $f(x)$ tends to 0 as x goes to 0. Denoting (abusively) by $\phi^{(3)}$ the right derivative of ϕ'' (which is well defined since $1/\phi''$ is concave) and by f' the right derivative of f, we have $f'(x) = x^2 \phi^{(3)}(x)$. Then $f'(x)$ is nonpositive because ϕ'' is nonincreasing. Hence f is nonincreasing. Since f tends to 0 at 0, this means that f is a nonpositive function and the same property holds true for the function ψ'', which completes the proof of the concavity of ψ.

5.3.1 ϕ-Sobolev and Moment Inequalities

Our aim is to derive exponential moments or moment inequalities from the tensorization inequality for ϕ-entropy via an adequate choice of the function ϕ. As compared to the quadratic case, the extra difficulty is that we shall not apply the tensorization inequality to the initial functional of interest Z but rather to a conveniently chosen transformation f of it. This is precisely our purpose now to understand how to couple ϕ and f to get interesting results on the exponential moments or the moments of Z. As a guideline let us recall from Chapter 1 how to derive Efron–Stein's inequality (and a variant of it) from the tensorization inequality for the variance, i.e., the ϕ-entropy when ϕ is defined on the whole real line as $\phi(x) = x^2$

$$\operatorname{Var}(Z) \leq \mathbb{E}\left[\sum_{i=1}^{n} \mathbb{E}\left[\left(Z - \mathbb{E}\left[Z \mid X^{(i)}\right]\right)^2 \mid X^{(i)}\right]\right].$$

We can either use a symmetrization device and introduce an independent copy X' of X. Defining

$$Z_i' = \zeta(X_1, \ldots, X_{i-1}, X_i', X_{i+1}, \ldots, X_n) \tag{5.8}$$

we use the property that, conditionally to $X^{(i)}$, Z_i' is an independent copy of Z to derive that

$$\mathbb{E}\left[\left(Z - \mathbb{E}\left[Z \mid X^{(i)}\right]\right)^2 \mid X^{(i)}\right] = \frac{1}{2}\mathbb{E}\left[(Z - Z_i')^2 \mid X^{(i)}\right]$$
$$= \mathbb{E}\left[(Z - Z_i')_+^2 \mid X^{(i)}\right].$$

This leads to Efron–Stein's inequality that we can write

$$\operatorname{Var}(Z) \leq \mathbb{E}\left[V^+\right] = \mathbb{E}\left[V^-\right],$$

where

$$V^+ = \mathbb{E}\left[\sum_{i=1}^{n}(Z - Z_i')_+^2 \mid X\right] \tag{5.9}$$

and

$$V^- = \mathbb{E}\left[\sum_{i=1}^{n}(Z - Z_i')_-^2 \mid X\right]. \tag{5.10}$$

A variant consists in using a variational argument, noticing that $\mathbb{E}\left[Z \mid X^{(i)}\right]$ is the best $X^{(i)}$-measurable approximation of Z in $\mathbb{L}_2$ which leads to

$$\operatorname{Var}(Z) \leq \sum_{i=1}^{n}\mathbb{E}\left[(Z - Z_i)^2\right]$$

for any family of square integrable random variables Z_i's such that Z_i is $X^{(i)}$-measurable. In other words one has

$$\mathrm{Var}\,(Z) \leq \sum_{i=1}^{n} \mathbb{E}\,[V]\,, \tag{5.11}$$

where

$$V = \sum_{i=1}^{n} (Z - Z_i)^2\,. \tag{5.12}$$

Our purpose is now to generalize these symmetrization and variational arguments. In what follows we shall keep the same notations as above. The quantities V^+, V^- and V (respectively defined by (5.9), (5.10) and (5.12)) will turn to play a crucial role in what follows, quite similar to the quadratic variation in the theory of martingales.

Symmetrization Inequalities

The following elementary Lemma provides symmetrization inequalities for ϕ-entropy which, though elementary, will turn to be extremely useful.

Lemma 5.5 *Let ϕ be some continuous and convex function on $\mathbb{R}_+$, $Z \in \mathbb{L}_1^+$ and Z' be some independent copy of Z'. Then, denoting by ϕ' the right derivative of ϕ one has*

$$H_\phi\,(Z) \leq \frac{1}{2}\mathbb{E}\,[(Z - Z')\,(\phi'\,(Z) - \phi'\,(Z'))]$$
$$\leq \mathbb{E}\,[(Z - Z')_+\,(\phi'\,(Z) - \phi'\,(Z'))] \tag{5.13}$$

If moreover $\psi : x \to (\phi\,(x) - \phi\,(0))\,/x$ is concave on $\mathbb{R}_+^*$ then

$$H_\phi\,(Z) \leq \frac{1}{2}\mathbb{E}\,[(Z - Z')\,(\psi\,(Z) - \psi\,(Z'))]$$
$$\leq \mathbb{E}\,[(Z - Z')_+\,(\psi\,(Z) - \psi\,(Z'))]\,. \tag{5.14}$$

Proof. Without loss of generality we assume that $\phi\,(0) = 0$. Since Z' is an independent copy of Z, we derive from (2.51) that

$$H_\phi\,(Z) \leq \mathbb{E}\,[\phi\,(Z) - \phi\,(Z') - (Z - Z')\,\phi'\,(Z')]$$
$$\leq \mathbb{E}\,[(Z - Z')\,\phi'\,(Z')]$$

and by symmetry

$$2H_\phi\,(Z) \leq \mathbb{E}\,[(Z' - Z)\,\phi'\,(Z)] + \mathbb{E}\,[(Z - Z')\,\phi'\,(Z')]\,,$$

which leads to (5.13). To prove (5.14), we simply note that

$$\frac{1}{2}\mathbb{E}\left[(Z-Z')(\psi(Z)-\psi(Z'))\right] - H_\phi(Z) = -\mathbb{E}[Z]\mathbb{E}[\psi(Z)] + \phi(\mathbb{E}[Z]).$$

But the concavity of ψ implies that $\mathbb{E}[\psi(Z)] \leq \psi(\mathbb{E}[Z]) = \phi(\mathbb{E}[Z])/\mathbb{E}[Z]$ and we derive from the previous identity that (5.14) holds. ∎

Note that by Proposition 5.4, we can apply (5.14) whenever $\phi \in \mathcal{LO}$. In particular, for our target example where $\phi(x) = x^p$, with $p \in [1,2]$, (5.14) improves on (5.13) within a factor p.

5.3.2 A Poissonian Inequality for Self-Bounding Functionals

As a first illustration of the method we intend to present here the extension to some nonnegative functionals of independent random variables due to [31] of a Poissonian bound for the supremum of nonnegative empirical processes established in [90] by using Ledoux's approach to concentration inequalities. The motivations for considering general nonnegative functionals of independent random variables came from random combinatorics. Several illustrations are given in [31] but we shall focus here on the case example of random combinatorial entropy since the corresponding concentration result will turn out to be very useful for designing data-driven penalties to solve the model selection problem for classification (see Section 8.2). Roughly speaking, under some self-bounding condition (SB) to be given below, we shall show that a nonnegative functional Z of independent variables concentrates around its expectation like a Poisson random variable with expectation $\mathbb{E}[Z]$ (this comparison being expressed in terms of moment generating function). This Poissonian inequality can be deduced from the integration of a differential inequality for the moment generating function of Z which derives from the combination of the tensorization inequality and the variational formula for entropy. Indeed applying (2.51) for entropy (conditionally to $X^{(i)}$) implies that for every positive measurable function $G^{(i)}$ of $X^{(i)}$ one has

$$\mathbb{E}^{(i)}[\Phi(G)] - \Phi\left(\mathbb{E}^{(i)}[G]\right) \leq \mathbb{E}^{(i)}\left[G\left(\ln G - \ln G^{(i)}\right) - \left(G - G^{(i)}\right)\right].$$

Hence, if Z is some measurable function of X and for every $i \in \{1, \ldots, n\}$, Z_i is some measurable function of $X^{(i)}$, applying the above inequality to the variables $G = e^{\lambda Z}$ and $G^{(i)} = e^{\lambda Z_i}$, one gets

$$\mathbb{E}^{(i)}[\Phi(G)] - \Phi\left(\mathbb{E}^{(i)}[G]\right) \leq \mathbb{E}^{(i)}\left[e^{\lambda Z}\varphi(-\lambda(Z - Z_i))\right],$$

where φ denotes the function $z \to \exp(z) - z - 1$. Therefore, we derive from (2.46), that

$$\lambda\mathbb{E}\left[Ze^{\lambda Z}\right] - \mathbb{E}\left[e^{\lambda Z}\right]\ln\mathbb{E}\left[e^{\lambda Z}\right] \leq \sum_{i=1}^{n}\mathbb{E}\left[e^{\lambda Z}\varphi(-\lambda(Z - Z_i))\right], \qquad (5.15)$$

for any λ such that $\mathbb{E}\left[e^{\lambda Z}\right] < \infty$. A very remarkable fact is that Han's inequality for Kullback–Leibler information is at the heart of the proof of this bound and is also deeply involved in the verification of condition (SB) below for combinatorial entropies.

A Poissonian Bound

We now turn to the main result of this section (due to [31]) which derives from (5.15).

Theorem 5.6 *Let $X_1, \ldots, X_n$ be independent random variables and define for every $i \in \{1, \ldots, n\}$ $X^{(i)} = (X_1, \ldots, X_{i-1}, X_{i+1}, \ldots, X_n)$. Let Z be some nonnegative and bounded measurable function of $X = (X_1, \ldots, X_n)$. Assume that for every $i \in \{1, \ldots, n\}$, there exists some measurable function Z_i of $X^{(i)}$ such that*

$$0 \leq Z - Z_i \leq 1. \tag{5.16}$$

Assume furthermore that

$$\sum_{i=1}^{n} (Z - Z_i) \leq Z. \tag{SB}$$

Defining h as $h(u) = (1 + u)\ln(1 + u) - u$, for $u \geq -1$, the following inequalities hold:

$$\ln \mathbb{E}\left[e^{\lambda(Z - \mathbb{E}[Z])}\right] \leq \mathbb{E}[Z]\,\varphi(\lambda) \quad \text{for every } \lambda \in \mathbb{R}. \tag{5.17}$$

and therefore

$$\mathbb{P}[Z \geq \mathbb{E}[Z] + x] \leq \exp\left[-\mathbb{E}[Z]\,h\left(\frac{x}{\mathbb{E}[Z]}\right)\right], \text{ for all } x > 0 \tag{5.18}$$

and

$$\mathbb{P}[Z \leq \mathbb{E}[Z] - x] \leq \exp\left[-\mathbb{E}[Z]\,h\left(-\frac{x}{\mathbb{E}[Z]}\right)\right], \text{ for } 0 < x \leq \mathbb{E}[Z]. \tag{5.19}$$

Proof. We know that (5.15) holds for any λ. Since the function φ is convex with $\varphi(0) = 0$, $\varphi(-\lambda u) \leq u\varphi(-\lambda)$ for any λ and any $u \in [0,1]$. Hence it follows from (5.16) that for every λ, $\varphi(-\lambda(Z - Z_i)) \leq (Z - Z_i)\varphi(-\lambda)$ and therefore we derive from (5.15) and (SB) that

$$\lambda \mathbb{E}\left[Ze^{\lambda Z}\right] - \mathbb{E}\left[e^{\lambda Z}\right]\ln\mathbb{E}\left[e^{\lambda Z}\right] \leq \mathbb{E}\left[\varphi(-\lambda)e^{\lambda Z}\sum_{i=1}^{n}(Z - Z_i)\right]$$

$$\leq \varphi(-\lambda)\mathbb{E}\left[Ze^{\lambda Z}\right].$$

We introduce $\widetilde{Z} = Z - \mathbb{E}[Z]$ and define for any λ, $F(\lambda) = \mathbb{E}\left[e^{\lambda \widetilde{Z}}\right]$. Setting $v = \mathbb{E}[Z]$, the previous inequality becomes

$$[\lambda - \varphi(-\lambda)] \frac{F'(\lambda)}{F(\lambda)} - \ln F(\lambda) \le v\varphi(-\lambda),$$

which in turn implies

$$\left(1 - e^{-\lambda}\right) \Psi'(\lambda) - \Psi(\lambda) \le v\varphi(-\lambda) \quad \text{with} \quad \Psi(\lambda) = \ln F(\lambda). \tag{5.20}$$

Now observe that $v\varphi$ is a solution of the ordinary differential equation $\left(1 - e^{-\lambda}\right) f'(\lambda) - f(\lambda) = v\varphi(-\lambda)$. In order to show that $\Psi \le v\varphi$, we set

$$\Psi(\lambda) = v\varphi(\lambda) + \left(e^{\lambda} - 1\right) g(\lambda), \tag{5.21}$$

for every $\lambda \ne 0$ and derive from (5.20) that

$$\left(1 - e^{-\lambda}\right) \left[e^{\lambda} g(\lambda) + \left(e^{\lambda} - 1\right) g'(\lambda)\right] - \left(e^{\lambda} - 1\right) g(\lambda) \le 0,$$

which yields

$$\left(1 - e^{-\lambda}\right)\left(e^{\lambda} - 1\right) g'(\lambda) \le 0.$$

We derive from this inequality that g' is nonpositive which means that g is nonincreasing. Now, since $\widetilde{Z}$ is centered at expectation $\Psi'(0) = \varphi(0) = 0$ and it comes from (5.21) that $g(\lambda)$ tends to 0 as λ goes to 0. This shows that g is nonnegative on $(-\infty, 0)$ and nonpositive on $(0, \infty)$ which in turn means by (5.21) that $\Psi \le v\varphi$ and we have proved that (5.17) holds. Thus, by Chernoff's inequality,

$$\mathbb{P}\left[Z - \mathbb{E}[Z] \ge x\right] \le \exp\left[-\sup_{\lambda > 0}\left(x\lambda - v\varphi(\lambda)\right)\right]$$

and

$$\mathbb{P}\left[Z - \mathbb{E}[Z] \le -x\right] \le \exp\left[-\sup_{\lambda < 0}\left(-x\lambda - v\varphi(\lambda)\right)\right].$$

The proof can be completed by using the easy to check (and well known) relations:

$$\sup_{\lambda > 0}\left[x\lambda - v\varphi(\lambda)\right] = vh(x/v)$$

for every $x > 0$ and

$$\sup_{\lambda < 0}\left[-x\lambda - v\varphi(\lambda)\right] = vh(-x/v)$$

for every $0 < x \le v$. ∎

The above inequalities are exactly the classical Cramér-Chernoff upper bounds for the Poisson distribution with mean $\mathbb{E}[Z]$ and in this sense this theorem establishes a comparison between the concentration around its mean of a nonnegative functional Z satisfying the above assumptions and that of the Poisson distribution with the same mean. Let us give some further comments.

- This theorem can typically be applied to the supremum of sums of nonnegative random variables. Indeed let $X_1, \ldots, X_n$ be independent $[0, 1]^N$-valued random variables and consider

$$Z = \sup_{1 \leq t \leq N} \sum_{i=1}^n X_{i,t}$$

with $Z_i = \sup_{1 \leq t \leq N} \sum_{j \neq i} X_{j,t}$ for all $i \leq n$. Then denoting by τ some random number such that $Z = \sum_{i=1}^n X_{i,\tau}$, one obviously has

$$0 \leq Z - Z_i \leq X_{i,\tau} \leq 1,$$

and therefore

$$\sum_{i=1}^n (Z - Z_i) \leq \sum_{i=1}^n X_{i,\tau} = Z,$$

which means that the self-bounding condition (SB) is satisfied. It is easy to see on this example that (5.18) and (5.19) are in some sense unimprovable. Indeed, if $N = 1$, and $X_1, \ldots, X_n$ are Bernoulli trials with parameter θ, then Z follows the binomial distribution $\mathcal{B}(n, \theta/n)$ and its asymptotic distribution is actually a Poisson distribution with mean θ.

- Inequality (5.19) readily implies the sub-Gaussian inequality

$$\mathbb{P}[Z \leq \mathbb{E}[Z] - x] \leq \exp\left[-\frac{x^2}{2\mathbb{E}[Z]}\right] \tag{5.22}$$

which holds for every $x > 0$. Indeed, (5.22) is trivial when $x > \mathbb{E}[Z]$ and follows from (5.19) otherwise since, for every $\varepsilon \in [0, 1]$ one has $h(-\varepsilon) \geq \varepsilon^2/2$.

Let us turn now to a somehow more subtle application of Theorem 5.6 to combinatorial entropy. Surprisingly, Han's inequality will be involved again to show that the combinatorial entropy satisfies condition (SB).

Application to Combinatorial Entropies

Let $\mathcal{F}$ be some class of measurable functions defined on some set $\mathcal{X}$ and taking their values in $\{1, \ldots, k\}$. We define the *combinatorial entropy* of $\mathcal{F}$ at point $x \in \mathcal{X}^n$ by

$$\zeta(x) = \ln_k |Tr(x)|,$$

where $Tr(x) = \{(f(x_1), \ldots, f(x_n)), f \in \mathcal{F}\}$ and $|Tr(x)|$ denotes the cardinality of $Tr(x)$. It is quite remarkable that, given some independent variables, $X_1, \ldots, X_n$, $Z = \zeta(X)$ satisfies to the assumptions of our Theorem 5.6. Indeed, let $Z_i = \zeta(X^{(i)})$ for every i. Obviously $0 \leq Z - Z_i \leq 1$ for all i. On the other hand, given $x \in \mathcal{X}^n$, let us consider some random variable Y with uniform distribution on the set $Tr(x)$. It comes from Han's inequality (see Corollary 2.23) that,

$$\ln|Tr(x)| = h_S(Y) \leq \frac{1}{n-1} \sum_{i=1}^{n} h_S\left(Y^{(i)}\right).$$

Now for every i, $Y^{(i)}$ takes its values in $Tr\left(x^{(i)}\right)$ and therefore by (2.48) we have $h_S\left(Y^{(i)}\right) \leq \ln|Tr\left(x^{(i)}\right)|$. Hence

$$\ln|Tr(x)| \leq \frac{1}{n-1} \sum_{i=1}^{n} \ln\left|Tr\left(x^{(i)}\right)\right|,$$

which means that

$$\zeta(x) \leq \frac{1}{n-1} \sum_{i=1}^{n} \zeta\left(x^{(i)}\right).$$

Thus the self-bounding condition (SB) is satisfied and Theorem 5.6 applies to the combinatorial entropy $\ln_k |Tr(X)|$, which, in particular implies that

$$\mathbb{P}\left[Z \leq \mathbb{E}[Z] - \sqrt{2\mathbb{E}[Z]x}\right] \leq e^{-x}, \text{ for all } x > 0. \tag{5.23}$$

This inequality has some importance in statistical learning theory as we shall see later. Another interesting example for statistical learning is the following.

Rademacher Conditional Means

We consider here some finite class $\{f_t, t \in T\}$ of measurable functions on $\mathcal{X}$, taking their values in $[0, 1]$ and $\varepsilon_1, \ldots, \varepsilon_n$ be i.i.d. Rademacher random variables. We define for every point $x \in \mathcal{X}^n$ the Rademacher mean by

$$\zeta(x) = \mathbb{E}\left[\sup_{t \in T} \sum_{i=1}^{n} \varepsilon_i f_t(x_i)\right].$$

Then for every integer $i \in [1, n]$ one has on the one hand

$$\zeta(x) = \mathbb{E}\left[\mathbb{E}\left[\left(\sup_{t \in T} \sum_{j=1}^{n} \varepsilon_j f_t(x_j)\right) \mid \varepsilon_i\right]\right]$$

$$\geq \mathbb{E}\left[\sup_{t \in T} \mathbb{E}\left[\sum_{j=1}^{n} \varepsilon_j f_t(x_j) \mid \varepsilon_i\right]\right] = \zeta\left(x^{(i)}\right)$$

and on the other hand, defining τ (depending on $\varepsilon_1, \ldots, \varepsilon_n$ and x) such that

$$\sup_{t \in T} \sum_{i=1}^{n} \varepsilon_i f_t(x_i) = \sum_{i=1}^{n} \varepsilon_i f_\tau(x_i)$$

one gets

$$\zeta(x) - \zeta\left(x^{(i)}\right) \leq \mathbb{E}\left[\varepsilon_i f_\tau(x_i)\right]. \tag{5.24}$$

We derive from (5.24) that

$$\zeta(x) - \zeta\left(x^{(i)}\right) \leq \mathbb{E}\left[|\varepsilon_i|\right] = 1$$

and one also has

$$\sum_{i=1}^n \zeta(x) - \zeta\left(x^{(i)}\right) \leq \sum_{i=1}^n \mathbb{E}\left[\varepsilon_i f_\tau(x_i)\right] = \zeta(x).$$

This means that if $(X_1, \ldots, X_n)$ are independent random variables, independent from $(\varepsilon_1, \ldots, \varepsilon_n)$, then the Rademacher conditional mean

$$Z = \mathbb{E}\left[\sup_{t \in T} \sum_{i=1}^n \varepsilon_i f_t(X_i) \mid (X_1, \ldots, X_n)\right]$$

satisfies to the assumptions of Theorem 5.6. Hence (5.23) is valid for the Rademacher conditional mean Z as well.

Of course Theorem 5.6 is designed for self-bounding nonnegative functionals and does not solve the problem of improving on Hoeffding type bounds for the supremum of a centered empirical process. It is one of our main tasks in what follows to produce such sharper bounds.

5.3.3 ϕ-Sobolev Type Inequalities

Our purpose is to derive from the tensorization inequality for ϕ-entropy and the variational formula or the symmetrization inequality above a bound on the ϕ-entropy of a conveniently chosen convex transformation f of the initial variable Z. The results will heavily depend on the monotonicity of the transformation f. We begin with the nondecreasing case. All along this section, the quantities V^+ and V^- are defined by (5.9) and (5.10) from the symmetrized variables $(Z_i')_{1 \leq i \leq n}$ and V is defined by (5.12) from $X^{(i)}$-measurable variables Z_i such that $Z \geq Z_i$, for every integer $1 \leq i \leq n$.

Theorem 5.7 *Let $X_1, \ldots, X_n$ be some independent random variables and Z be some $(X_1, \ldots, X_n)$-measurable taking its values in some interval $\mathcal{I}$. Let ϕ belong to $\mathcal{LO}$ and f be some nonnegative and differentiable convex function on $\mathcal{I}$. Let ψ denote the function $x \to (\phi(x) - \phi(0))/x$. Under the assumption that f is nondecreasing, one has*

$$H_\phi(f(Z)) \leq \frac{1}{2}\mathbb{E}\left[V f'^2(Z) \phi''(f(Z))\right], \tag{5.25}$$

whenever $\phi' \circ f$ is convex, while if $\psi \circ f$ is convex one has

$$H_\phi(f(Z)) \leq \mathbb{E}\left[V^+ f'^2(Z) \psi'(f(Z))\right]. \tag{5.26}$$

Proof. We first assume f to be nondecreasing and fix $x < y$. Under the assumption that $g = \phi' \circ f$ is convex, we notice that

$$\phi\left(f\left(y\right)\right) - \phi\left(f\left(x\right)\right) - \left(f\left(y\right) - f\left(x\right)\right)\phi'\left(f\left(x\right)\right) \le \frac{1}{2}\left(y - x\right)^2 f'^2\left(y\right)\phi''\left(f\left(y\right)\right).$$
(5.27)

Indeed, setting

$$h\left(t\right) = \phi\left(f\left(y\right)\right) - \phi\left(f\left(t\right)\right) - \left(f\left(y\right) - f\left(t\right)\right)g\left(t\right),$$

we have

$$h'\left(t\right) = -g'\left(t\right)\left(f\left(y\right) - f\left(t\right)\right).$$

But for every $t \le y$, the monotonicity and convexity assumptions on f and g yield

$$0 \le g'\left(t\right) \le g'\left(y\right) \text{ and } 0 \le f\left(y\right) - f\left(t\right) \le \left(y - t\right)f'\left(y\right),$$

hence

$$-h'\left(t\right) \le \left(y - t\right)f'\left(y\right)g'\left(y\right).$$

Integrating this inequality with respect to t on $[x, y]$ leads to (5.27). Under the assumption that $\psi \circ f$ is convex we notice that

$$0 \le f\left(y\right) - f\left(x\right) \le \left(y - x\right)f'\left(y\right)$$

and

$$0 \le \psi\left(f\left(y\right)\right) - \psi\left(f\left(x\right)\right) \le \left(y - x\right)f'\left(y\right)\psi'\left(f\left(y\right)\right),$$

which leads to

$$\left(f\left(y\right) - f\left(x\right)\right)\left(\psi\left(f\left(y\right)\right) - \psi\left(f\left(x\right)\right)\right) \le \left(x - y\right)^2 f'^2\left(y\right)\psi'\left(f\left(y\right)\right).$$
(5.28)

Now the tensorization inequality combined with (2.51) and (5.27) leads to

$$H_\phi\left(f\left(Z\right)\right) \le \frac{1}{2}\sum_{i=1}^{n}\mathbb{E}\left[\left(Z - Z_i\right)^2 f'^2\left(Z\right)\phi''\left(f\left(Z\right)\right)\right]$$

and therefore to (5.25), while we derive from the tensorization inequality via (5.14) and (5.28) that

$$H_\phi\left(f\left(Z\right)\right) \le \sum_{i=1}^{n}\mathbb{E}\left[\left(Z - Z_i'\right)_+^2 f'^2\left(Z\right)\psi'\left(f\left(Z\right)\right)\right],$$

which means that (5.26) holds. ∎

We are now dealing with the case where f is nonincreasing.

Theorem 5.8 *Let* $X_1, \ldots, X_n$ *be some independent random variables and* Z *be some* $\left(X_1, \ldots, X_n\right)$-*measurable taking its values in some interval* $\mathcal{I}$. *Let* ϕ *belong to* $\mathcal{LO}$ *and* f *be some nonnegative and differentiable convex function* on $\mathcal{I}$. Let ψ denote the function $x \to (\phi(x) - \phi(0))/x$. Under the assumption that f is nonincreasing, for any random variable $\widetilde{Z} \leq \min_{1 \leq i \leq n} Z_i$, one has

$$H_\phi(f(Z)) \leq \frac{1}{2} \mathbb{E}\left[V f'^2\left(\widetilde{Z}\right) \phi''\left(f\left(\widetilde{Z}\right)\right)\right], \tag{5.29}$$

whenever $\phi' \circ f$ is convex, while if $\psi \circ f$ is convex, for any random variable $\widetilde{Z} \leq \min_{1 \leq i \leq n} Z_i'$ one has

$$H_\phi(f(Z)) \leq \mathbb{E}\left[V^+ f'^2\left(\widetilde{Z}\right) \psi'\left(f\left(\widetilde{Z}\right)\right)\right]. \tag{5.30}$$

or

$$H_\phi(f(Z)) \leq \mathbb{E}\left[V^- f'^2(Z) \psi'(f(Z))\right]. \tag{5.31}$$

Proof. We proceed exactly as in the proof of Theorem 5.7. Fixing $\widetilde{y} \leq x \leq y$, under the assumption that $g = \phi' \circ f$ is convex, we notice that this time

$$\phi(f(y)) - \phi(f(x)) - (f(y) - f(x))\phi'(f(x)) \leq \frac{1}{2}(y-x)^2 f'^2(\widetilde{y})\phi''(f(\widetilde{y})). \tag{5.32}$$

Indeed, still denoting by h the function

$$h(t) = \phi(f(y)) - \phi(f(t)) - (f(y) - f(t))g(t),$$

we have

$$h'(t) = -g'(t)(f(y) - f(t)).$$

But for every $t \leq y$, the monotonicity and convexity assumptions on f and g yield

$$0 \leq -g'(t) \leq -g'(\widetilde{y}) \quad \text{and} \quad 0 \leq -(f(y) - f(t)) \leq -(y-t)f'(\widetilde{y}),$$

hence

$$-h'(t) \leq (y-t)f'(\widetilde{y})g'(\widetilde{y}).$$

Integrating this inequality with respect to t on $[x, y]$ leads to (5.32). Under the assumption that $\psi \circ f$ is convex we notice that

$$0 \leq -(f(y) - f(x)) \leq -(y-x)f'(\widetilde{y})$$

and

$$0 \leq -(\psi(f(y)) - \psi(f(x))) \leq -(y-x)f'(\widetilde{y})\psi'(f(\widetilde{y})),$$

which leads to

$$(f(y) - f(x))(\psi(f(y)) - \psi(f(x))) \leq (x-y)^2 f'^2(y)\psi'(f(y)). \tag{5.33}$$

The tensorization inequality again, combined with (2.51) and (5.32) leads to

$$H_\phi(f(Z)) \leq \frac{1}{2} \sum_{i=1}^n \mathbb{E}\left[(Z - Z_i)^2 f'^2\left(\widetilde{Z}\right)\phi''\left(f\left(\widetilde{Z}\right)\right)\right]$$

and therefore to (5.29), while we derive from the tensorization inequality (5.14) and (5.33) that

$$H_\phi\left(f\left(Z\right)\right) \le \sum_{i=1}^n \mathbb{E}\left[\left(Z - Z_i'\right)_+^2 f'^2\left(\widetilde{Z}\right)\psi'\left(f\left(\widetilde{Z}\right)\right)\right],$$

which means that (5.30) holds. In order to prove (5.31) we simply define $\widetilde{f}\left(x\right) = f\left(-x\right)$ and $\widetilde{Z} = -Z$. Then $\widetilde{f}$ is nondecreasing and convex and we can use (5.26) to bound $H_\phi\left(f\left(Z\right)\right) = H_\phi\left(\widetilde{f}\left(\widetilde{Z}\right)\right)$, which gives

$$H_\phi\left(\widetilde{f}\left(\widetilde{Z}\right)\right) \le \sum_{i=1}^n \mathbb{E}\left[\left(\widetilde{Z} - \widetilde{Z}_i'\right)_+^2 \widetilde{f}'^2\left(\widetilde{Z}\right)\psi'\left(\widetilde{f}\left(\widetilde{Z}\right)\right)\right]$$

$$\le \sum_{i=1}^n \mathbb{E}\left[\left(Z - Z_i'\right)_-^2 f'^2\left(Z\right)\psi'\left(f\left(Z\right)\right)\right]$$

completing the proof of the result. ∎

As an exercise, we can derive from Theorem 5.7 and Theorem 5.8 some Logarithmic Sobolev type inequalities. Indeed, taking $f\left(z\right) = \exp\left(\lambda z\right)$ and $\phi\left(x\right) = x\ln\left(x\right)$ leads to

$$H\left(\exp\left(\lambda Z\right)\right) \le \lambda^2 \mathbb{E}\left[V^+ \exp\left(\lambda Z\right)\right], \tag{5.34}$$

if $\lambda \ge 0$, while provided that $Z - Z_i' \le 1$, one has

$$H\left(\exp\left(-\lambda Z\right)\right) \le e^\lambda \lambda^2 \mathbb{E}\left[V^+ \exp\left(-\lambda Z\right)\right]. \tag{5.35}$$

We shall see in the next section how to derive exponential moment bounds by integrating this inequality. Applying, this time Theorem 5.7 with $f\left(z\right) = \left(z - \mathbb{E}\left[Z\right]\right)_+^\alpha$ and $\phi\left(x\right) = x^{q/\alpha}$, with $1 \le q/2 \le \alpha \le q - 1$ leads to

$$\mathbb{E}\left[\left(Z - \mathbb{E}\left[Z\right]\right)_+^q\right] \le \mathbb{E}\left[\left(Z - \mathbb{E}\left[Z\right]\right)_+^\alpha\right]^{q/\alpha} + \frac{q\left(q - \alpha\right)}{2}\mathbb{E}\left[V\left(Z - \mathbb{E}\left[Z\right]\right)_+^{q-2}\right]. \tag{5.36}$$

and

$$\mathbb{E}\left[\left(Z - \mathbb{E}\left[Z\right]\right)_+^q\right] \le \mathbb{E}\left[\left(Z - \mathbb{E}\left[Z\right]\right)_+^\alpha\right]^{q/\alpha} + \alpha\left(q - \alpha\right)\mathbb{E}\left[V^+\left(Z - \mathbb{E}\left[Z\right]\right)_+^{q-2}\right], \tag{5.37}$$

Keeping the same definition for ϕ, if $f\left(z\right) = \left(z - \mathbb{E}\left[Z\right]\right)_-^\alpha$, we can apply this time Theorem 5.8 to get the exact analogue of (5.37)

$$\mathbb{E}\left[\left(Z - \mathbb{E}\left[Z\right]\right)_-^q\right] \le \mathbb{E}\left[\left(Z - \mathbb{E}\left[Z\right]\right)_-^\alpha\right]^{q/\alpha} + \alpha\left(q - \alpha\right)\mathbb{E}\left[V^-\left(Z - \mathbb{E}\left[Z\right]\right)_-^{q-2}\right]. \tag{5.38}$$

If we can warrant that the increments $Z - Z_i$ or $Z - Z'_i$ remain bounded by some positive random variable M, then we may also use the alternative bounds for the lower deviations stated in Theorem 5.8 to derive that either

$$\mathbb{E}\left[(Z - \mathbb{E}[Z])^q_-\right] \leq \mathbb{E}\left[(Z - \mathbb{E}[Z])^\alpha_-\right]^{q/\alpha}$$
$$+ \frac{q(q-\alpha)}{2}\mathbb{E}\left[V(Z - \mathbb{E}[Z] - M)^{q-2}_-\right] \tag{5.39}$$

or

$$\mathbb{E}\left[(Z - \mathbb{E}[Z])^q_-\right] \leq \mathbb{E}\left[(Z - \mathbb{E}[Z])^\alpha_-\right]^{q/\alpha}$$
$$+ \alpha(q-\alpha)\mathbb{E}\left[V^+(Z - \mathbb{E}[Z] - M)^{q-2}_-\right] \tag{5.40}$$

These inequalities will lead to moment inequalities by induction on the order of the moment. This will be done in Section 5.3.5.

5.3.4 From Efron–Stein to Exponential Inequalities

We are now in position to prove the main result of [32].

Theorem 5.9 *For every positive real numbers θ and λ such that $\theta\lambda < 1$ and* $\mathbb{E}\left[\exp\left(\lambda V^+/\theta\right)\right] < \infty$, *one has*

$$\ln\mathbb{E}\left[\exp\left(\lambda(Z - \mathbb{E}[Z])\right)\right] \leq \frac{\lambda\theta}{1 - \lambda\theta}\ln\mathbb{E}\left[\exp\left(\frac{\lambda V^+}{\theta}\right)\right] \tag{5.41}$$

while if $Z - Z'_i \leq 1$ for every $1 \leq i \leq n$ and $0 \leq \lambda < 1/2$

$$\ln\mathbb{E}\left[\exp\left(-\lambda(Z - \mathbb{E}[Z])\right)\right] \leq \frac{2\lambda}{1 - 2\lambda}\ln\mathbb{E}\left[\exp\left(\lambda V^+\right)\right] \tag{5.42}$$

Proof. Starting from (5.34), we need to decouple the right-hand side. To do this, we use the decoupling device proposed in [90]. Notice from the duality formula (2.25) that for any random variable W such that $\mathbb{E}\left[\exp\left(\lambda V^+/\theta\right)\right] < \infty$ one has

$$\mathbb{E}\left[(W - \ln\mathbb{E}\left[e^W\right])e^{\lambda Z}\right] \leq H\left(e^{\lambda Z}\right)$$

or equivalently

$$\mathbb{E}\left[We^{\lambda Z}\right] \leq \ln\mathbb{E}\left[e^W\right]\mathbb{E}\left[e^{\lambda Z}\right] + H\left(e^{\lambda Z}\right).$$

Setting $W = \lambda V^+/\theta$ and combining this inequality with (5.34) yields

$$H\left(e^{\lambda Z}\right) \leq \lambda\theta\left[\ln\mathbb{E}\left[e^{\lambda V^+/\theta}\right]\mathbb{E}\left[e^{\lambda Z}\right] + H\left(e^{\lambda Z}\right)\right]$$

and therefore, setting for every positive x, $\rho(x) = \ln\mathbb{E}\left[e^{xV^+}\right]$,

$$(1 - \lambda\theta)H\left(e^{\lambda Z}\right) \leq \lambda\theta\rho(\lambda/\theta)\mathbb{E}\left[e^{\lambda Z}\right].$$

Let $F(\lambda) = \mathbb{E}\left[\exp\left(\lambda\left(Z - \mathbb{E}\left[Z\right]\right)\right)\right]$. The previous inequality can be re-written as

$$\lambda F'(\lambda) - F(\lambda)\ln F(\lambda) \leq \frac{\lambda\theta}{1 - \lambda\theta}\rho\left(\lambda/\theta\right)F(\lambda),$$

and it remains to integrate this differential inequality. We proceed as in the proof of Proposition 2.14. Dividing each side by $\lambda^2 F(\lambda)$, we get

$$\frac{1}{\lambda}\frac{F'(\lambda)}{F(\lambda)} - \frac{1}{\lambda^2}\ln F(\lambda) \leq \frac{\theta\rho\left(\lambda/\theta\right)}{\lambda\left(1 - \lambda\theta\right)}$$

and setting $G(\lambda) = \lambda^{-1}\ln F(\lambda)$, we see that the differential inequality becomes

$$G'(\lambda) \leq \frac{\theta\rho\left(\lambda/\theta\right)}{\lambda\left(1 - \lambda\theta\right)}$$

which in turn implies since $G(\lambda)$ tends to 0 as λ tends to 0,

$$G(\lambda) \leq \int_0^\lambda \frac{\theta\rho\left(u/\theta\right)}{u\left(1 - u\theta\right)}du.$$

Now $\rho(0) = 0$ and the convexity of ρ implies that $\rho\left(u/\theta\right)/u\left(1 - u\theta\right)$ is a nondecreasing function, therefore

$$G(\lambda) \leq \frac{\theta\rho\left(\lambda/\theta\right)}{\left(1 - \lambda\theta\right)}$$

and (5.41) follows. The proof of (5.42) is quite similar. We start this time from (5.35) and notice that since $\lambda < 1/2$, one has $e^\lambda < 2$. Hence

$$H\left(\exp\left(-\lambda Z\right)\right) \leq 2\lambda^2\mathbb{E}\left[V^+\exp\left(-\lambda Z\right)\right]$$

and we use the same decoupling device as above to derive that

$$-\lambda F'(-\lambda) - F(-\lambda)\ln F(-\lambda) \leq \frac{2\lambda}{1 - 2\lambda}\rho(\lambda)F(-\lambda).$$

Integrating this differential inequality and using again the convexity of ρ leads to (5.42). ∎

This inequality should be viewed as an analogue of Efron–Stein's inequality in the sense that it relates the exponential moments of $Z - \mathbb{E}\left[Z\right]$ to those of V^+ while Efron–Stein's inequality does the same for the second order moments. Several examples of applications of this inequality are given in [32]. Here, we will focus on the derivation of concentration inequalities for empirical processes. Let us start by the simpler example of Rademacher processes and complete the results obtained in Chapter 1.

Rademacher Processes

Let $Z = \sup_{t \in T} \sum_{i=1}^{n} \varepsilon_i \alpha_{i,t}$, where T is a finite set, $(\alpha_{i,t})$ are real numbers and $\varepsilon_1, \ldots, \varepsilon_n$ are independent random signs. We can now derive from Theorem 5.9 an analogue for exponential moments of what we got from Efron–Stein's inequality in Chapter 1. Taking $\varepsilon_1', \ldots, \varepsilon_n'$ as an independent copy of $\varepsilon_1, \ldots, \varepsilon_n$ we set for every $i \in [1, n]$

$$Z_i' = \sup_{t \in T} \left[\left(\sum_{\substack{j=1 \\ j \neq i}}^{n} \varepsilon_j \alpha_{j,t} \right) + \varepsilon_i' \alpha_{i,t} \right].$$

Considering t^* such that $\sup_{t \in T} \sum_{j=1}^{n} \varepsilon_j \alpha_{j,t} = \sum_{j=1}^{n} \varepsilon_j \alpha_{j,t^*}$ we have for every $i \in [1, n]$

$$Z - Z_i' \leq (\varepsilon_i - \varepsilon_i') \alpha_{i,t^*}$$

which yields

$$(Z - Z_i')_+^2 \leq (\varepsilon_i - \varepsilon_i')^2 \alpha_{i,t^*}^2$$

and therefore by independence of ε_i' from $\varepsilon_1, \ldots, \varepsilon_n$

$$\mathbb{E}\left[(Z - Z_i')_+^2 \mid \varepsilon \right] \leq \mathbb{E}\left(1 + \varepsilon_i^2\right) \alpha_{i,t^*}^2 \leq 2\alpha_{i,t^*}^2.$$

Hence

$$V^+ \leq 2\sigma^2,$$

where $\sigma^2 = \sup_{t \in T} \sum_{i=1}^{n} \alpha_{i,t}^2$. Plugging this upper bound in (5.41) and letting θ tend to 0, we finally recover an inequality which is due to Ledoux (see [77])

$$\ln \mathbb{E}\left[\exp\left(\lambda\left(Z - \mathbb{E}[Z]\right)\right) \right] \leq 2\lambda^2 \sigma^2.$$

As compared to what we had derived from the coupling approach, we see that this time we have got the expected order for the variance i.e., $\sup_{t \in T} \sum_{i=1}^{n} \alpha_{i,t}^2$ instead of $\sum_{i=1}^{n} \sup_{t \in T} \alpha_{i,t}^2$, at the price of loosing a factor 4 in the constants. Of course we immediately derive from this control for the moment generating function that, for every positive x

$$\mathbb{P}\left[Z - \mathbb{E}[Z] \geq x\right] \leq \exp\left(-\frac{x^2}{8\sigma^2}\right). \tag{5.43}$$

Talagrand's Inequalities for Empirical Processes

In [113] (see Theorem 4.1), Talagrand obtained some striking concentration inequality for the supremum of an empirical process which is an infinite dimensional analogue of Bernstein's inequality. Using the tools that we have developed above it is not difficult to prove this inequality. Let us consider some finite set T and some independent random vectors $X_1, \ldots, X_n$ (not necessarily i.i.d.), taking their values in $\mathbb{R}^T$. Assume that

$$\mathbb{E}\left[X_{i,t}\right] = 0 \text{ and } |X_{i,t}| \leq 1 \text{ for every } 1 \leq i \leq n \text{ and } t \in T \tag{5.44}$$

and define Z as

$$\text{either } \sup_{t \in T} \sum_{i=1}^{n} X_{i,t} \text{ or } \sup_{t \in T} \left| \sum_{i=1}^{n} X_{i,t} \right|.$$

Defining τ as a function of $X_1, \ldots, X_n$ such that

$$\text{either } Z = \sum_{i=1}^{n} X_{i,\tau} \text{ or } Z = \left| \sum_{i=1}^{n} X_{i,\tau} \right|,$$

we easily see that

$$Z - Z'_i \leq \sum_{j=1}^{n} X_{j,\tau} - \sum_{j \neq i}^{n} X_{j,\tau} - X'_{i,t} \leq X_{i,\tau} - X'_{i,\tau}$$

and therefore, setting $\sigma_{i,t}^2 = \mathbb{E}\left[X_{i,t}^2\right] = \mathbb{E}\left[X_{i,t}'^2\right]$

$$V^+ \leq \sum_{i=1}^{n} \mathbb{E}\left[\left(X_{i,\tau} - X'_{i,\tau}\right)^2 \mid X\right] = \sum_{i=1}^{n} X_{i,\tau}^2 + \sum_{i=1}^{n} \sigma_{i,\tau}^2.$$

Let $v = \sup_{t \in T} \sum_{i=1}^{n} \sigma_{i,t}^2$ and $W = \sup_{t \in T} \sum_{i=1}^{n} X_{i,t}^2$, then

$$V^+ \leq W + v.$$

We may apply Theorem 5.6 to W, hence, combining inequality (5.17) for W and (5.41) with $\theta = 1$ leads to

$$\ln \mathbb{E}\left[\exp\left(\lambda\left(Z - \mathbb{E}\left[Z\right]\right)\right)\right] \leq \frac{\lambda}{1 - \lambda}\left(\lambda v + \left(e^\lambda - 1\right)\mathbb{E}\left[W\right]\right), \text{ for every } \lambda \in (0,1).$$

Using the elementary remark that $\left(e^\lambda - 1\right)\left(1 - \lambda\right) \leq \lambda$, the previous inequality implies that for every $\lambda \in (0, 1/2)$

$$\ln \mathbb{E}\left[\exp\left(\lambda\left(Z - \mathbb{E}\left[Z\right]\right)\right)\right] \leq \frac{\lambda^2}{(1 - \lambda)^2}\left(v + \mathbb{E}\left[W\right]\right) \leq \frac{\lambda^2}{(1 - 2\lambda)}\left(v + \mathbb{E}\left[W\right]\right).$$

Using the calculations of Chapter 1, this evaluation for the moment generating function readily implies the following pleasant form (with explicit constants) for Talagrand's inequality. For every positive x, one has

$$\mathbb{P}\left[Z - \mathbb{E}\left[Z\right] \geq 2\sqrt{(v + \mathbb{E}\left[W\right])x} + 2x\right] \leq e^{-x}. \tag{5.45}$$

As for the left tail, using this time (5.42) we get the same kind of inequality (with slightly worse constants)

$$\mathbb{P}\left[-Z + \mathbb{E}\left[Z\right] \geq 2\sqrt{2\left(v + \mathbb{E}\left[W\right]\right)x} + 4x\right] \leq e^{-x} \tag{5.46}$$

Actually, these inequalities are variants of those proved in [90]. Of course, by monotone convergence the assumption that T is finite may be relaxed and one can assume in fact T to be countable.

In order to use an inequality like (5.45), it is desirable to get a more tractable formulation of it, involving just v instead of v and $\mathbb{E}[W]$. This can be done for centered empirical processes at the price of additional technicalities related to classical symmetrization and contraction inequalities as described in [79]. One indeed has (see [90] for more details)

$$\mathbb{E}[W] \leq v + 16\mathbb{E}\left[\sup_{t \in T}\left|\sum_{i=1}^{n} X_{i,t}\right|\right]. \tag{5.47}$$

Hence, in the case where $Z = \sup_{t \in T}|\sum_{i=1}^{n} X_{i,t}|$, one derives from (5.45) that

$$\mathbb{P}\left[Z - \mathbb{E}[Z] \geq 2\sqrt{(2v + 16\mathbb{E}[Z])x} + 2x\right] \leq e^{-x}$$

The above inequalities in particular applies to empirical processes, since if we consider n independent and identically distributed random variables $\xi_1, \ldots, \xi_n$ and a countable class of functions $\{f_t, t \in T\}$ such that

$$|f_t - \mathbb{E}[f_t(\xi_1)]| \leq 1 \text{ for every } t \in T \tag{5.48}$$

we can use the previous result by setting $X_{i,t} = f_t(\xi_i) - \mathbb{E}[f_t(\xi_i)]$. For these i.i.d. empirical processes further refinements of the entropy method due to Rio (see [103]) and then Bousquet (see [33]) are possible which lead to even better constants (and indeed optimal constants as far as Bousquet'sresult is concerned) and applies to the one sided suprema $Z = \sup_{t \in T}\sum_{i=1}^{n}(f_t(\xi_i) - \mathbb{E}[f_t(\xi_i)])$ (instead of the two-sided suprema as for the previous one). It is proved in [33] that under the uniform boundedness assumption $|f_t - \mathbb{E}[f_t(\xi_1)]| \leq b$ for every $t \in T$, one has for every positive x

$$\mathbb{P}\left[Z - \mathbb{E}[Z] \geq \sqrt{2(v + 2b\mathbb{E}[Z])x} + \frac{bx}{3}\right] \leq e^{-x}. \tag{5.49}$$

The very nice and remarkable feature of Bousquet's inequality is that it exactly gives Bernstein's inequality in the one dimensional situation where T is reduced to a single point! Using the simple upper bound

$$\sqrt{2(v + 2b\mathbb{E}[Z])x} \leq \sqrt{2vx} + 2\sqrt{b\mathbb{E}[Z]x} \leq \sqrt{2vx} + \varepsilon\mathbb{E}[Z] + b\varepsilon^{-1}x$$

for every positive ε, we derive from Bousquet's version of Talagrand's inequality the following upper bound which will be very useful in the applications that we have in view

$$\mathbb{P}\left[Z \geq (1 + \varepsilon)\mathbb{E}[Z] + \sqrt{2vx} + b\left(\frac{1}{3} + \varepsilon^{-1}\right)x\right] \leq \exp(-x). \tag{5.50}$$

A very simple example of application is the study of chi-square statistics which was the initial motivation for advocating the use of concentration inequalities in [20].

A First Application to Chi-Square Statistics

One very remarkable feature of the concentration inequalities stated above is that, despite of their generality, they turn out to be sharp when applied to the particular and apparently simple problem of getting nonasymptotic exponential bounds for chi-square statistics. Let $X_1, \ldots, X_n$ be i.i.d. random variables with common distribution P and define the *empirical probability measure* P_n by

$$P_n = \sum_{i=1}^{n} \delta_{X_i}.$$

Following [20], we denote by ν_n the centered empirical measure $P_n - P$, given some finite set of bounded functions $\{\varphi_I\}_{I \in m}$, we can indeed write

$$\sqrt{\sum_{I \in m} \nu_n^2 (\varphi_I)} = \sup_{|a|_2 = 1} \nu_n \left[\sum_{I \in m} a_I \varphi_I \right], \quad \text{where } |a|_2^2 = \sum_{I \in m} a_I^2.$$

Let $Z^2 = n \sum_{I \in m} \nu_n^2 (\varphi_I)$. Applying (5.50) to the countable class of functions

$$\left\{ \sum_{I \in m} a_I \varphi_I : a \in \mathcal{S}'_m \right\},$$

where $\mathcal{S}'_m$ denotes some countable and dense subset of the unit sphere $\mathcal{S}_m$ in $\mathbb{R}^m$, one derives that, for every positive numbers ε and x,

$$\mathbb{P}\left[Z \geq (1+\varepsilon) \sqrt{\mathbb{E}[Z]} + \sqrt{2x v_m} + \kappa(\varepsilon) \sqrt{\frac{\|\Phi_m\|_\infty}{n}} x \right] \leq \exp(-x), \quad (5.51)$$

where $\kappa(\varepsilon) = 2 \left(\frac{1}{3} + \varepsilon^{-1} \right)$

$$\Phi_m = \sum_{I \in m} \varphi_I^2 \quad \text{and} \quad v_m = \sup_{a \in \mathcal{S}_m} \left[\mathrm{Var} \left(\sum_{I \in m} a_I \varphi_I(\xi_1) \right) \right].$$

Moreover by Cauchy–Schwarz inequality

$$\mathbb{E}[Z] \leq \sqrt{n \sum_{I \in m} \mathbb{E}[\nu_n^2(\varphi_I)]} \leq \sqrt{\sum_{I \in m} \mathrm{Var}(\varphi_I(\xi_1))}. \quad (5.52)$$

Let us now turn to the case example of "classical" chi-square statistics, which is of special interest by itself and also in view of the application to the problem of histogram selection. Let us take m to be some finite partition of $[0,1]$ which elements are intervals and define for every interval $I \in m$

$$\varphi_I = P(I)^{-1/2} \, \mathbb{1}_I,$$

then, the resulting functional Z^2 is the chi-square statistics

$$\chi_n^2 (m) = \sum_{I \in m} \frac{n \left[P_n (I) - P (I) \right]^2}{P (I)}. \tag{5.53}$$

In this case, we derive from (5.52) that

$$\mathbb{E} \left[\chi_n (m) \right] \leq \sqrt{\sum_{I \in m} (1 - P (I))} \leq \sqrt{D_m},$$

where $1 + D_m$ denotes the number of pieces of m. We also notice that $v_m \leq 1$ and setting $\delta_m = \sup_{I \in m} P (I)^{-\frac{1}{2}}$, that

$$\| \Phi_m \|_\infty \leq \delta_m^2.$$

Therefore (5.51) becomes

$$\mathbb{P} \left[\chi_n (m) \geq (1 + \varepsilon) \sqrt{D_m} + \sqrt{2x} + \kappa (\varepsilon) \frac{\delta_m}{\sqrt{n}} x \right] \leq \exp (-x). \tag{5.54}$$

We do not know of any other way for deriving this inequality while on the other hand there already exists some deviation inequality on the right tail, obtained by Mason and Van Zwet. As compared to Mason and Van Zwet's inequality in [89], (5.54) is sharper but however not sharp enough for our needs in the problem of optimal histogram selection since for irregular partitions the linear term $\delta_m x / \sqrt{n}$ can become too large. To do better, the above argument needs to be substantially refined and this is precisely what we shall perform.

5.3.5 Moment Inequalities

The main results of this section are derived from (5.36), (5.37) and (5.38) or (5.39) and (5.40) by using induction on the order of the moment. Bounding the moments of the lower deviation $(Z - \mathbb{E} [Z])_-$ will require slightly more work than bounding the moments of the upper deviation $(Z - \mathbb{E} [Z])_+$.

Upper Deviations

Our main result has a similar flavor as Burkholder's inequality but involves better constants with respect to q, namely the dependency is of order $\sqrt{q}$ instead of q. This means that in some sense, for functionals of independent variables, the quantities V or V^+ are maybe nicer than the quadratic variation.

Theorem 5.10 *For any real number $q \geq 2$, let us define*

$$\kappa_q = \frac{1}{2} \left(1 - \left(1 - \frac{1}{q} \right)^{q/2} \right)^{-1}, \tag{5.55}$$

then (κ_q) increases to $\kappa = \sqrt{e} \left(2\left(\sqrt{e} - 1\right)\right)^{-1} < 1.271$ as q goes to infinity. One has for any real number $q \geq 2$

$$\left\| (Z - \mathbb{E}[Z])_+ \right\|_q \leq \sqrt{\left(1 - \frac{1}{q}\right) 2\kappa_q q \left\| V^+ \right\|_{q/2}} \leq \sqrt{2\kappa q \left\| V^+ \right\|_{q/2}}, \qquad (5.56)$$

and similarly

$$\left\| (Z - \mathbb{E}[Z])_- \right\|_q \leq \sqrt{\left(1 - \frac{1}{q}\right) 2\kappa_q q \left\| V^- \right\|_{q/2}} \leq \sqrt{2\kappa q \left\| V^- \right\|_{q/2}}. \qquad (5.57)$$

Moreover

$$\left\| (Z - \mathbb{E}[Z])_+ \right\|_q \leq \sqrt{\kappa_q q \left\| V \right\|_{q/2}} \leq \sqrt{\kappa q \left\| V \right\|_{q/2}}. \qquad (5.58)$$

Proof. It is enough to prove (5.56) and (5.58) since (5.57) derives from (5.56) by changing Z into $-Z$. It follows from Efron–Stein's inequality and its variant (5.11) respectively that for every $q \in [1, 2]$

$$\left\| (Z - \mathbb{E}[Z])_+ \right\|_q \leq \left\| Z - \mathbb{E}[Z] \right\|_2 \leq \sqrt{\left\| V_+ \right\|_1} \qquad (5.59)$$

and

$$\left\| (Z - \mathbb{E}[Z])_+ \right\|_q \leq \left\| Z - \mathbb{E}[Z] \right\|_2 \leq \sqrt{\left\| V \right\|_1} \qquad (5.60)$$

We intend to prove by induction on k that for every $q \in [k, k+1)$ one has

$$\left\| (Z - \mathbb{E}[Z])_+ \right\|_q \leq \sqrt{c_q q} \qquad (5.61)$$

where either $c_q = 2\left(1 - 1/q\right)\kappa_q \left\| V_+ \right\|_{q/2}$ or $c_q = \kappa_q \left\| V \right\|_{q/2}$ for $q \geq 2$ and either $c_q = \left\| V_+ \right\|_1$ or $c_q = \left\| V \right\|_1$ for $q \in [1, 2)$. $\kappa'_q = \kappa_q = 1$ for $q \in [1, 2)$. For $k = 1$, (5.61) follows either from (5.59) or (5.60). We assume now that (5.61) holds for some integer $k \geq 1$ and every $q \in [k, k+1)$. Let q be some real number belonging to $[k+1, k+2)$. We want to prove that (5.61) holds. Hölder's inequality implies that for every nonnegative random variable Y

$$\mathbb{E}\left[Y \left(Z - \mathbb{E}[Z]\right)_+^{q-2}\right] \leq \left\| Y \right\|_{q/2} \left\| (Z - \mathbb{E}[Z])_+ \right\|_q^{q-2},$$

hence, using either (5.37) or (5.36) with $\alpha = q - 1$ we get

$$\left\| (Z - \mathbb{E}[Z])_+ \right\|_q^q \leq \left\| (Z - \mathbb{E}[Z])_+ \right\|_{q-1}^q + \frac{q}{2\kappa_q} c_q \left\| (Z - \mathbb{E}[Z])_+ \right\|_q^{q-2}.$$

Defining for every real number $p \geq 1$

$$x_p = \left\| (Z - \mathbb{E}[Z])_+ \right\|_p^p \left(p c_p\right)^{-p/2},$$

we aim at proving that $x_q \leq 1$ and the previous inequality becomes

$$x_q q^{q/2} c_q^{q/2} \leq x_{q-1}^{q/q-1} (q-1)^{q/2} c_{q-1}^{q/2} + \frac{1}{2} x_q^{1-2/q} q^{q/2} c_q^{q/2} \kappa_q^{-1},$$

from which we derive since $c_{q-1} \leq c_q$

$$x_q \leq x_{q-1}^{q/q-1} \left(1 - \frac{1}{q}\right)^{q/2} + \frac{1}{2\kappa_q} x_q^{1-2/q}. \tag{5.62}$$

Knowing by induction that $x_{q-1} \leq 1$, we derive from (5.62) that

$$x_q \leq \left(1 - \frac{1}{q}\right)^{q/2} + \frac{1}{2\kappa_q} x_q^{1-2/q}.$$

Now the function

$$f_q : x \to \left(1 - \frac{1}{q}\right)^{q/2} + \frac{1}{2\kappa_q} x^{1-2/q} - x$$

is strictly concave on $\mathbb{R}_+$ and positive at point $x = 0$. Hence, since $f_q(1) = 0$ (because of our choice of κ_q) and $f_q(x_q) \geq 0$ we derive that $x_q \leq 1$, which means that (5.61) holds, achieving the proof of the result. ∎

The next result establishes a link between the Poissonian bound established in the previous section for self-bounding processes and moment inequalities.

Corollary 5.11 *If we assume that*

$$\sum_{i=1}^{n} (Z - Z_i) \leq AZ, \text{ for some constant } A \geq 1, \tag{5.63}$$

then on the one hand, for every integer $q \geq 1$

$$\|Z\|_q \leq \mathbb{E}[Z] + \frac{A(q-1)}{2} \tag{5.64}$$

and on the other hand, for every real number $q \geq 2$

$$\left\| (Z - \mathbb{E}[Z])_+ \right\|_q \leq \sqrt{\kappa} \left[\sqrt{Aq\mathbb{E}[Z]} + \frac{Aq}{2} \right],$$

where κ stands for the absolute constant of Theorem 5.10 ($\kappa < 1.271$).

Proof. Applying Theorem 5.7 with $f(z) = z^{q-1}$ and $\phi(x) = x^{q/q-1}$ leads to

$$\|Z\|_q^q \leq \|Z\|_{q-1}^q + \frac{q}{2} \mathbb{E}\left[VZ^{q-2}\right].$$

But, under assumption (5.63), we have $V \leq AZ$, hence

$$\|Z\|_q^q \leq \|Z\|_{q-1}^q + \frac{qA}{2} \|Z\|_{q-1}^{q-1}$$

$$\leq \|Z\|_{q-1}^q \left[1 + \frac{qA}{2\|Z\|_{q-1}} \right].$$

Now, we notice that for every nonnegative real number u, one has $1 + uq \leq (1+u)^q$ and therefore

$$\|Z\|_q^q \leq \|Z\|_{q-1}^q \left(1 + \frac{A}{2\|Z\|_{q-1}} \right)^q$$

or equivalently

$$\|Z\|_q \leq \|Z\|_{q-1} + \frac{A}{2}.$$

Hence by induction $\|Z\|_q \leq \|Z\|_1 + (A/2)(q-1)$ which means (5.64). By Theorem 5.10 and (5.63) we have

$$\left\| (Z - \mathbb{E}[Z])_+ \right\|_q \leq \sqrt{\kappa q \|V\|_{q/2}} \leq \sqrt{\kappa q A \|Z\|_{q/2}}.$$

Let s be the smallest integer such that $q/2 \leq s$, then (5.64) yields

$$\|Z\|_{q/2} \leq \mathbb{E}[Z] + \frac{A(s-1)}{2} \leq \mathbb{E}[Z] + \frac{Aq}{4}$$

$$\left\| (Z - \mathbb{E}[Z])_+ \right\|_q \leq \sqrt{\kappa} \left[\sqrt{qA\mathbb{E}[Z] + \frac{q^2 A^2}{4}} \right]$$

$$\leq \sqrt{\kappa} \left[\sqrt{qA\mathbb{E}[Z]} + \frac{qA}{2} \right]$$

and the result follows. ∎

Lower Deviations

We intend to provide lower deviation results under boundedness assumptions on some increments of the functional Z.

Theorem 5.12 *We assume that either for some positive random variable M*

$$(Z - Z_i')_+ \leq M, \text{ for every } 1 \leq i \leq n \tag{5.65}$$

or that for some $X^{(i)}$-measurable random variables Z_i's one has

$$0 \leq Z - Z_i \leq M, \text{ for every } 1 \leq i \leq n. \tag{5.66}$$

Then, there exists some universal constants C_1 and C_2 ($C_1 < 4.16$ and $C_2 < 2.42$) such that for every real number $q \geq 2$ one has, under assumption (5.65)

$$\left\| (Z - \mathbb{E}[Z])_- \right\|_q \leq \sqrt{C_1 q \left(\|V_+\|_{q/2} \vee q \|M\|_q^2 \right)}, \tag{5.67}$$

while under assumption (5.66)

$$\left\| (Z - \mathbb{E}[Z])_- \right\|_q \leq \sqrt{C_2 q \left(\|V\|_{q/2} \vee q \|M\|_q^2 \right)}. \tag{5.68}$$

Moreover if (5.66) holds with $M = 1$ and

$$\sum_{i=1}^n (Z - Z_i) \leq AZ, \text{ for some constant } A \geq 1, \tag{5.69}$$

then for every integer $q \geq 1$

$$\left\| (Z - \mathbb{E}[Z])_- \right\|_q \leq \sqrt{CqA\mathbb{E}[Z]}, \tag{5.70}$$

where C is some universal constant smaller than 1.131.

Proof. In the sequel, we use the notation $m_q = \left\| (Z - \mathbb{E}[Z])_- \right\|_q$. We first prove (5.67) and (5.68). For $a > 0$, the continuous function

$$x \to e^{-1/2} + \frac{1}{ax} e^{1/\sqrt{x}} - 1$$

decreases from $+\infty$ to $e^{-1/2} - 1 < 0$ on $(0, +\infty)$. Let us define C_a as the unique zero of this function. According to whether we are dealing with assumption (5.65) or assumption (5.66), we set either $a = 1$ and $c_q = \|V_+\|_{q/2} \vee q \|M\|_q^2$ if $q \geq 2$ ($c_q = \|V_+\|_1$ if $q \in [1, 2)$) or $a = 2$ and $c_q = \|V\|_{q/2} \vee q \|M\|_q^2$ if $q \geq 2$ ($c_q = \|V\|_1$ if $q \in [1, 2)$). Our aim is to prove by induction on k that for every $q \in [k, k+1)$, the following inequality holds for $a = 1, 2$

$$m_q \leq \sqrt{C_a q c_q}. \tag{5.71}$$

Since C_1 and C_2 are larger than 1, if $k = 1$, Efron–Stein's inequality or its variant (5.11) imply that (5.71) holds for every $q \in [1, 2)$ and we have an even better result in terms of constants since then

$$m_q \leq \sqrt{c_q} \tag{5.72}$$

Now let $k \geq 2$ be some integer and consider some real number $q \in [k, k+1)$. By induction we assume that $m_{q-1} \leq \sqrt{C_a (q-1) c_{q-1}}$. Then we use either (5.40) or (5.39) with $\alpha = q - 1$, which gives either

$$m_q^q \leq m_{q-1}^q + q\mathbb{E}\left[V_+ \left((Z - \mathbb{E}[Z])_- + M \right)^{q-2} \right] \tag{5.73}$$

or

$$m_q^q \leq m_{q-1}^q + \frac{q}{2} \mathbb{E}\left[V\left((Z - \mathbb{E}[Z])_- + M\right)^{q-2}\right]. \tag{5.74}$$

From the convexity of $x \to x^{q-2}$ if $k \geq 3$ or subadditivity if $k = 2$, we derive that for every $\theta \in (0, 1)$

$$((Z - \mathbb{E}[Z])_- + 1)^{q-2} \leq \theta^{-(q-3)_+} M^{q-2} + (1 - \theta)^{-(q-3)_+} (Z - \mathbb{E}[Z])_-^{q-2}. \tag{5.75}$$

Using Hölder's inequality we get from (5.73) (or (5.74)) and (5.75) either

$$m_q^q \leq m_{q-1}^q + q\theta^{-(q-3)_+} \|M\|_q^{q-2} \|V_+\|_{q/2} + q(1 - \theta)^{-(q-3)_+} \|V_+\|_{q/2} m_q^{q-2}$$

or

$$m_q^q \leq m_{q-1}^q + \frac{q}{2}\theta^{-(q-3)_+} \|M\|_q^{q-2} \|V\|_{q/2} + \frac{q}{2}(1 - \theta)^{-(q-3)_+} \|V\|_{q/2} m_q^{q-2}. \tag{5.76}$$

Let us first deal with the case $k \geq 3$. Since $m_{q-1} \leq \sqrt{C_a(q-1)c_{q-1}}$ and $c_{q-1} \leq c_q$, we derive that

$$m_q^q \leq C_a^{q/2}(q-1)^{q/2} c_q^{q/2} + \frac{1}{a}q^{-q+2}\theta^{-(q-3)_+} q^{q/2} c_q^{q/2}$$
$$+ \frac{1}{a}q(1 - \theta)^{-(q-3)_+} c_q m_q^{q-2}.$$

Let $x_q = C_a^{-q/2} m_q^q (qc_q)^{-q/2}$, then

$$x_q \leq \left(1 - \frac{1}{q}\right)^{q/2}$$
$$+ \frac{1}{aC_a}\left(\theta^{-(q-3)_+}\left(\sqrt{C_a}q\right)^{-q+2} + (1 - \theta)^{-(q-3)_+} x_q^{1-2/q}\right). \tag{5.77}$$

Let us choose θ minimizing

$$g(\theta) = \theta^{-q+3}\left(\sqrt{C_a}q\right)^{-q+2} + (1 - \theta)^{-q+3},$$

i.e., $\theta = 1/\left(\sqrt{C_a}q + 1\right)$. Since for this value of θ one has

$$g(\theta) = \left(1 + \frac{1}{\sqrt{C_a}q}\right)^{q-2},$$

(5.77) becomes

$$x_q \leq \left(1 - \frac{1}{q}\right)^{q/2} + \frac{1}{aC_a}\left(\left(1 + \frac{1}{\sqrt{C_a}q}\right)^{q-2} + (1 - \theta)^{-q+3}\left(x_q^{1-2/q} - 1\right)\right).$$

Hence, using the elementary inequalities

$$\left(1 - \frac{1}{q}\right)^{q/2} \le e^{-1/2}$$

$$\left(1 + \frac{1}{\sqrt{C_a}q}\right)^{q-2} \le e^{1/\sqrt{C_a}}$$

which derive from the well known upper bound $\ln(1 + u) \le u$, we get

$$x_q \le e^{-1/2} + \frac{1}{aC_a}\left(e^{1/\sqrt{C_a}} + (1 - \theta)^{-q+3}\left(x_q^{1-2/q} - 1\right)\right).$$

Now since $e^{1/\sqrt{C_a}} \ge g(\theta) > (1 - \theta)^{-q+3}$, the function

$$f_q : x \to e^{-1/2} + \frac{1}{aC_a}\left(e^{1/\sqrt{C_a}} + (1 - \theta)^{-q+3}\left(x^{1-2/q} - 1\right)\right) - x$$

is positive at 0 and strictly concave on $\mathbb{R}_+$. So, noticing that C_a has been defined in such a way that $f_q(1) = 0$, the function f_q can be nonnegative at point x_q only if $x_q \le 1$ which proves (5.71). To treat the case where $k = 2$, we note that in this case we can use (5.72) which ensures that $m_{q-1} \le \sqrt{c_{q-1}} \le \sqrt{c_q}$ so that (5.76) becomes

$$m_q^q \le c_q^{q/2} + \frac{q}{2}\|M\|_q^{q-2}\|V\|_{q/2} + \frac{q}{2}\|V\|_{q/2}m_q^{q-2}.$$

Let $x_q = C_a^{-q/2}m_q^q(qc_q)^{-q/2}$, then

$$x_q \le \left(\frac{1}{C_a q}\right)^{q/2} + \frac{1}{aC_a}\left(\left(\sqrt{C_a}q\right)^{-q+2} + x_q^{1-2/q}\right)$$

and therefore, since $q \ge 2$ and $C_a \ge 1$

$$x_q \le \frac{1}{2C_a} + \frac{1}{aC_a}\left(1 + x_q^{1-2/q}\right).$$

The function

$$g_q : x \to \frac{1}{2C_a} + \frac{1}{aC_a}\left(1 + x^{1-2/q}\right) - x$$

is strictly concave on $\mathbb{R}_+$ and positive at 0. Furthermore

$$g_q(1) = \frac{(4 + a)}{2aC_a} - 1 < 0,$$

since $C_a > (4 + a)/2a$. Hence g_q can be nonnegative at point x_q only if $x_q \le 1$ yielding (5.71), achieving the proof of (5.67) and (5.68). In order to prove (5.70) we first define C as the unique positive root of the equation

$$e^{-1/2} + \frac{1}{2C}e^{-1+1/C} - 1 = 0.$$

We derive from the upper bound $V \leq AZ$ and (5.11) that

$$(\mathbb{E}\,|Z - \mathbb{E}\,[Z]|)^2 \leq \mathbb{E}\left[(Z - \mathbb{E}\,[Z])^2\right] \leq A\mathbb{E}\,[Z]$$

which allows to deal with the cases where $q = 1$ and $q = 2$ since $C > 1$. Then, for $q \geq 3$, we assume by induction that $m_k \leq \sqrt{CkA\mathbb{E}\,[Z]}$, for $k = q - 2$ and $k = q - 1$ and use $V \leq AZ$ together with (5.39) for $\alpha = q - 1$. This gives

$$m_q^q \leq m_{q-1}^q + \frac{q}{2}A\mathbb{E}\left[Z\left((Z - \mathbb{E}\,[Z])_- + 1\right)^{q-2}\right]$$

which in turn implies since $z \to \left((z - \mathbb{E}\,[Z])_- + 1\right)^{q-2}$ decreases while $z \to z$ increases

$$m_q^q \leq m_{q-1}^q + \frac{q}{2}A\mathbb{E}\,[Z]\,\mathbb{E}\left[\left((Z - \mathbb{E}\,[Z])_- + 1\right)^{q-2}\right].$$

By the triangle inequality $\left\|(Z - \mathbb{E}\,[Z])_- + 1\right\|_{q-2} \leq 1 + \left\|(Z - \mathbb{E}\,[Z])_-\right\|_{q-2}$, this inequality becomes

$$m_q^q \leq m_{q-1}^q + \frac{q}{2}A\mathbb{E}\,[Z]\,(1 + m_{q-2})^{q-2}$$

and therefore our induction assumption yields

$$m_q^q \leq \left(1 - \frac{1}{q}\right)^{q/2}(CqA\mathbb{E}\,[Z])^{q/2}$$

$$+ \frac{(CqA\mathbb{E}\,[Z])^{q/2}}{2C}\left[\frac{1}{\sqrt{CqA\mathbb{E}\,[Z]}} + \sqrt{1 - \frac{2}{q}}\right]^{q-2}. \qquad (5.78)$$

At this stage of the proof we can use the fact that we know a crude upper bound on m_q deriving from the nonnegativity of Z, namely $m_q \leq \mathbb{E}\,[Z]$. Hence, we can always assume that $CqA \leq \mathbb{E}\,[Z]$, since otherwise (5.70) is implied by this crude upper bound. Combining this inequality with $A > 1$, leads to

$$\frac{1}{\sqrt{CqA\mathbb{E}\,[Z]}} \leq \frac{1}{Cq},$$

so that plugging this inequality in (5.78) and setting $x_q = m_q^q\,(CqA\mathbb{E}\,[Z])^{-q/2}$, we derive that

$$x_q \leq \left(1 - \frac{1}{q}\right)^{q/2} + \frac{1}{2C}\left(\frac{1}{Cq} + \sqrt{1 - \frac{2}{q}}\right)^{q-2}.$$

We claim that

$$\left(\frac{1}{Cq} + \sqrt{1 - \frac{2}{q}}\right)^{q-2} \le e^{-1+1/C}. \tag{5.79}$$

Indeed (5.79) can be checked numerically for $q = 3$, while for $q \ge 4$, combining

$$\sqrt{1 - 2/q} \le 1 - \frac{1}{q} - \frac{1}{2q^2}$$

with $\ln(1 + u) \le u$ leads to

$$\ln\left[\left(\frac{1}{Cq} + \sqrt{1 - \frac{2}{q}}\right)^{q-2}\right] \le -1 + 1/C + \frac{1}{q}\left(\frac{3}{2} + \frac{1}{q} - \frac{2}{C}\right)$$

$$\le -1 + 1/C + \frac{1}{q}\left(\frac{7}{4} - \frac{2}{C}\right)$$

which, since $C < 8/7$, implies (5.79). Hence

$$x_q \le \left(1 - \frac{1}{q}\right)^{q/2} + \frac{1}{2C}e^{-1+1/C} \le e^{-1/2} + \frac{1}{2C}e^{-1+1/C},$$

which, by definition of C means that $x_q \le 1$, achieving the proof of the result. ∎

Application to Suprema of Unbounded Empirical Processes

Let us consider again some finite set T and some independent random vectors $X_1, \ldots, X_n$ (not necessarily i.i.d.), taking their values in $\mathbb{R}^T$. Assume this time that

$$\mathbb{E}[X_{i,t}] = 0 \text{ for every } 1 \le i \le n \text{ and } t \in T$$

and for some $q \ge 2$

$$M = \sup_{i,t} |X_{i,t}| \in \mathbb{L}_q.$$

Define Z as

$$\text{either } \sup_{t \in T} \sum_{i=1}^{n} X_{i,t} \text{ or } \sup_{t \in T}\left|\sum_{i=1}^{n} X_{i,t}\right|.$$

Arguing exactly as in the proof of Talagrand's inequality, one has

$$V^+ \le W + v,$$

where $v = \sup_{t \in T} \sum_{i=1}^{n} \sigma_{i,t}^2$ and $W = \sup_{t \in T} \sum_{i=1}^{n} X_{i,t}^2$. It comes from (5.56) that

$$\left\|(Z - \mathbb{E}[Z])_+\right\|_q \le \sqrt{2\kappa q \left\|V^+\right\|_{q/2}} \le \sqrt{2\kappa q v} + \sqrt{2\kappa q}\left\|\sqrt{W}\right\|_q.$$

In order to control $\left\|\sqrt{W}\right\|_q$, we use this time (5.58). Defining

$$W_i = \sup_{t \in T} \sum_{j \neq i}^{n} X_{j,t}^2$$

we have $W \geq W_i$ and

$$\sum_{i=1}^{n} (W - W_i)^2 \leq W M^2$$

and therefore

$$\sum_{i=1}^{n} \left(\sqrt{W} - \sqrt{W_i}\right)^2 \leq M^2.$$

Hence, (5.58) yields

$$\left\|\sqrt{W}\right\|_q \leq \mathbb{E}\left[\sqrt{W}\right] + \sqrt{\kappa q} \left\|M\right\|_q$$

and since $\mathbb{E}\left[\sqrt{W}\right] \leq \sqrt{\mathbb{E}[W]}$ with $v \leq \mathbb{E}[W]$, collecting the above inequalities we get

$$\left\|(Z - \mathbb{E}[Z])_+\right\|_q \leq 2\sqrt{2\kappa q \mathbb{E}[W]} + \sqrt{2}\kappa q \left\|M\right\|_q. \tag{5.80}$$

This provides a version with explicit constants (as functions of the order q of the moment) of an inequality due to Baraud (see [8]) and which turns out to be the main tool used by Baraud to prove model selection theorems for regression on a fixed design. Note that even in the one-dimensional case, getting the factor $\sqrt{q}$ in the first term in the right-hand side of (5.80) is far from being trivial since straightforward proofs rather lead to some factor q. In this one-dimensional situation we recover, up to absolute numerical constants the inequality due to Pinelis (see [97]).

6

Maximal Inequalities

The main issue of this chapter is to provide exponential bounds for suprema of empirical processes. Thanks to the powerful concentrations tools developed in the previous chapter, this amounts (at least for bounded empirical processes) to controls in expectation which can be obtained from the maximal inequalities for random vectors given in Chapter 2 through chaining arguments, exactly as in the Gaussian case (see Chapter 3).

All along this chapter we consider independent random variables $\xi_1,...,\xi_n$ defined on a probability space $(\Omega, \mathcal{A}, \mathbb{P})$ with values in some measurable space Ξ. We denote by P_n the empirical probability measure associated with $\xi_1,...,\xi_n$, which means that for any measurable function $f : \Xi \to \mathbb{R}$

$$P_n(f) = \frac{1}{n} \sum_{i=1}^{n} f(\xi_i).$$

If furthermore $f(\xi_i)$ is integrable for all $i \le n$, we set

$$S_n(f) = \sum_{i=1}^{n} (f(\xi_i) - \mathbb{E}f(\xi_i)) \text{ and } \nu_n(f) = \frac{S_n(f)}{n}.$$

Given a collection of functions $\mathcal{F}$ our purpose is to control $\sup_{f \in \mathcal{F}} \nu_n(f)$ under appropriate assumptions on $\mathcal{F}$. Since we are mainly interested by sub-Gaussian type inequalities, the classes of interest are Donsker classes. It is known (see [57] for instance) that universal entropy and entropy with bracketing are appropriate ways of measuring the massiveness of $\mathcal{F}$ and roughly speaking play the same role in the empirical process theory as metric entropy for the study of Gaussian processes. Hence, it is not surprising that such conditions will used below. We begin with the simplest case of set indexed i.i.d. empirical processes.

6.1 Set-Indexed Empirical Processes

The purpose of this section is to provide maximal inequalities for set-indexed empirical processes under either the Vapnik–Chervonenkis condition or an entropy with bracketing assumption. All along this section we assume the variables $\xi_1, \dots, \xi_n$ to be i.i.d. with common distribution P. We need some basic maximal inequalities for random vectors and also Rademacher processes since we shall sometimes use symmetrization techniques.

6.1.1 Random Vectors and Rademacher Processes

Lemma 2.3 ensures that if $(Z_f)_{f \in \mathcal{F}}$ is a finite family of real valued random variables and ψ is some convex and continuously differentiable function on $[0, b)$ with $0 < b \leq +\infty$ such that $\psi(0) = \psi'(0) = 0$ and for every $\lambda \in (0, b)$ and $f \in \mathcal{F}$, one has

$$\ln \mathbb{E}\left[\exp\left(\lambda Z_f\right)\right] \leq \psi(\lambda) \tag{6.1}$$

then, if N denotes the cardinality of $\mathcal{F}$ we have

$$\mathbb{E}\left[\sup_{f \in \mathcal{F}} Z_f\right] \leq \psi^{*-1}\left(\ln(N)\right).$$

In particular, if for some nonnegative number σ one has $\psi(\lambda) = \lambda^2 v/2$ for every $\lambda \in (0, +\infty)$, then

$$\mathbb{E}\left[\sup_{f \in \mathcal{F}} Z_f\right] \leq \sqrt{2v \ln(N)},$$

while, if $\psi(\lambda) = \lambda^2 v/\left(2\left(1 - c\lambda\right)\right)$ for every $\lambda \in (0, 1/c)$, one has

$$\mathbb{E}\left[\sup_{f \in \mathcal{F}} Z_f\right] \leq \sqrt{2v \ln(N)} + c\ln(N). \tag{6.2}$$

The two situations where we shall apply this Lemma in order to derive chaining bounds are the following.

- $\mathcal{F}$ is a finite subset of $\mathbb{R}^n$ and $Z_f = \sum_{i=1}^{n} \varepsilon_i f_i$, where $(\varepsilon_1, \dots, \varepsilon_n)$ are independent Rademacher variables. Then, setting $v = \sup_{f \in \mathcal{F}} \sum_{i=1}^{n} f_i^2$, it is well known that (6.1) is fulfilled with $\psi(\lambda) = \lambda^2 v/2$ and therefore

$$\mathbb{E}\left[\sup_{f \in \mathcal{F}} \sum_{i=1}^{n} \varepsilon_i f_i\right] \leq \sqrt{2v \ln(N)}. \tag{6.3}$$

- $\mathcal{F}$ is a finite set of functions f such that $\|f\|_\infty \leq 1$ and $Z_f = \sum_{i=1}^n f(\xi_i)$
  $- \mathbb{E}[f(\xi_i)]$, where $\xi_1, \dots, \xi_n$ are independent random variables. Then, setting $v = \sup_{f \in \mathcal{F}} \sum_{i=1}^n \mathbb{E}[f^2(\xi_i)]$, as a by-product of the proof of Bernstein's inequality, assumption (6.1) is fulfilled with $\psi(\lambda) = \lambda^2 v / (2(1 - \lambda/3))$ and therefore

$$\mathbb{E}\left[\sup_{f \in \mathcal{F}} Z_f\right] \leq \sqrt{2v \ln(N)} + \frac{1}{3}\ln(N) \qquad (6.4)$$

We are now ready to prove a maximal inequality for Rademacher processes which will be useful to analyze symmetrized empirical processes.

Let $\mathcal{F}$ be some bounded subset of $\mathbb{R}^n$ equipped with the usual Euclidean norm defined by

$$\|z\|_2^2 = \sum_{i=1}^n z_i^2$$

and let for any positive δ, $H_2(\delta, \mathcal{F})$ denote the logarithm of the maximal number of points $\{f^{(1)}, \dots, f^{(N)}\}$ belonging to $\mathcal{F}$ such that $\left\|f^{(j)} - f^{(j')}\right\|_2^2 > \delta^2$ for every $j \neq j'$. It is easy to derive from the maximal inequality (6.3) the following chaining inequality which is quite standard (see [79] for instance). The proof being short we present it for the sake of completeness.

Lemma 6.1 *Let $\mathcal{F}$ be some bounded subset of $\mathbb{R}^n$ and $(\varepsilon_1, \dots, \varepsilon_n)$ be independent Rademacher variables. We consider the Rademacher process defined by $Z_f = \sum_{i=1}^n \varepsilon_i f_i$ for every $f \in \mathcal{F}$. Let δ such that $\sup_{f \in \mathcal{F}} \|f\|_2 \leq \delta$, then*

$$\mathbb{E}\left[\sup_{f \in \mathcal{F}} Z_f\right] \leq 3\delta \sum_{j=0}^\infty 2^{-j}\sqrt{H_2(2^{-j-1}\delta, \mathcal{F})} \qquad (6.5)$$

Proof. Since $\mathcal{F} \to H_2(., \mathcal{F})$ is nondecreasing with respect to the inclusion ordering, if we can prove that (6.5) holds true when $\mathcal{F}$ is a finite set, then for any finite subset $\mathcal{F}'$ of $\mathcal{F}$ one has

$$\mathbb{E}\left[\sup_{f \in \mathcal{F}'} Z_f\right] \leq 3\delta \sum_{j=0}^\infty 2^{-j}\sqrt{H_2(2^{-j-1}\delta, \mathcal{F})}$$

which leads to (6.5) by continuity of $f \to Z_f$ and separability of $\mathcal{F}$. So we can assume $\mathcal{F}$ to be finite. For any integer j, we set $\delta_j = \delta 2^{-j}$. By definition of $H_2(., \mathcal{F})$, for any integer $j \geq 1$ we can define some mapping Π_j from $\mathcal{F}$ to $\mathcal{F}$ such that

$$\ln|\Pi_j(\mathcal{F})| \leq H_2(\delta_j, \mathcal{F}) \qquad (6.6)$$

and

$$d(f, \Pi_j f) \leq \delta_j \quad \text{for all } f \in \mathcal{F}. \qquad (6.7)$$

For $j = 0$, we choose Π_0 to be identically equal to 0. For this choice, (6.7) and (6.6) are also satisfied by definition of δ. Since $\mathcal{F}$ is finite, there exists some integer J such that for all $f \in \mathcal{F}$

$$Z_f = \sum_{j=0}^{J} Z_{\Pi_{j+1}f} - Z_{\Pi_j f},$$

from which we deduce that

$$\mathbb{E}\left[\sup_{f \in \mathcal{F}} Z_f\right] \leq \sum_{j=0}^{J} \mathbb{E}\left[\sup_{f \in \mathcal{F}} Z_{\Pi_{j+1}f} - Z_{\Pi_j f}\right].$$

Since for any integer j, $(\Pi_j f, \Pi_{j+1}f)$ ranges in a set with cardinality not larger than $\exp\left(2H_2\left(\delta_{j+1}, \mathcal{F}\right)\right)$ when f varies and

$$d\left(\Pi_j f, \Pi_{j+1}f\right) \leq \frac{3}{2}\delta_j$$

for all $f \in \mathcal{F}$, by (6.3) we get

$$\sum_{j=0}^{J} \mathbb{E}\left[\sup_{f \in \mathcal{F}} Z_{\Pi_{j+1}f} - Z_{\Pi_j f}\right] \leq 3\sum_{j=0}^{J} \delta_j \sqrt{H_2\left(\delta_{j+1}, \mathcal{F}\right)}$$

and therefore

$$\mathbb{E}\left[\sup_{f \in \mathcal{F}} Z_f\right] \leq 3\sum_{j=0}^{J} \delta_j \sqrt{H_2\left(\delta_{j+1}, \mathcal{F}\right)},$$

which implies (6.5). ∎

We turn now to maximal inequalities for set-indexed empirical processes. The VC case will be treated via symmetrization by using the previous bounds for Rademacher processes while the bracketing case will be studied via some convenient chaining argument.

6.1.2 Vapnik–Chervonenkis Classes

The notion of Vapnik–Chervonenkis dimension (*VC-dimension*) for a class of sets $\mathcal{B}$ is very important. It is one way of defining a proper notion of finite dimension for the nonlinear set $\{1\!\!1_B, B \in \mathcal{B}\}$. As we shall see in Chapter 8, the VC-classes may be typically used for defining proper models of classifiers. Let us recall the basic definitions and properties of VC-classes.

Definition 6.2 *Let $\mathcal{B}$ be some class of subsets of Ξ. For every integer n, let*

$$m_n\left(\mathcal{B}\right) = \sup_{A \subset \Xi, |A| = n} |\{A \cap B, B \in \mathcal{B}\}|.$$

Let us define the VC-dimension of $\mathcal{B}$ by

$$V = \sup\left\{n \geq 0, m_n\left(\mathcal{B}\right) = 2^n\right\}.$$

If $V < \infty$, then $\mathcal{B}$ is called a Vapnik–Chervonenkis class (VC-class).

A basic example of a VC-class is the class of half spaces of $\mathbb{R}^d$ which has VC-dimension $d + 1$ according to Radon's Theorem (see [57] for instance). One of the main striking results on VC-classes is Sauer's lemma that we recall below (for a proof of Sauer's lemma, we refer to [57]).

Lemma 6.3 (Sauer's lemma) *Let $\mathcal{B}$ be a VC-class with VC-dimension V. Then for every integer $n \geq V$*

$$m_n(\mathcal{B}) \leq \sum_{j=0}^{V} \binom{n}{j}.$$

There exists at least two ways of measuring the "size" of a VC-class $\mathcal{B}$. The first one is directly based on the combinatorial properties of $\mathcal{B}$. Let $H_\mathcal{B}$ denote the (random) combinatorial entropy of $\mathcal{B}$

$$H_\mathcal{B} = \ln |\{B \cap \{\xi_1, \ldots, \xi_n\}, B \in \mathcal{B}\}|. \tag{6.8}$$

Note that Sauer's lemma implies via (2.9) that for a VC-class $\mathcal{B}$ with dimension V, one has

$$H_\mathcal{B} \leq V \ln\left(\frac{en}{V}\right). \tag{6.9}$$

The second one is more related to $\mathbb{L}_2(Q)$-covering metric properties of $S = \{\mathbb{1}_B : B \in \mathcal{B}\}$ with respect to any discrete probability measure Q. For any positive number δ, and any probability measure Q, let $N(\delta, \mathcal{B}, Q)$ denote the maximal number of indicator functions $\{t_1, \ldots, t_N\}$ such that $E_Q[t_i - t_j]^2 > \delta^2$ for every $i \neq j$. The universal δ-metric entropy of $\mathcal{B}$ is then defined by

$$H(\delta, \mathcal{B}) = \sup_Q \ln N(\delta, \mathcal{B}, Q) \tag{6.10}$$

where the supremum is extended to the set of all discrete probability measures. The following upper bound for the universal entropy of a VC-class is due to Haussler (see [66]). For some absolute constant κ, one has for every positive δ

$$H(\delta, \mathcal{B}) \leq \kappa V \left(1 + \ln\left(\delta^{-1} \vee 1\right)\right). \tag{6.11}$$

These two different ways of measuring the massiveness of $\mathcal{B}$ lead to the following maximal inequalities for VC-classes due to [92].

Lemma 6.4 *Let $\mathcal{B}$ be some countable VC-class of measurable subsets of Ξ with VC-dimension not larger than $V \geq 1$ and assume that $\sigma > 0$ is such that*

$$P[B] \leq \sigma^2, \text{ for every } B \in \mathcal{B}.$$

Let

$$W_\mathcal{B}^+ = \sup_{B \in \mathcal{B}} \nu_n[B] \text{ and } W_\mathcal{B}^- = \sup_{B \in \mathcal{B}} -\nu_n[B],$$

and $H_\mathcal{B}$ be the *combinatorial entropy* of $\mathcal{B}$ defined by (6.8). Then there exists some absolute constant K such that the following inequalities hold

$$\sqrt{n}\left(\mathbb{E}\left[W_\mathcal{B}^-\right] \vee \mathbb{E}\left[W_\mathcal{B}^+\right]\right) \leq \frac{K}{2}\sigma\sqrt{\mathbb{E}\left[H_\mathcal{B}\right]} \tag{6.12}$$

provided that $\sigma \geq K\sqrt{\mathbb{E}\left[H_\mathcal{B}\right]/n}$ *and*

$$\sqrt{n}\left(\mathbb{E}\left[W_\mathcal{B}^-\right] \vee \mathbb{E}\left[W_\mathcal{B}^+\right]\right) \leq \frac{K}{2}\sigma\sqrt{V\left(1 + \ln\left(\sigma^{-1}\vee 1\right)\right)} \tag{6.13}$$

provided that $\sigma \geq K\sqrt{V\left(1 + |\ln\sigma|\right)/n}$.

Proof. We take some independent copy $\xi' = (\xi_1', \ldots, \xi_n')$ of $\xi = (\xi_1, \ldots, \xi_n)$ and consider the corresponding copy P_n' of P_n. Then, by Jensen's inequality, for any countable class $\mathcal{F}$ of uniformly bounded functions

$$\mathbb{E}\left[\sup_{f\in\mathcal{F}}(P_n - P)(f)\right] \leq \mathbb{E}\left[\sup_{f\in\mathcal{F}}\left(P_n - P_n'\right)(f)\right],$$

so that, given independent random signs $(\varepsilon_1, \ldots, \varepsilon_n)$, independent of ξ, one derives the following symmetrization inequality

$$\mathbb{E}\left[\sup_{f\in\mathcal{F}}(P_n - P)(f)\right] \leq \frac{1}{n}\mathbb{E}\left[\sup_{f\in\mathcal{F}}\sum_{i=1}^{n}\varepsilon_i\left(f(\xi_i) - f(\xi_i')\right)\right]$$

$$\leq \frac{2}{n}\mathbb{E}\left[\sup_{f\in\mathcal{F}}\sum_{i=1}^{n}\varepsilon_i f(\xi_i)\right]. \tag{6.14}$$

Applying this symmetrization inequality to the class $\mathcal{F} = \{\mathbb{1}_B, B \in \mathcal{B}\}$ and the sub-Gaussian inequalities for suprema of Rademacher processes which are stated above, namely (6.3) or (6.5) and setting $\delta_n^2 = [\sup_{B\in\mathcal{B}} P_n(B)] \vee \sigma^2$ we get either

$$\mathbb{E}\left[W_\mathcal{B}^+\right] \leq 2\sqrt{\frac{2}{n}}\mathbb{E}\sqrt{H_\mathcal{B}\delta_n^2} \tag{6.15}$$

or if $H(.,\mathcal{B})$ denotes the *universal entropy* of $\mathcal{B}$ as defined by (6.10)

$$\mathbb{E}\left[W_\mathcal{B}^+\right] \leq \frac{6}{\sqrt{n}}\mathbb{E}\left[\sqrt{\delta_n^2}\sum_{j=0}^{\infty}2^{-j}\sqrt{H\left(2^{-j-1}\delta_n, \mathcal{B}\right)}\right]. \tag{6.16}$$

Then by Cauchy–Schwarz inequality, on the one hand (6.15) becomes

$$\mathbb{E}\left[W_\mathcal{B}^+\right] \leq 2\sqrt{\frac{2}{n}}\sqrt{\mathbb{E}\left[H_\mathcal{B}\right]\mathbb{E}\left[\delta_n^2\right]} \tag{6.17}$$

so that

$$\mathbb{E}\left[W_{\mathcal{B}}^{+}\right] \leq 2\sqrt{\frac{2}{n}}\sqrt{\mathbb{E}\left[H_{\mathcal{B}}\right]\left(\sigma^2 + \mathbb{E}\left[W_{\mathcal{B}}^{+}\right]\right)}$$

and on the other hand since $H\left(.,\mathcal{B}\right)$ is nonincreasing, we derive from (6.16) that

$$\mathbb{E}\left[W_{\mathcal{B}}^{+}\right] \leq \frac{6}{\sqrt{n}}\sqrt{\mathbb{E}\left[\delta_n^2\right]}\sum_{j=0}^{\infty}2^{-j}\sqrt{H\left(2^{-j-1}\sigma,\mathcal{B}\right)}.$$

So that by Haussler's bound (6.11), one derives the following alternative upper bound for $\mathbb{E}\left[W_{\mathcal{B}}^{+}\right]$

$$\frac{6}{\sqrt{n}}\sqrt{\sigma^2 + \mathbb{E}\left[W_{\mathcal{B}}^{+}\right]}\sqrt{\kappa V}\sum_{j=0}^{\infty}2^{-j}\sqrt{(j+1)\ln(2) + \ln\left(\sigma^{-1}\vee 1\right) + 1}.$$

Setting either $D = C^2\mathbb{E}\left[H_{\mathcal{B}}\right]$ or $D = C^2 V\left(1 + \ln\left(\sigma^{-1}\vee 1\right)\right)$, where C is some conveniently chosen absolute constant C ($C = 2$ in the first case and $C = 6\sqrt{\kappa}\left(1+\sqrt{2}\right)$ in the second case), the following inequality holds in both cases

$$\mathbb{E}\left[W_{\mathcal{B}}^{+}\right] \leq \sqrt{\frac{D}{n}}\sqrt{2\left[\sigma^2 + \mathbb{E}\left[W_{\mathcal{B}}^{+}\right]\right]}$$

or equivalently

$$\mathbb{E}\left[W_{\mathcal{B}}^{+}\right] \leq \sqrt{\frac{D}{n}}\left[\sqrt{\frac{D}{n}} + \sqrt{\frac{D}{n} + 2\sigma^2}\right].$$

which whenever $\sigma \geq 2\sqrt{3}\sqrt{D/n}$, implies that

$$\mathbb{E}\left[W_{\mathcal{B}}^{+}\right] \leq \sqrt{3}\sigma\sqrt{\frac{D}{n}}. \tag{6.18}$$

The control of $\mathbb{E}\left[W_{\mathcal{B}}^{-}\right]$ is very similar. This time we apply the symmetrization inequality (6.14) to the class $\mathcal{F} = \{-\mathbb{1}_B, B \in \mathcal{B}\}$ and derive by the same arguments as above that

$$\mathbb{E}\left[W_{\mathcal{B}}^{-}\left(\sigma\right)\right] \leq \sqrt{\frac{D}{n}}\sqrt{2\left[\sigma^2 + \mathbb{E}\left[W_{\mathcal{B}}^{+}\right]\right]}.$$

Hence, provided that $\sigma \geq 2\sqrt{3}\sqrt{D/n}$, (6.18) implies that

$$\mathbb{E}\left[W_{\mathcal{B}}^{+}\right] \leq \frac{\sigma^2}{2}$$

which in turn yields

$$\mathbb{E}\left[W_{\mathcal{B}}^{-}\right] \leq \sqrt{\frac{D}{n}}\sqrt{3\sigma^2},$$

completing the proof of the Lemma. ∎

The case of entropy with bracketing can be treated via some direct chaining argument.

6.1.3 $\mathbb{L}_1$-Entropy with Bracketing

We recall that if f_1 and f_2 are measurable functions such that $f_1 \leq f_2$, the collection of measurable functions f such that $f_1 \leq f \leq f_2$ is denoted by $[f_1, f_2]$ and called bracket with lower extremity f_1 and upper extremity f_2. The $\mathbb{L}_1(P)$-diameter of a bracket $[f_1, f_2]$ is given by $P(f_2) - P(f_1)$. The $\mathbb{L}_1(P)$-entropy with bracketing of $\mathcal{F}$ is defined for every positive δ, as the logarithm of the minimal number of brackets with $\mathbb{L}_1(P)$-diameter not larger than δ which are needed to cover S and is denoted by $H_{[\cdot]}(\delta, \mathcal{F}, P)$.

We can now prove a maximal inequality via some classical chaining argument. Note that the same kind of result is valid for $\mathbb{L}_2(P)$-entropy with bracketing conditions but the chaining argument involves adaptive truncations which are not needed for $\mathbb{L}_1(P)$-entropy with bracketing. Since $\mathbb{L}_1(P)$-entropy with bracketing suffice to our needs for set-indexed processes, for the sake of simplicity we begin first by considering this notion here, postponing to the next section the results involving $\mathbb{L}_2(P)$-entropy with bracketing conditions.

Lemma 6.5 *Let $\mathcal{F}$ be some countable collection of measurable functions such that $0 \leq f \leq 1$ for every $f \in \mathcal{F}$, f_0 be some measurable function such that $0 \leq f_0 \leq 1$, δ be some positive number such that $P(|f - f_0|) \leq \delta^2$ and assume $\delta \to H_{[\cdot]}^{1/2}(\delta^2, \mathcal{F}, P)$ to be integrable at 0. Then, setting*

$$\varphi(\delta) = \int_0^\delta H_{[\cdot]}^{1/2}(u^2, \mathcal{F}, P)\, du,$$

the following inequality is available

$$\sqrt{n}\left(\mathbb{E}\left[\sup_{f \in \mathcal{F}} \nu_n(f_0 - f)\right] \vee \mathbb{E}\left[\sup_{f \in \mathcal{F}} \nu_n(f - f_0)\right]\right) \leq 12\varphi(\delta)$$

provided that $4\varphi(\delta) \leq \delta^2 \sqrt{n}$.

Proof. We perform first the control of $\mathbb{E}\left[\sup_{f \in \mathcal{F}} \nu_n(f_0 - f)\right]$. For any integer j, we set $\delta_j = \delta 2^{-j}$ and $H_j = H_{[\cdot]}(\delta_j^2, \mathcal{F}, P)$. By definition of $H_{[\cdot]}(., \mathcal{F}, P)$, for any integer $j \geq 1$ we can define some mapping Π_j from $\mathcal{F}$ to some finite collection of functions such that

$$\ln \#\Pi_j \mathcal{F} \leq H_j \tag{6.19}$$

and

$$\Pi_j f \leq f \text{ with } P(f - \Pi_j f) \leq \delta_j^2 \text{ for all } f \in \mathcal{F}. \tag{6.20}$$

For $j = 0$, we choose Π_0 to be identically equal to 0. For this choice of Π_0, we still have

$$P(|f - \Pi_0 f|) \leq \delta_0^2 \tag{6.21}$$

for every $f \in \mathcal{F}$. Furthermore, since we may always assume that the extremities of the brackets used to cover $\mathcal{F}$ take their values in $[0,1]$, we also have for every integer j that

$$0 \leq \Pi_j f \leq 1$$

Noticing that since $H_{[\cdot]}\left(\delta_j^2, \mathcal{F}, P\right)$ is nonincreasing,

$$H_1 \leq \delta_1^{-2} \varphi^2\left(\delta\right)$$

and under the condition $4\varphi\left(\delta\right) \leq \delta^2 \sqrt{n}$, one has $H_1 \leq \delta_1^2 n$. Thus, since $j \rightarrow H_j \delta_j^{-2}$ increases to infinity, the set $\left\{j \geq 0 : H_j \leq \delta_j^2 n\right\}$ is a nonvoid interval of the form

$$\left\{j \geq 0 : H_j \leq \delta_j^2 n\right\} = [0, J]$$

with $J \geq 1$. For every $f \in \mathcal{F}$, starting from the decomposition

$$-\nu_n\left(f\right) = \sum_{j=0}^{J-1} \nu_n\left(\Pi_j f\right) - \nu_n\left(\Pi_{j+1} f\right) + \nu_n\left(\Pi_J f\right) - \nu_n\left(f\right)$$

we derive since $\Pi_J\left(f\right) \leq f$ and $P\left(f - \Pi_J\left(f\right)\right) \leq \delta_J^2$ that

$$-\nu_n\left(f\right) \leq \sum_{j=0}^{J-1} \nu_n\left(\Pi_j f\right) - \nu_n\left(\Pi_{j+1} f\right) + \delta_J^2$$

and therefore

$$\mathbb{E}\left[\sup_{f \in \mathcal{F}}\left[-\nu_n\left(f\right)\right]\right] \leq \sum_{j=0}^{J-1} \mathbb{E}\left[\sup_{f \in \mathcal{F}}\left[\nu_n\left(\Pi_j f\right) - \nu_n\left(\Pi_{j+1} f\right)\right]\right] + \delta_J^2. \qquad (6.22)$$

Now, it comes from (6.20) and (6.21) that for every integer j and every $f \in \mathcal{F}$, one has

$$P\left[|\Pi_j f - \Pi_{j+1} f|\right] \leq \delta_j^2 + \delta_{j+1}^2 = 5\delta_{j+1}^2$$

and therefore, since $|\Pi_j f - \Pi_{j+1} f| \leq 1$,

$$P\left[\left|\Pi_j f - \Pi_{j+1} f\right|^2\right] \leq 5\delta_{j+1}^2.$$

Moreover (6.19) ensures that the number of functions of the form $\Pi_j f - \Pi_{j+1} f$ when f varies in $\mathcal{F}$ is not larger than $\exp\left(H_j + H_{j+1}\right) \leq \exp\left(2H_{j+1}\right)$. Hence, we derive from (6.25) that

$$\sqrt{n}\mathbb{E}\left[\sup_{f \in \mathcal{F}}\left[\nu_n\left(\Pi_j f\right) - \nu_n\left(\Pi_{j+1} f\right)\right]\right] \leq 2\left[\delta_{j+1}\sqrt{5H_{j+1}} + \frac{1}{3\sqrt{n}}H_{j+1}\right]$$

and (6.22) becomes

$$\sqrt{n}\mathbb{E}\left[\sup_{f\in\mathcal{F}}\left[-\nu_n\left(f\right)\right]\right] \leq 2\sum_{j=1}^{J}\left[\delta_j\sqrt{5H_j} + \frac{1}{3\sqrt{n}}H_j\right] + 4\sqrt{n}\delta_{J+1}^2. \qquad (6.23)$$

It comes from the definition of J that on the one hand, for every $j \leq J$

$$\frac{1}{3\sqrt{n}}H_j \leq \frac{1}{3}\delta_j\sqrt{H_j}$$

and on the other hand

$$4\sqrt{n}\delta_{J+1}^2 \leq 4\delta_{J+1}\sqrt{H_{J+1}}.$$

Hence, plugging these inequalities in (6.23) yields

$$\sqrt{n}\mathbb{E}\left[\sup_{f\in\mathcal{F}}\left[-\nu_n\left(f\right)\right]\right] \leq 6\sum_{j=1}^{J+1}\left[\delta_j\sqrt{H_j}\right]$$

and the result follows. The control of $\mathbb{E}\left[\sup_{f\in\mathcal{F}}\nu_n\left(f - f_0\right)\right]$ can be performed analogously, changing lower into upper approximations in the dyadic approximation scheme described above. ■

6.2 Function-Indexed Empirical Processes

In order to develop some chaining argument for empirical processes, the following Lemma is absolutely fundamental.

Lemma 6.6 *Let $\mathcal{G}$ be some class of real valued measurable functions on Ξ. Assume that there exists some positive numbers v and c such that for all $g \in \mathcal{G}$ and all integers $k \geq 2$*

$$\sum_{i=1}^{n}\mathbb{E}\left[\left|g\left(\xi_i\right)\right|^k\right] \leq \frac{k!}{2}vc^{k-2}.$$

If $\mathcal{G}$ has finite cardinality N then, for all measurable set A with $\mathbb{P}\left[A\right] > 0$

$$\mathbb{E}^A\left[\sup_{g\in\mathcal{G}}S_n\left(g\right)\right] \leq \sqrt{2v\ln\left(\frac{N}{\mathbb{P}\left[A\right]}\right)} + c\ln\left(\frac{N}{\mathbb{P}\left[A\right]}\right). \qquad (6.24)$$

Proof. We know from the proof of Bernstein's inequality (more precisely from (2.21)) that for any positive λ and all $g \in \mathcal{G}$

$$\ln\mathbb{E}\left[\exp\left(\lambda S_n\left(g\right)\right)\right] \leq \frac{\lambda^2 v}{2\left(1 - c\lambda\right)}.$$

Since

$$\sup_{\lambda > 0} \left(z\lambda - \frac{\lambda^2 v}{2\left(1 - c\lambda\right)} \right) = \frac{v}{c^2} h_1 \left(\frac{cz}{v} \right)$$

where $h_1(u) = 1 + u - \sqrt{1 + 2u}$ for any positive u, we can apply Lemma 2.3 with $h = \frac{v}{c^2} h_1$. Since $h_1^{-1}(u) = u + \sqrt{2u}$ for $u > 0$, the conclusion of Lemma 2.3 leads to Lemma 6.6. ∎

Remark 6.7 *Lemma 6.6 typically applies for a class of uniformly bounded functions. Indeed, if we assume that there exists some positive numbers v and a such that for all $g \in \mathcal{G}$*

$$\sum_{i=1}^{n} \mathbb{E}\left[g^2\left(\xi_i\right) \right] \leq v \text{ and } \|g\|_\infty \leq a,$$

then the assumptions of Lemma 6.6 are fulfilled with $c = a/3$.

An exponential bound under $\mathbb{L}_2$-bracketing assumptions
We are in position to state

Theorem 6.8 *Let $\mathcal{F}$ be some countable class of real valued and measurable functions on Ξ. Assume that there exists some positive numbers σ and b such that for all $f \in \mathcal{F}$ and all integers $k \geq 2$*

$$\frac{1}{n} \sum_{i=1}^{n} \mathbb{E}\left[|f\left(\xi_i\right)|^k \right] \leq \frac{k!}{2} \sigma^2 b^{k-2}.$$

Assume furthermore that for any positive number δ, there exists some finite set $\mathcal{B}_\delta$ of brackets covering $\mathcal{F}$ such that for any bracket $[g_1, g_2] \in \mathcal{B}_\delta$

$$\frac{1}{n} \sum_{i=1}^{n} \mathbb{E}\left[\left(g_2 - g_1\right)^k \left(\xi_i\right) \right] \leq \frac{k!}{2} \delta^2 b^{k-2}$$

all integers $k \geq 2$. Let $e^{H(\delta)}$ denote the minimal cardinality of such a covering. There exists some absolute constant κ such that, for any $\varepsilon \in \,]0, 1]$ and any measurable set A with $\mathbb{P}[A] > 0$, we have

$$\mathbb{E}^A \left[\sup_{f \in \mathcal{F}} S_n(f) \right] \leq E + (1 + 6\varepsilon)\, \sigma \sqrt{2n \ln\left(\frac{1}{\mathbb{P}[A]} \right)} + 2b \ln\left(\frac{1}{\mathbb{P}[A]} \right), \quad (6.25)$$

where

$$E = \frac{\kappa}{\varepsilon} \sqrt{n} \int_0^{\varepsilon\sigma} \sqrt{H(u) \wedge n}\, du + 2\left(b + \sigma\right) H(\sigma)$$

($\kappa = 27$ works).

Applying Lemma 2.4 with $\varphi(x) = E + (1 + 6\varepsilon)\sigma\sqrt{2nx} + bx$ immediately yields the following exponential inequality.

Corollary 6.9 *Under the assumptions of Theorem 6.8 we have that for any $\varepsilon \in \,]0, 1]$ and all positive number x*

$$\mathbb{P}\left[\sup_{f \in \mathcal{F}} S_n(f) \geq E + (1 + 6\varepsilon)\sigma\sqrt{2nx} + 2bx\right] \leq \exp(-x),$$

where for some absolute constant κ

$$E = \frac{\kappa}{\varepsilon}\sqrt{n}\int_0^{\varepsilon\sigma}\sqrt{H(u) \wedge n}\,du + 2(b + \sigma)H(\sigma)$$

($\kappa = 27$ works).

We now turn to the proof of Theorem 6.8.

Proof. We write p for $\mathbb{P}[A]$ for short. For any positive δ, we consider some covering of $\mathcal{F}$ by a set of brackets $\mathcal{B}_\delta$ with cardinality $e^{H(\delta)}$ such that for any bracket $[g_1, g_2] \in \mathcal{B}_\delta$

$$\frac{1}{n}\sum_{i=1}^n\mathbb{E}\left[(g_2 - g_1)^k(\xi_i)\right] \leq \frac{k!}{2}\delta^2 b^{k-2}$$

all integers $k \geq 2$. For any integer j we set $\delta_j = \varepsilon\sigma 2^{-j}$ and consider for any function $f \in \mathcal{F}$, some bracket $[f_j^L, f_j^U] \in \mathcal{B}_{\delta_j}$ containing f. Moreover we define the cumulative function

$$\mathbf{H}(\delta) = \sum_{\delta_j \geq \delta}H(\delta_j),$$

for all $\delta \leq \delta_0$. We then set $\Pi_j f = f_j^U$ and $\Delta_j f = f_j^U - f_j^L$, so that the bracketing assumptions imply that for all $f \in \mathcal{F}$ and all integers j,

$$0 \leq \Pi_j f - f \leq \Delta_j f \tag{6.26}$$

$$\frac{1}{n}\sum_{i=1}^n\mathbb{E}\left[(\Delta_j f)^2(\xi_i)\right] \leq \delta_j^2 \tag{6.27}$$

and for all integers $k \geq 2$

$$\frac{1}{n}\sum_{i=1}^n\mathbb{E}\left[|f(\xi_i)|^k\right] \leq \frac{k!}{2}\sigma^2 b^{k-2}, \tag{6.28}$$

$$\frac{1}{n}\sum_{i=1}^n\mathbb{E}\left[\Delta_0^k f(\xi_i)\right] \leq \frac{k!}{2}\delta_0^2 b^{k-2}. \tag{6.29}$$

We have in view to use an adaptive truncation argument. Towards this aim, we introduce for all integer j

$$a_j = \delta_j \sqrt{\frac{3n}{\mathbf{H}(\delta_{j+1}) - \ln(p)}} \qquad (6.30)$$

and define

$$\tau f = \min\{j \geq 0 : \Delta_j f > a_j\} \wedge J,$$

where J is some given integer to be chosen later. We notice that for any integer τ and all $f \in \mathcal{F}$

$$f = \Pi_0 f + (f - \Pi_\tau f \wedge \Pi_{\tau-1}f) + (\Pi_\tau f \wedge \Pi_{\tau-1}f - \Pi_{\tau-1}f)$$
$$+ \sum_{j=1}^{\tau-1} (\Pi_j f - \Pi_{j-1}f),$$

where $\Pi_{-1} = \Pi_0$ and the summation extended from 1 to $\tau - 1$ is taken to be 0 whenever $\tau = 0$ or $\tau = 1$. We use this decomposition with $\tau = \tau f$. In that case we can furthermore write

$$f - \Pi_{\tau f} f \wedge \Pi_{\tau f-1}f = \sum_{j=0}^{J} (f - \Pi_j f \wedge \Pi_{j-1}f)\, \mathbb{1}_{\tau f=j}$$

and

$$\Pi_{\tau f} f \wedge \Pi_{\tau f-1}f - \Pi_{\tau f-1}f = \sum_{j=1}^{J} (\Pi_j f \wedge \Pi_{j-1}f - \Pi_{j-1}f)\, \mathbb{1}_{\tau f=j},$$

from which we derive that,

$$\mathbb{E}^A \left[\sup_{f \in \mathcal{F}} S_n(f) \right] \leq \mathbb{E}_1 + \mathbb{E}_2 + \mathbb{E}_3 \qquad (6.31)$$

where

$$\mathbb{E}_1 = \mathbb{E}^A \left[\sup_{f \in \mathcal{F}} S_n(\Pi_0 f) \right]$$

$$\mathbb{E}_2 = \sum_{j=0}^{J} \mathbb{E}^A \left[\sup_{f \in \mathcal{F}} S_n((f - \Pi_j f \wedge \Pi_{j-1}f)\, \mathbb{1}_{\tau f=j}) \right]$$

and either $\mathbb{E}_3 = 0$ whenever $J = 0$ or otherwise

$$\mathbb{E}_3 = \sum_{j=1}^{J} \mathbb{E}^A \left[\sup_{f \in \mathcal{F}} S_n(\rho_j f) \right],$$

where for every $j \leq J$

$$\rho_j f = (\Pi_j f \wedge \Pi_{j-1}f - \Pi_{j-1}f)\, \mathbb{1}_{\tau f=j} + (\Pi_j f - \Pi_{j-1}f)\, \mathbb{1}_{\tau f>j}.$$

It remains to control these three quantities.

Control of $\mathbb{E}_1$

We note that Π_0 ranges in a set of functions with cardinality bounded by $\exp(H(\delta_0))$. Moreover, by Minkovski's inequality we have

$$\left(\frac{1}{n}\sum_{i=1}^{n}\mathbb{E}\left[|\Pi_0 f(\xi_i)|^k\right]\right)^{\frac{1}{k}} \leq \left(\frac{1}{n}\sum_{i=1}^{n}\mathbb{E}\left[|f(\xi_i)|^k\right]\right)^{\frac{1}{k}} + \left(\frac{1}{n}\sum_{i=1}^{n}\mathbb{E}\left[\Delta_0^k f(\xi_i)\right]\right)^{\frac{1}{k}}.$$

Hence, it comes from inequalities (6.28) and (6.29) that

$$\left(\frac{1}{n}\sum_{i=1}^{n}\mathbb{E}\left[|\Pi_0 f(\xi_i)|^k\right]\right)^{\frac{1}{k}} \leq \left(\sigma^{2/k} + \delta_0^{2/k}\right)\left(\frac{k!}{2}b^{k-2}\right)^{\frac{1}{k}}$$

from which we derive by using the concavity of the function $x \to x^{2/k}$

$$\frac{1}{n}\sum_{i=1}^{n}\mathbb{E}\left[|\Pi_0 f(\xi_i)|^k\right] \leq \frac{k!}{2}(\sigma + \delta_0)^2 (2b)^{k-2}.$$

Therefore we get by applying Lemma 6.6

$$\mathbb{E}_1 \leq (1+\varepsilon)\left(\sigma\sqrt{2n(H(\delta_0) - \ln(p))} + 2b(H(\delta_0) - \ln(p))\right)$$

$$\leq 2\sqrt{2}\sigma\sqrt{n}\sqrt{H(\delta_0)} + 2bH(\delta_0) + (1+\varepsilon)\sigma\sqrt{2n\ln\left(\frac{1}{p}\right)}$$

$$+ 2b\ln\left(\frac{1}{p}\right).$$

Control of $\mathbb{E}_2$

By (6.26) we have $0 \leq \Pi_j f \wedge \Pi_{j-1} f - f \leq \Delta_j f$ for all $j \in \mathbb{N}$ and $f \in \mathcal{F}$, which implies that

$$\mathbb{E}_2 \leq \sum_{j=0}^{J}\sup_{f\in\mathcal{F}}\sum_{i=1}^{n}\mathbb{E}\left[(\Delta_j f \mathbb{1}_{\tau f=j})(\xi_i)\right].$$

For all integer $j < J$, $\tau f = j$ implies that $\Delta_j f > a_j$, hence by inequality (6.27)

$$\sum_{i=1}^{n}\mathbb{E}\left[(\Delta_j f \mathbb{1}_{\tau f=j})(\xi_i)\right] \leq \sum_{i=1}^{n}\mathbb{E}\left[(\Delta_j f \mathbb{1}_{\Delta_j f>a_j})(\xi_i)\right] \leq n\frac{\delta_j^2}{a_j}.$$

Moreover by Cauchy–Schwarz inequality and inequality (6.27) again we have

$$\sum_{i=1}^{n}\mathbb{E}\left[(\Delta_J f \mathbb{1}_{\tau f=J})(\xi_i)\right] \leq \sum_{i=1}^{n}\mathbb{E}\left[(\Delta_J f)(\xi_i)\right] \leq n\delta_J.$$

Collecting these bounds and using the definition of the truncation levels given by (6.30) we get if $J > 0$

$$\mathbb{E}_2 \leq \sqrt{n} \sum_{j=0}^{J-1} \delta_j \sqrt{\mathbf{H}(\delta_{j+1})} + \sqrt{n} \sqrt{\ln\left(\frac{1}{p}\right)} \sum_{j=0}^{J-1} \delta_j + n\delta_J$$

$$\leq \frac{2}{\sqrt{3}} \sqrt{n} \sum_{j=1}^{J} \delta_j \sqrt{\mathbf{H}(\delta_j)} + n\delta_J + \sqrt{\frac{2}{3}} \varepsilon\sigma \sqrt{2n \ln\left(\frac{1}{p}\right)},$$

or

$$\mathbb{E}_2 \leq n\delta_0$$

whenever $J = 0$.

Control of $\mathbb{E}_3$

We may assume that $J > 0$ otherwise $\mathbb{E}_3 = 0$.

- We note that for all $j < J$

$$(\tau f > j) = (\Delta_0 f \leq a_0, \ldots, \Delta_j f \leq a_j)$$

and

$$(\tau f = J) = (\Delta_0 f \leq a_0, \ldots, \Delta_J f \leq a_J).$$

Hence for any integer j in $[1, J]$, the cardinality of the set of functions

$$\rho_j f = (\Pi_j f \wedge \Pi_{j-1} f - \Pi_{j-1} f)\, \mathbb{1}_{\tau f = j} + (\Pi_j f - \Pi_{j-1} f)\, \mathbb{1}_{\tau f > j}$$

when f varies in $\mathcal{F}$, is bounded by $\exp\left(\mathbf{H}(\delta_j)\right)$. Given j in $[1, J]$, our aim is now to bound $|\rho_j f|$. We consider two situations.

- If $\Pi_j f \leq \Pi_{j-1} f$, then $\rho_j f = (\Pi_j f - \Pi_{j-1} f)\, \mathbb{1}_{\tau f \geq j}$ and by (6.26) it implies that

$$0 \leq -\rho_j f \leq \Delta_{j-1} f \mathbb{1}_{\tau f \geq j}.$$

- If $\Pi_j f > \Pi_{j-1} f$, then $\rho_j f = (\Pi_j f - \Pi_{j-1} f)\, \mathbb{1}_{\tau f > j}$ and therefore we get via (6.26) again that

$$0 \leq \rho_j f \leq \Delta_j f \mathbb{1}_{\tau f > j}.$$

It follows that in any case

$$|\rho_j f| \leq \max\left(\Delta_{j-1} f \mathbb{1}_{\tau f \geq j}, \Delta_j f \mathbb{1}_{\tau f > j}\right),$$

from which we deduce that, on the one hand $\|\rho_j f\|_\infty \leq \max\left(a_{j-1}, a_j\right) \leq a_{j-1}$ and on the other hand by (6.27) and Minkovski's inequality

$$\left(\frac{1}{n} \sum_{i=1}^{n} \mathbb{E}\left[(\rho_j f)^2 (\xi_i)\right]\right)^{\frac{1}{2}} \leq \delta_j + \delta_{j-1} \leq 3\delta_j.$$

We can now apply Lemma 6.6 together with Remark 6.7 and get

$$\mathbb{E}_3 \le \sum_{j=1}^{J} 3\delta_j \sqrt{2n\left(\mathbf{H}\left(\delta_j\right) - \ln\left(p\right)\right)} + \frac{a_{j-1}}{3}\left(\mathbf{H}\left(\delta_j\right) - \ln\left(p\right)\right).$$

Using the definition of the levels of truncation (6.30) this becomes

$$\mathbb{E}_3 \le \sqrt{n}\sum_{j=1}^{J}\left(3\sqrt{2}\delta_j + \frac{1}{\sqrt{3}}\delta_{j-1}\right)\sqrt{\left(\mathbf{H}\left(\delta_j\right) - \ln\left(p\right)\right)}$$

$$\le \sqrt{n}\sum_{j=1}^{J}\left(3\sqrt{2} + \frac{2}{\sqrt{3}}\right)\delta_j\left(\sqrt{\mathbf{H}\left(\delta_j\right)} + \sqrt{\ln\frac{1}{p}}\right)$$

and therefore

$$\mathbb{E}_3 \le \sqrt{n}\left(3\sqrt{2} + \frac{2}{\sqrt{3}}\right)\left(\sum_{j=1}^{J}\delta_j\sqrt{\mathbf{H}\left(\delta_j\right)}\right)$$

$$+ \left(3 + \sqrt{\frac{2}{3}}\right)\varepsilon\sigma\sqrt{2n\ln\left(\frac{1}{p}\right)}.$$

End of the proof

It remains to collect the inequalities above and to choose J properly. If $J > 0$ we derive from (6.31) that

$$\mathbb{E}^A\left[\sup_{f\in\mathcal{F}} S_n\left(f\right)\right] \le \frac{\sqrt{n}}{\varepsilon}\left(3\sqrt{2} + \frac{4}{\sqrt{3}}\right)\left(\sum_{j=0}^{J}\delta_j\sqrt{\mathbf{H}\left(\delta_j\right)}\right) + 2bH\left(\delta_0\right)$$

$$+ n\delta_J + \left(1 + \left(4 + 2\sqrt{\frac{2}{3}}\right)\varepsilon\right)\sigma\sqrt{2n\ln\left(\frac{1}{p}\right)}$$

$$+ 2b\ln\left(\frac{1}{p}\right). \tag{6.32}$$

Now

$$\sum_{j=0}^{J}\delta_j\sqrt{\mathbf{H}\left(\delta_j\right)} \le \sum_{j=0}^{J}\delta_j\left(\sum_{k=0}^{j}\sqrt{H\left(\delta_k\right)}\right) \le \sum_{k=0}^{J}\sqrt{H\left(\delta_k\right)}\left(\sum_{j=k}^{J}\delta_j\right)$$

$$\le 2\sum_{k=0}^{J}\delta_k\sqrt{H\left(\delta_k\right)}$$

and therefore using the monotonicity of the function H we get

$$\sum_{j=0}^{J} \delta_j \sqrt{\mathbf{H}(\delta_j)} \le 4 \int_{\delta_{J+1}}^{\delta_0} \sqrt{H(x) \wedge H(\delta_J)} dx. \tag{6.33}$$

Let us consider the set $\mathcal{J} = \{j \in \mathbb{N} : H(\delta_j) \le n\}$. If it is not bounded this means that $H(x) \le n$ for all positive x. We can therefore derive inequality (6.25) and derive from (6.32) and (6.33) by letting J goes to infinity. We can now assume the set $\mathcal{J}$ to be bounded. Then either $\mathcal{J}$ is empty and we set $J = 0$ or it is not empty and we define J to be the largest element of $\mathcal{J}$. The point is that according to this definition of J we have $H(\delta_{J+1}) > n$ and therefore

$$\sqrt{n} \int_0^{\delta_{J+1}} \sqrt{H(x) \wedge n} dx \ge n\delta_{J+1} \ge \frac{n\delta_J}{2}. \tag{6.34}$$

It remains to consider two cases.

- If $J = 0$, then

$$\mathbb{E}_1 + \mathbb{E}_2 + \mathbb{E}_3 \le 2\sigma\sqrt{n}\sqrt{2H(\delta_0)} + 2bH(\delta_0) + n\delta_0$$
$$+ (1 + \varepsilon)\sigma\sqrt{2n\ln\left(\frac{1}{p}\right)} + 2b\ln\left(\frac{1}{p}\right)$$

and since $2\sqrt{n}\sqrt{2H(\delta_0)} \le n + 2H(\delta_0)$,

$$\mathbb{E}_1 + \mathbb{E}_2 + \mathbb{E}_3 \le 2(b + \sigma)H(\delta_0) + n(\delta_0 + \sigma)$$
$$+ (1 + \varepsilon)\sigma\sqrt{2n\ln\left(\frac{1}{p}\right)} + 2b\ln\left(\frac{1}{p}\right).$$

We derive from inequality (6.34) that

$$2\sqrt{n} \int_0^{\delta_0} \sqrt{H(x) \wedge n} dx \ge n\delta_0$$

and we can easily conclude that inequality (6.25) holds via (6.31).
- If $J > 0$, then it comes from the definition of J that $H(\delta_J) \le n$. Hence we derive from (6.33) that

$$\sum_{j=0}^{J} \delta_j \sqrt{\mathbf{H}(\delta_j)} \le 4 \int_{\delta_{J+1}}^{\delta_0} \sqrt{H(x) \wedge n} dx$$

which combined to (6.34) yields inequality (6.25) thanks to (6.32). This achieves the proof of Theorem 6.8. ∎

Density Estimation via Model Selection

7.1 Introduction and Notations

One of the most studied problem in nonparametric statistics is density estimation. Suppose that one observes n independent random variables $X_1, \ldots, X_n$ valued in some measurable space $(\mathbb{X}, \mathcal{X})$ with distribution P and assume that P has an unknown density with respect to some given and known positive measure μ. Our purpose is to estimate this density with as few prior information as possible (especially on the smoothness of the density). We wish to generalize the approach developed in the Gaussian case. The first problem that we have to face with is that in the Gaussian case the least squares criterion was exactly equivalent to the maximum likelihood criterion. In the density estimation framework, this is no longer true. This is the reason why we shall indeed study two different model selection penalized criteria, either penalized least squares or penalized log-likelihood. For least squares we shall translate the Gaussian model selection theorem for linear models in the density estimation context, while for maximum likelihood, we shall do the same for histograms and follow an other path (more general but less precise) to deal with arbitrary models. The main sources of inspiration for this chapter are [20], [34] and [12]. As it is well known Kullback–Leibler information and Hellinger distance between two probability measures P and Q can be computed from the densities of P and Q with respect to μ and these calculations do not depend on the dominating measure μ. All along this chapter we shall therefore abusively use the following notations. Setting $f = dP/d\mu$ and $g = dQ/d\mu$ we shall write $\mathbf{K}(f, g)$ or $\mathbf{K}(P, Q)$ indifferently for the Kullback–Leibler information

$$\int f \ln\left(\frac{f}{g}\right) d\mu, \text{ whenever } P \ll Q$$

and $\mathbf{h}^2(f, g)$ or $\mathbf{h}^2(P, Q)$ indifferently for the squared Hellinger distance

$$\frac{1}{2} \int \left(\sqrt{f} - \sqrt{g}\right)^2 d\mu.$$

More generally when a quantity of the form

$$\int \psi(f, g) \, d\mu$$

does not depend on the dominating measure μ, we shall note it

$$\int \psi(dP, dQ).$$

From a technical point of view, the major difference with respect to the Gaussian case is that we shall face to boundedness. This will be true both for least squares and maximum likelihood. This is not surprising for least squares since using this loss function for estimating a density structurally leads to integrability problems. This is also true for maximum likelihood. One reason is that the Kullback–Leibler loss which is naturally linked to this procedure behaves nicely when it is conveniently connected to the square Hellinger loss and this holds true only under appropriate boundedness assumptions. Another reason is that the log-likelihood itself appears as the sum of not necessarily bounded random variables.

7.2 Penalized Least Squares Model Selection

We write this unknown density as $f = s_0 + s$, where s_0 is a given and known bounded function (typically we shall take $s_0 \equiv 0$ or $s_0 \equiv 1$) and $\int s \, s_0 \, d\mu = 0$. Our purpose is to estimate s by using as few prior information on s as possible. Moreover all along Section 7.2, the measure μ will be assumed to be a *probability* measure. The approach that we wish to develop has a strong analogy with the one used in Chapter 4 in the Gaussian framework. Basically, we consider some countable collection of models $\{S_m\}_{m \in \mathcal{M}}$, some least squares criterion $\gamma_n : \mathbb{L}_2(\mu) \to \mathbb{R}$ and some penalty function pen $: \mathcal{M} \to \mathbb{R}_+$. Note that pen possibly depends on the observations $X_1, \ldots, X_n$ but not on s. We then consider for every $m \in \mathcal{M}$, the least squares estimator (LSE) within model S_m (if it exists!)

$$\widehat{s}_m = \mathrm{argmin}_{t \in S_m} \gamma_n(t)$$

and define the model selection procedure

$$\widehat{m} = \mathrm{argmin}_{m \in \mathcal{M}} (\gamma_n(\widehat{s}_m) + \mathrm{pen}(m)).$$

We finally estimate s by the *penalized LSE*

$$\widetilde{s} = \widehat{s}_{\widehat{m}}.$$

Before defining the empirical contrast that we intend to study, let us fix some notations that we shall use all along this chapter.

Definition 7.1 *We denote by* $\langle \cdot, \cdot \rangle$, *respectively by* $\|\cdot\|$, *the scalar product, respectively the norm in* $\mathbb{L}_2(\mu)$. *We consider the empirical probability measure* P_n *associated with the sample* $X_1, \ldots, X_n$ *and the centered empirical measure* $\nu_n = P_n - P$. *For any* P-*integrable function* t *on* $(\mathbb{X}, \mathcal{X})$, *one therefore has*

$$\nu_n(t) = P_n(t) - P(t) = \frac{1}{n} \sum_{i=1}^{n} t(X_i) - \int_{\mathbb{X}} t(x) \, dP(x).$$

The adequate way of defining the empirical criterion γ_n as a least squares criterion for density estimation is to set for every $t \in \mathbb{L}_2(\mu)$

$$\gamma_n(t) = \|t\|^2 - 2P_n(t).$$

Similarly to the Gaussian least squares studied in Chapter 4, we see that whenever S_m is a linear finite dimensional model orthogonal to s_0 in $\mathbb{L}_2(\mu)$, the corresponding LSE $\widehat{s}_m$ does exist and is merely the usual *empirical projection* estimator on S_m. Indeed if we consider S_m to be a linear space of dimension D_m with orthonormal basis (relatively to the scalar product in $\mathbb{L}_2(\mu)$) $\{\varphi_{\lambda,m}\}_{\lambda \in \Lambda_m}$, one can write $\widehat{s}_m$ as

$$\widehat{s}_m = \sum_{\lambda \in \Lambda_m} \widehat{\beta}_{\lambda,m} \varphi_{\lambda,m} \quad \text{with} \quad \widehat{\beta}_{\lambda,m} = P_n(\varphi_{\lambda,m}) \quad \text{for all } \lambda \in \Lambda_m. \tag{7.1}$$

Once a penalty function is given, we shall call penalized LSE the corresponding penalized estimator. More precisely we state.

Definition 7.2 *Let* $X_1, \ldots, X_n$ *be a sample of the distribution* $P = (s_0 + s)\mu$ *where* s_0 *is a given and known bounded function and* $\int s \, s_0 \, d\mu = 0$. *Denoting by* P_n *the empirical probability measure associated with* $X_1, \ldots, X_n$, *we define the least squares criterion on* $\mathbb{L}_2(\mu)$ *as*

$$\gamma_n(t) = \|t\|^2 - 2P_n(t) \quad \text{for all } t \in \mathbb{L}_2(\mu).$$

Given some countable collection of linear finite dimensional models $\{S_m\}_{m \in \mathcal{M}}$, *each model being orthogonal to* s_0, *and some penalty function* pen $: \mathcal{M} \to \mathbb{R}_+$, *the penalized LSE* $\widetilde{s}$ *associated with the collection of models* $\{S_m\}_{m \in \mathcal{M}}$ *and the penalty function* pen *is defined by*

$$\widetilde{s} = \widehat{s}_{\widehat{m}},$$

where, for all $m \in \mathcal{M}$, $\widehat{s}_m = \text{argmin}_{t \in S_m} \gamma_n(t)$ *and* $\widehat{m}$ *minimizes the penalized least squares criterion*

$$\gamma_n(\widehat{s}_m) + \text{pen}(m).$$

Our purpose is to study the properties of the penalized LSE. In particular we intend to show that adequate choices of the penalty function allow to interpret some well known adaptive estimators as penalized LSE. The classical estimation procedures that we have in mind are cross-validation or hard thresholding of the empirical coefficients related to some orthonormal basis. The main issue will be to establish risk bounds and discuss various strategies for choosing the penalty function or the collection of models. We will also show that these risk bounds imply that the penalized LSE are adaptive in the minimax sense on various collections of parameter spaces which directly depend on the collection of models $\{S_m\}_{m \in \mathcal{M}}$. It is worth recalling that the least squares estimation criterion is no longer equivalent (unlike in the Gaussian framework) to a maximum likelihood criterion. The next section will be devoted to a specific study of the penalized MLEs which can also be defined in analogy to the penalized LSE in the Gaussian framework.

7.2.1 The Nature of Penalized LSE

We focus on penalized LSE defined from a collection of finite dimensional linear models. Note that if S_m is a linear finite dimensional model, then one derives from 7.1 that

$$\gamma_n(\widehat{s}_m) = -\sum_{\lambda \in \Lambda_m} \widehat{\beta}_{\lambda,m}^2 = -\|\widehat{s}_m\|^2$$

so that, given some penalty function pen : $\mathcal{M} \to \mathbb{R}_+$, our model selection criterion can be written as

$$\widehat{m} = \mathrm{argmin}_{m \in \mathcal{M}} \left(-\|\widehat{s}_m\|^2 + \mathrm{pen}(m) \right). \tag{7.2}$$

We see that this expression is very simple. It allows to interpret some well known adaptive estimators as penalized LSE.

Connection with Other Adaptive Density Estimation Procedures

Several adaptive density estimators are indeed defined from data-driven selection procedures within a given collection of LSEs. Some of these procedures can be interpreted as model selection via penalized least squares procedures. This is what we want to show for two important examples: unbiased cross-validation and hard thresholding.

Cross-Validation

Unbiased cross-validation has been introduced by Rudemo (see [104]) for histogram or kernel estimation. In the context of least squares estimation, it provides a data-driven method for selecting the order of an expansion. To put it in a model selection language, let us consider some nested collection of finite dimensional linear models $\{S_m\}_{m \in \mathcal{M}}$, which means that the mapping $m \to D_m$ is one to one and that $D_m < D_{m'}$ implies that $S_m \subset S_{m'}$. The unbiased cross-validation method is based upon the following heuristics. An ideal model should minimize $\|s - \widehat{s}_m\|^2$ or equivalently $\|\widehat{s}_m\|^2 - 2\int \widehat{s}_m s d\mu$

with respect to $m \in \mathcal{M}$. Since this quantity involves the unknown s, it has to be estimated and the unbiased cross-validation method defines $\widehat{m}$ as the minimizer with respect to $m \in \mathcal{M}$ of

$$\|\widehat{s}_m\|^2 - \frac{2}{n(n-1)} \sum_{i \neq i'} \sum_{\lambda \in \Lambda_m} \varphi_{\lambda,m}(X_i)\,\varphi_{\lambda,m}(X_{i'}),$$

where $\{\varphi_{\lambda,m}\}_{\lambda \in \Lambda_m}$ is an orthonormal basis of S_m. The cross-validated estimator of s is then defined as $\widehat{s}_{\widehat{m}}$. Since

$$\|\widehat{s}_m\|^2 = \frac{1}{n^2} \sum_{i,i'} \sum_{\lambda \in \Lambda_m} \varphi_{\lambda,m}(X_i)\,\varphi_{\lambda,m}(X_{i'})$$

one finds $\widehat{m}$ as the minimizer of

$$-\frac{n+1}{n-1} \|\widehat{s}_m\|^2 + \frac{2}{(n-1)} P_n \left(\sum_{\lambda \in \Lambda_m} \varphi_{\lambda,m}^2 \right)$$

or equivalently of

$$-\|\widehat{s}_m\|^2 + \frac{2}{(n+1)} P_n \left(\sum_{\lambda \in \Lambda_m} \varphi_{\lambda,m}^2 \right).$$

If we introduce the function

$$\Phi_m = \sup \left\{ t^2 : t \in S_m,\ \|t\|_2 = 1 \right\}$$

the quantity $\sum_{\lambda \in \Lambda_m} \varphi_{\lambda,m}^2$ is easily seen to be exactly equal to Φ_m. Therefore it depends only on S_m and not on a particular choice of the orthonormal basis $\{\varphi_{\lambda,m}\}_{\lambda \in \Lambda_m}$. Now

$$\widehat{m} = \operatorname{argmin}_{m \in \mathcal{M}} \left(-\|\widehat{s}_m\|^2 + \frac{2}{(n+1)} P_n(\Phi_m) \right),$$

and it follows from 7.2 that the cross-validated estimator of s is a penalized LSE with penalty function

$$\operatorname{pen}(m) = \frac{2}{(n+1)} P_n(\Phi_m). \tag{7.3}$$

Hard Thresholding Estimators

Let $\{\varphi_\lambda\}_{\lambda \in \Lambda_n}$ be a finite orthonormal system in $\mathbb{L}_2(\mu)$ with $|\Lambda_n| = N_n$. Denoting by $\widehat{\beta}_\lambda$ the empirical coefficient $P_n(\varphi_\lambda)$ for $\lambda \in \Lambda_n$, the hard thresholding estimator introduced in the density estimation context for wavelet basis by Donoho, Johnstone, Kerkyacharian and Picard (see [54]) is defined as

$$\sum_{\lambda \in \Lambda_0} \widehat{\beta}_\lambda \varphi_\lambda + \sum_{\lambda \notin \Lambda_0} \widehat{\beta}_\lambda \mathbb{1}_{\left\{ \widehat{\beta}_\lambda^2 > \mathcal{L}(n)/n \right\}} \varphi_\lambda, \tag{7.4}$$

where Λ_0 is a given finite subset of Λ_n (with cardinality not depending on n) and $\mathcal{L}(n)$ is an adequate (possibly random) function of n which is typically taken as some constant times $\ln n$ in [54]. To put it in a model selection framework, we simply take $\mathcal{M}$ as the collection of all subsets m of Λ_n such that $\Lambda_0 \subseteq m$. Taking S_m as the linear span of $\{\varphi_\lambda\}_{\lambda \in m}$ and defining the penalty function as $\mathrm{pen}(m) = \mathcal{L}(n)\,|m|\,/n$, we derive from 7.2 that

$$\gamma_n(\widehat{s}_m) + \mathrm{pen}(m) = -\sum_{\lambda \in m} \widehat{\beta}_\lambda^2 + \frac{\mathcal{L}(n)\,|m|}{n}$$

$$= -\sum_{\lambda \in \Lambda_0} \widehat{\beta}_\lambda^2 + \frac{\mathcal{L}(n)\,|\Lambda_0|}{n} - \sum_{\lambda \in m/\Lambda_0} \left(\widehat{\beta}_\lambda^2 - \frac{\mathcal{L}(n)}{n}\right).$$

From the latter identity it is clear that a minimizer $\widehat{m}$ of $\gamma_n(\widehat{s}_m) + \mathrm{pen}(m)$ over $\mathcal{M}$ is given by

$$\widehat{m} = \Lambda_0 \cup \left\{\lambda \in \Lambda_n/\Lambda_0 : \widehat{\beta}_\lambda^2 - \frac{\mathcal{L}(n)}{n} > 0\right\}$$

which means that the penalized LSE $\widehat{s}_{\widehat{m}}$ is exactly equal to the hard thresholding estimator defined by 7.4.

Risk Bounds and Choice of the Penalty Function?

Given some countable collection of linear finite dimensional models $\{S_m\}_{m \in \mathcal{M}}$, each of them being orthogonal to s_0, and some penalty function $\mathrm{pen} : \mathcal{M} \to \mathbb{R}_+$, the study of the risk of the corresponding penalized LSE $\widetilde{s}$ can be analyzed thanks to a fundamental though elementary inequality. Indeed, by definition of $\widetilde{s}$ we have whatever $m \in \mathcal{M}$ and $s_m \in S_m$

$$\gamma_n(\widetilde{s}) + \mathrm{pen}(\widehat{m}) \leq \gamma_n(\widehat{s}_m) + \mathrm{pen}(m) \leq \gamma_n(s_m) + \mathrm{pen}(m).$$

Assuming that $s \in \mathbb{L}_2(\mu)$, we note that for all $t \in \mathbb{L}_2(\mu)$ which is orthogonal to s_0

$$\gamma_n(t) = \|t\|^2 - 2\langle s, t\rangle - 2\nu_n(t) = \|s - t\|^2 - \|s\|^2 - 2\nu_n(t).$$

This leads to the following interesting control on the distance between s and $\widetilde{s}$

$$\|s - \widetilde{s}\|^2 \leq \|s - s_m\|^2 + 2\nu_n(\widetilde{s} - s_m) - \mathrm{pen}(\widehat{m}) + \mathrm{pen}(m), \qquad (7.5)$$

for all $m \in \mathcal{M}$ and $s_m \in S_m$. Our approach consists in, starting from inequality 7.5, to choose a penalty function in a way that it contributes to dominate the fluctuation of the variable $\nu_n(\widetilde{s} - s_m)$. The trouble is that this variable is not that easy to control. This is the reason why we need to use empirical processes techniques. More precisely, we first write

$$\nu_n(\widetilde{s} - s_m) \leq \|\widetilde{s} - s_m\| \sup_{t \in S_m + S_{\widehat{m}}} \frac{\nu_n(t - s_m)}{\|t - s_m\|},$$

so that, setting

$$\chi_n\left(m, \widehat{m}\right) = \sup_{t \in S_m + S_{\widehat{m}}} \frac{\nu_n\left(t\right)}{\|t\|}$$

we get

$$\|s - \widetilde{s}\|^2 \le \|s - s_m\|^2 + 2\|\widetilde{s} - s_m\| \chi_n\left(m, \widehat{m}\right) - \mathrm{pen}\left(\widehat{m}\right) + \mathrm{pen}\left(m\right).$$

Using twice the inequality

$$2ab \le \theta^{-1}a^2 + \theta b^2 \qquad (7.6)$$

for adequate values of the positive number θ (successively $\theta = \varepsilon/2$ and $\theta = 1 + (\varepsilon/2)$), we derive that for any arbitrary positive number ε

$$2\left(a + a'\right)b \le \frac{2}{\varepsilon}a^2 + \frac{2}{2+\varepsilon}a'^2 + (1 + \varepsilon)b^2.$$

Hence, the triangle inequality $\|\widetilde{s} - s_m\| \le \|\widetilde{s} - s\| + \|s - s_m\|$ yields

$$2\|\widetilde{s} - s_m\| \chi_n\left(m, \widehat{m}\right) \le \frac{2}{\varepsilon}\|s - s_m\|^2 + \frac{2}{2+\varepsilon}\|\widetilde{s} - s\|^2$$
$$+ (1 + \varepsilon)\chi_n^2\left(m, \widehat{m}\right)$$

and therefore

$$\frac{\varepsilon}{2 + \varepsilon}\|\widetilde{s} - s\|^2 \le \left(1 + \frac{2}{\varepsilon}\right)\|s - s_m\|^2 + (1 + \varepsilon)\chi_n^2\left(m, \widehat{m}\right)$$
$$- \mathrm{pen}\left(\widehat{m}\right) + \mathrm{pen}\left(m\right). \qquad (7.7)$$

The model selection problem that we want to consider is subset selection within a given basis. In fact if we consider a given finite orthonormal family of functions $\{\varphi_\lambda\}_{\lambda \in \Lambda_n}$, we can define for any subset m of Λ_n, S_m to be the linear span of $\{\varphi_\lambda\}_{\lambda \in m}$. In such a case $\mathcal{M}$ is taken as a collection of subsets of Λ_n and we note by Cauchy–Schwarz inequality that

$$\chi_n\left(m, \widehat{m}\right) = \sup_{a \in \mathbb{R}^{m \cup \widehat{m}}, a \ne 0} \frac{\sum_{\lambda \in m \cup \widehat{m}} a_\lambda \nu_n\left(\varphi_\lambda\right)}{\left[\sum_{\lambda \in m \cup \widehat{m}} a_\lambda^2\right]^{1/2}} \le \left[\sum_{\lambda \in m \cup \widehat{m}} \nu_n^2\left(\varphi_\lambda\right)\right]^{1/2},$$

which yields

$$\frac{\varepsilon}{2 + \varepsilon}\|\widetilde{s} - s\|^2 \le \left(1 + \frac{2}{\varepsilon}\right)\|s - s_m\|^2 + (1 + \varepsilon)\sum_{\lambda \in m \cup \widehat{m}} \nu_n^2\left(\varphi_\lambda\right)$$
$$- \mathrm{pen}\left(\widehat{m}\right) + \mathrm{pen}\left(m\right). \qquad (7.8)$$

We see that the main issue is to choose a penalty function which is large enough to essentially annihilate the chi-square type statistics

$$\sum_{\lambda \in m \cup \widehat{m}} \nu_n^2\left(\varphi_\lambda\right).$$

Achieving this program leads to a risk bound for $\tilde{s}$ as expected but requires to control the chi-square type statistics $\sum_{\lambda \in m \cup m'} \nu_n^2 (\varphi_\lambda)$ for all values of m' in $\mathcal{M}$ simultaneously. The probabilistic tools needed for this are presented in the next section. From inequality 7.8 we see that the penalty function pen (m') should be chosen at least as big as the expectation of $(1+\varepsilon) \sum_{\lambda \in m'} \nu_n^2 (\varphi_\lambda)$ which is equal to $(1+\varepsilon) \sum_{\lambda \in m'} \mathrm{Var}(\varphi_\lambda) /n$. The dependency with respect to the unknown underlying density s makes a difference with what we dealt with in the Gaussian framework. This explains why in our forthcoming model selection theorems, we propose several structures for the penalty function. We either take deterministic penalty functions which dominate this expectation term or data-driven penalty functions which overestimate it. But again, as in the Gaussian case, we need exponential bounds.

Exponential Bounds for Chi-Square Type Statistics

More precisely it is our purpose now to derive exponential bounds for chi-square type statistics from Bousquet's version of Talagrand's inequality. A first possibility is to use a straightforward way.

Proposition 7.3 *Let $X_1, \ldots, X_n$ be independent and identically distributed random variables valued in some measurable space $(\mathbb{X}, \mathcal{X})$. Let P denote their common distribution and ν_n be the corresponding centered empirical measure (see Definition 7.1). Let $\{\varphi_\lambda\}_{\lambda \in \Lambda}$ be a finite family of measurable and bounded functions on $(\mathbb{X}, \mathcal{X})$. Let*

$$\Phi_\Lambda = \sum_{\lambda \in \Lambda} \varphi_\lambda^2 \quad and \quad V_\Lambda = \mathbb{E} \left[\Phi_\Lambda (X_1) \right].$$

Moreover, let $\mathcal{S}_\Lambda = \left\{ a \in \mathbb{R}^\Lambda : \sum_{\lambda \in \Lambda} a_\lambda^2 = 1 \right\}$ and

$$M_\Lambda = \sup_{a \in \mathcal{S}_\Lambda} \left[\mathrm{Var} \left(\sum_{\lambda \in \Lambda} a_\lambda \varphi_\lambda (X_1) \right) \right].$$

Then, the following inequality holds for all positive x and ε

$$\mathbb{P} \left[\left(\sum_{\lambda \in \Lambda} \nu_n^2 (\varphi_\lambda) \right)^{1/2} \geq (1+\varepsilon) \sqrt{\frac{V_\Lambda}{n}} + \sqrt{\frac{2x M_\Lambda}{n}} + \kappa (\varepsilon) \frac{\sqrt{\|\Phi_\Lambda\|_\infty}}{n} x \right]$$

$$\leq \exp (-x), \tag{7.9}$$

where $\kappa (\varepsilon) = 2 \left(\varepsilon^{-1} + 1/3 \right)$

Proof. We know that

$$\chi_n (\Lambda) = \left[\sum_{\lambda \in \Lambda} \nu_n^2 (\varphi_\lambda) \right]^{1/2} = \sup_{a \in \mathcal{S}_\Lambda} \left| \nu_n \left[\sum_{\lambda \in \Lambda} a_\lambda \varphi_\lambda \right] \right|.$$

Hence, if $\mathcal{S}'_\Lambda$ is a dense subset of $\mathcal{S}_\Lambda$ we also have

$$\chi_n\left(\Lambda\right) = \sup_{a\in\mathcal{S}'_\Lambda}\left|\nu_n\left[\sum_{\lambda\in\Lambda}a_\lambda\varphi_\lambda\right]\right|.$$

We can apply inequality (5.50) to the countable set of functions

$$\left\{\sum_{\lambda\in\Lambda}a_\lambda\varphi_\lambda, a\in\mathcal{S}'_\Lambda\right\}$$

and get by (7.11) and by Cauchy–Schwarz inequality since for every $a\in\mathcal{S}_\Lambda$, $\left\|\sum_{\lambda\in\Lambda}a_\lambda\varphi_\lambda\right\|_\infty \leq \sqrt{\|\Phi_\Lambda\|_\infty}$

$$\mathbb{P}\left[\chi_n\left(\Lambda\right) \geq (1+\varepsilon)\,\mathbb{E}\left[\chi_n\left(\Lambda\right)\right] + \sqrt{\frac{2M_\Lambda x}{n}} + \kappa\left(\varepsilon\right)\frac{\sqrt{\|\Phi_\Lambda\|_\infty}}{n}x\right]$$
$$\leq \exp\left(-x\right),$$

for all positive x. Now

$$\mathbb{E}\left[\chi_n^2\left(\Lambda\right)\right] = \frac{1}{n}\sum_{\lambda\in\Lambda}\mathrm{Var}\left(\varphi_\lambda\left(X_1\right)\right) \leq \frac{V_\Lambda}{n},$$

so by Jensen's inequality

$$\mathbb{E}\left[\chi_n\left(\Lambda\right)\right] \leq \left(\mathbb{E}\left[\chi_n^2\left(\Lambda\right)\right]\right)^{1/2} \leq \sqrt{\frac{V_\Lambda}{n}} \qquad (7.10)$$

which implies 7.9.

By refining somehow the previous arguments it is possible to derive from (5.50) another inequality which is a little more subtle and especially well fitted to the purpose of controlling many chi-square type statistics simultaneously...

Proposition 7.4 *Let $X_1,\ldots,X_n$ be independent and identically distributed random variables valued in some measurable space $(\mathbb{X},\mathcal{X})$. Let P denote their common distribution and ν_n be the corresponding centered empirical measure (see Definition 7.1). Let $\{\varphi_\lambda\}_{\lambda\in\Lambda}$ be a finite family of measurable and bounded functions on $(\mathbb{X},\mathcal{X})$. Let $\varepsilon > 0$ be given. Consider the sets*

$$\mathcal{S}_\Lambda = \left\{a\in\mathbb{R}^\Lambda : \sum_{\lambda\in\Lambda}a_\lambda^2 = 1\right\} \quad and \quad \mathcal{C}_\Lambda = \left\{a\in\mathbb{R}^\Lambda : \sup_{\lambda\in\Lambda}|a_\lambda| = 1\right\}$$

and define

$$M = \sup_{a\in\mathcal{S}_\Lambda}\left[\mathrm{Var}\left(\sum_{\lambda\in\Lambda}a_\lambda\varphi_\lambda\left(X_1\right)\right)\right] \quad, \quad B = \sup_{a\in\mathcal{C}_\Lambda}\left[\left\|\sum_{\lambda\in\Lambda}a_\lambda\varphi_\lambda\right\|_\infty\right]. \qquad (7.11)$$

Let moreover for any subset m of Λ

$$\Phi_m = \sum_{\lambda \in m} \varphi_\lambda^2 \ \ and \ \ V_m = \mathbb{E}\left[\Phi_m(X_1)\right].$$

Then, setting $\kappa(\varepsilon) = 2\left(\varepsilon^{-1} + 1/3\right)$, there exists some event Ω_n (depending on ε) such that on the one hand

$$\mathbb{P}\left[\Omega_n^c\right] \le 2\left|\Lambda\right| \exp\left(-\eta(\varepsilon)\frac{nM}{B^2}\right)$$

where

$$\eta(\varepsilon) = \frac{2\varepsilon^2}{\kappa(\varepsilon)\left(\kappa(\varepsilon) + \frac{2\varepsilon}{3}\right)}$$

and, on the other hand for any subset m of Λ and any positive x

$$\mathbb{P}\left[\left(\sum_{\lambda \in m} \nu_n^2(\varphi_\lambda)\right)^{1/2} \mathbb{1}_{\Omega_n} \ge (1+\varepsilon)\left(\sqrt{\frac{V_m}{n}} + \sqrt{\frac{2Mx}{n}}\right)\right] \le e^{-x}. \quad (7.12)$$

Proof. Let θ, z be positive numbers to be chosen later and Ω_n be defined by

$$\Omega_n = \left\{\sup_{\lambda \in \Lambda}\left|\nu_n(\varphi_\lambda)\right| \le \theta\right\}.$$

For any $a \in \mathcal{S}_m$, it follows from Cauchy–Schwarz inequality that

$$\chi_n(m) = \left[\sum_{\lambda \in m} \nu_n^2(\varphi_\lambda)\right]^{1/2} \ge \left|\sum_{\lambda \in m} a_\lambda \nu_n(\varphi_\lambda)\right|$$

with equality when $a_\lambda = \nu_n(\varphi_\lambda)\left(\chi_n(m)\right)^{-1}$ for all $\lambda \in m$. Hence, defining A to be the set of those elements $a \in \mathcal{S}_m$ satisfying $\left|a\right|_2 = 1$ and $\sup_{\lambda \in \Lambda}\left|a_\lambda\right| \le \theta/z$, we have that on the event $\Omega_n \cap \left\{\chi_n(m) \ge z\right\}$

$$\chi_n(m) = \sup_{a \in A}\left|\sum_{\lambda \in m} a_\lambda \nu_n(\varphi_\lambda)\right| = \sup_{a \in A}\left|\nu_n\left(\sum_{\lambda \in m} a_\lambda \varphi_\lambda\right)\right|. \quad (7.13)$$

Taking A' to be a countable and dense subset of A we have

$$\sup_{a \in A'}\left|\nu_n\left[\sum_{\lambda \in m} a_\lambda \varphi_\lambda\right]\right| = \sup_{a \in A}\left|\nu_n\left(\sum_{\lambda \in m} a_\lambda \varphi_\lambda\right)\right|.$$

We can apply inequality (5.50) to the countable set of functions

$$\left\{\sum_{\lambda \in m} a_\lambda \varphi_\lambda, a \in A'\right\}$$

and get by (7.11)

$$\mathbb{P}\left[\sup_{a\in A}\left|\nu_n\left(\sum_{\lambda\in m}a_\lambda\varphi_\lambda\right)\right| \geq (1+\varepsilon)E + \sqrt{\frac{2Mx}{n}} + \kappa(\varepsilon)\frac{B\theta}{nz}x\right] \leq e^{-x} \quad (7.14)$$

for all positive x, where by (7.10)

$$E = \mathbb{E}\left[\sup_{a\in A'}\left|\nu_n\left(\sum_{\lambda\in m}a_\lambda\varphi_\lambda\right)\right|\right] \leq \mathbb{E}\left(\chi_n\left(m\right)\right) \leq \sqrt{\frac{V_m}{n}}.$$

Hence we get by (7.13) and (7.14)

$$\mathbb{P}\left[\chi_n\left(m\right)\mathbb{1}_{\Omega_n\cap\{\chi_n(m)\geq z\}} \geq (1+\varepsilon)\sqrt{\frac{V_m}{n}} + \sqrt{\frac{2Mx}{n}} + \kappa(\varepsilon)\frac{B\theta}{nz}x\right] \leq e^{-x}.$$

If we now choose $z = \sqrt{2Mx/n}$ and $\theta = 2\varepsilon M\left[B\kappa(\varepsilon)\right]^{-1}$, we get

$$\mathbb{P}\left[\chi_n\left(m\right)\mathbb{1}_{\Omega_n} \geq (1+\varepsilon)\left(\sqrt{\frac{V_m}{n}} + \sqrt{\frac{2Mx}{n}}\right)\right] \leq e^{-x}.$$

On the other hand, Bernstein's inequality (2.24) ensures that

$$\mathbb{P}\left[|\nu_n\left(\varphi_\lambda\right)| \geq \theta\right] \leq 2\exp\left(-\frac{n\theta^2}{2\left(M+B\theta/3\right)}\right) \leq 2\exp\left(-\eta\left(\varepsilon\right)\frac{nM}{B^2}\right)$$

where

$$\eta\left(\varepsilon\right) = \frac{2\varepsilon^2}{\kappa\left(\varepsilon\right)\left(\kappa\left(\varepsilon\right) + \frac{2\varepsilon}{3}\right)},$$

which leads to the required bound on $\mathbb{P}\left[\Omega_n^c\right]$ since

$$\mathbb{P}\left[\Omega_n^c\right] \leq |\Lambda|\sup_{\lambda\in\Lambda}\mathbb{P}\left[|\nu_n\left(\varphi_\lambda\right)| \geq \theta\right].$$

This achieves the proof of the theorem. ∎

Each of the two exponential inequalities for chi-square type statistics that we have established in Proposition 7.3 and Proposition 7.4 above can be applied to penalized least squares. Proposition 7.3 is well suited for rather small collection of models (typically nested) while Proposition 7.4 can be fruitfully used as a sub-Gaussian inequality when dealing with subset model selection within a conveniently localized basis (typically a wavelet basis).

7.2.2 Model Selection for a Polynomial Collection of Models

We consider a polynomial collection of model $\{S_m\}_{m\in\mathcal{M}}$ in the sense that for some nonnegative constants Γ and R

$$|\{m\in\mathcal{M}: D_m = D\}| \leq \Gamma D^R \quad \text{for all } D\in\mathbb{N} \quad (7.15)$$

where, for every $m \in \mathcal{M}$, D_m denotes the dimension of the linear space S_m. This situation typically occurs when the collection of models is nested that is totally ordered for inclusion. Indeed in this case the collection is polynomial with $\Gamma = 1$ and $R = 0$.

Deterministic Penalty Functions

Our first model selection theorem for LSE deals with deterministic penalty functions.

Theorem 7.5 *Let $X_1, \ldots, X_n$ be a sample of the distribution $P = (s_0 + s)\,\mu$ where s_0 is a given function in $\mathbb{L}_2(\mu)$ and $s \in \mathbb{L}_2(\mu)$ with $\int s \, s_0 \, d\mu = 0$. Let $\{\varphi_\lambda\}_{\lambda \in \Lambda}$ be some orthonormal system of $\mathbb{L}_2(\mu)$, orthogonal to s_0. Let $\{S_m\}_{m \in \mathcal{M}}$ be a collection of linear subspaces of $\mathbb{L}_2(\mu)$ such that for every $m \in \mathcal{M}$, S_m is spanned by $\{\varphi_\lambda\}_{\lambda \in \Lambda_m}$ where Λ_m is a subset of Λ with cardinality $D_m \leq n$. Assume that the polynomial cardinality condition 7.15 holds. Let for every $m \in \mathcal{M}$*

$$\Phi_m(x) = \sup\{t^2(x) : t \in S_m, \ \|t\|_2 = 1\} \quad \text{for all } x \in \mathbb{X}.$$

Assume that for some positive constant Φ the following condition holds

$$\|\Phi_m\|_\infty \leq \Phi D_m \quad \text{for all } m \in \mathcal{M}. \tag{7.16}$$

Let $\varepsilon > 0$ be given. Take the penalty function as

$$\text{pen}(m) = (1 + \varepsilon)^6 \, \frac{\Phi D_m}{n} \quad \text{for all } m \in \mathcal{M}. \tag{7.17}$$

Let $\tilde{s}$ be the penalized LSE associated with the collection of models $\{S_m\}_{m \in \mathcal{M}}$ and the penalty function pen according to Definition 7.2. Then,

$$\mathbb{E}_s\left[\|\tilde{s} - s\|^2\right] \leq C(\varepsilon, \Phi) \inf_{m \in \mathcal{M}} \left[d^2(s, S_m) + \frac{D_m}{n}\right]$$
$$+ \frac{C'(\varepsilon, \Phi, \Gamma)\left(1 \vee \|s\|^{4+2R}\right)}{n}. \tag{7.18}$$

Proof. Let $m \in \mathcal{M}$ and $m' \in \mathcal{M}$ be given. We have in view to apply Proposition 7.3 to the variable

$$\chi_n^2(m, m') = \sup_{t \in S_m + S_{m'}} \frac{\nu_n^2(t)}{\|t\|^2}.$$

Towards this aim we recall that S_m and $S_{m'}$ are spanned by the orthonormal bases $\{\varphi_\lambda\}_{\lambda \in \Lambda_m}$ and $\{\varphi_\lambda\}_{\lambda \in \Lambda_{m'}}$ respectively, so that $\{\varphi_\lambda\}_{\lambda \in \Lambda_m \cup \Lambda_{m'}}$ is a basis of $S_m + S_{m'}$. Then

$$\chi_n^2(m, m') = \sum_{\lambda \in \Lambda_m \cup \Lambda_{m'}} \varphi_\lambda^2$$

and we can therefore apply Proposition 7.3. Noticing that

$$\sum_{\lambda \in \Lambda_m \cup \Lambda_{m'}} \varphi_\lambda^2 \leq \sum_{\lambda \in \Lambda_m} \varphi_\lambda^2 + \sum_{\lambda \in \Lambda_{m'}} \varphi_\lambda^2 \leq \Phi_m + \Phi_{m'}$$

we derive from inequality (7.9) that for all positive $x_{m'}$

$$\sqrt{n} \chi_n (m, m') \leq (1 + \varepsilon) \sqrt{V_m + V_{m'}} + \sqrt{2 x_{m'} M_{m,m'}}$$
$$+ \kappa (\varepsilon) x_{m'} \sqrt{\frac{\|\Phi_m\|_\infty + \|\Phi_{m'}\|_\infty}{n}},$$

except on a set of probability less than $\exp(-x_{m'})$, where

$$M_{m,m'} = \sup \left\{ \int_{\mathbb{X}} s(x) t^2(x) d\mu(x) : t \in S_m + S_{m'} \text{ and } \|t\| = 1 \right\}.$$

In order to bound $M_{m,m'}$, we note that any $t \in S_m + S_{m'}$ can be written as

$$t = \sum_{\lambda \in \Lambda_m \cup \Lambda_{m'}} a_\lambda \varphi_\lambda$$

and so by Cauchy–Schwarz

$$\|t\|_\infty \leq \|t\| \sqrt{\left\| \sum_{\lambda \in \Lambda_m \cup \Lambda_{m'}} \varphi_\lambda^2 \right\|_\infty} \leq \|t\| \sqrt{\|\Phi_m\|_\infty + \|\Phi_{m'}\|_\infty}.$$

Also, by Cauchy–Schwarz again we have

$$\int_{\mathbb{X}} s(x) t^2(x) d\mu(x) \leq \|t\|_\infty \|s\| \|t\|$$

and therefore

$$M_{m,m'} \leq \|s\| \sqrt{\|\Phi_m\|_\infty + \|\Phi_{m'}\|_\infty}. \tag{7.19}$$

Moreover, we derive from assumption (7.16) that

$$\|\Phi_m\|_\infty + \|\Phi_{m'}\|_\infty \leq \Phi (D_m + D_{m'}) \leq 2\Phi n, \tag{7.20}$$

hence, if we consider some positive constants ρ, ε' and z to be chosen later, by taking $x_{m'} = z + \rho^2 \left(\sqrt{\Phi} (\|s\| \vee 1) \right)^{-1} \sqrt{D_{m'}}$ and setting

$$\Sigma = \sum_{m' \in \mathcal{M}} \exp \left[-\rho^2 \left(\sqrt{\Phi} (\|s\| \vee 1) \right)^{-1} \sqrt{D_{m'}} \right],$$

we get by summing over m' in $\mathcal{M}$ the probability bounds above and using (7.19) together with (7.20)

$$\sqrt{n}\chi_n\left(m, m'\right) \leq (1 + \varepsilon)\left(\sqrt{V_m} + \sqrt{V_{m'}}\right)$$
$$+ 2\sqrt{2x_{m'}\,\|s\|\,\sqrt{\Phi}\left(\sqrt{D_m} + \sqrt{D_{m'}}\right)} + \kappa\left(\varepsilon\right)x_{m'}\sqrt{2\Phi}$$

for all $m' \in \mathcal{M}$, except on a set of probability less than Σe^{-z}. Let ρ be some positive constant to be chosen later. By inequality (7.6) we have

$$2\sqrt{x_{m'}\,\|s\|\,\sqrt{\Phi}\left(\sqrt{D_m} + \sqrt{D_{m'}}\right)} \leq \rho^{-1}\,\|s\|\,\sqrt{\Phi}x_{m'} + \rho\left(\sqrt{D_m} + \sqrt{D_{m'}}\right),$$

hence, except on a set of probability less than Σe^{-z} the following inequality holds

$$\sqrt{n}\chi_n\left(m, \widehat{m}\right) \leq W_m\left(z\right) + (1 + \varepsilon)\sqrt{V_{\widehat{m}}} + \sqrt{2D_{\widehat{m}}\left[2\rho + \kappa\left(\varepsilon\right)\rho^2\right]}$$

where since $V_m \leq \Phi D_m$, one can take

$$W_m\left(z\right) = (1 + \varepsilon)\sqrt{\Phi D_m} + \sqrt{2}\left[\rho\sqrt{D_m} + (\|s\| \vee 1)\sqrt{\Phi}z\left(\kappa\left(\varepsilon\right) + \frac{1}{\rho}\right)\right].$$
$$(7.21)$$

Choosing ρ such that

$$2\rho + \kappa\left(\varepsilon\right)\rho^2 = \frac{\varepsilon'}{\sqrt{2}}$$

yields

$$\sqrt{n}\chi_n\left(m, \widehat{m}\right) \leq W_m\left(z\right) + (1 + \varepsilon)\sqrt{V_{\widehat{m}}} + \varepsilon'\sqrt{D_{\widehat{m}}} \qquad (7.22)$$

except on a set of probability less than Σe^{-z}. We finish the proof by plugging this inequality in (7.7). Indeed by (7.6) we derive from (7.22) that, except on a set of probability less than Σe^{-z}

$$\left[\frac{n}{(1 + \varepsilon)^2}\right]\chi_n^2\left(m, \widehat{m}\right) \leq \frac{W_m^2\left(z\right)}{\varepsilon\left(1 + \varepsilon\right)} + (1 + \varepsilon)\left(\sqrt{V_{\widehat{m}}} + \varepsilon'\sqrt{D_{\widehat{m}}}\right)^2$$

which in turn, by choosing $\varepsilon' = \varepsilon\sqrt{\Phi}$ implies since $V_{\widehat{m}} \leq \Phi D_{\widehat{m}}$

$$(1 + \varepsilon)\chi_n^2\left(m, \widehat{m}\right) \leq \frac{(1 + \varepsilon)^2\,W_m^2\left(z\right)}{\varepsilon n} + \text{pen}\left(\widehat{m}\right).$$

So that we get by (7.7)

$$\frac{\varepsilon}{2 + \varepsilon}\,\|\widetilde{s} - s\|^2 \leq \left(1 + \frac{2}{\varepsilon}\right)\|s - s_m\|^2 + \frac{(1 + \varepsilon)^2\,W_m^2\left(z\right)}{\varepsilon n} + \text{pen}\left(m\right) \quad (7.23)$$

except on a set of probability less than Σe^{-z} and it remains to integrate this probability bound with respect to z. By (7.21) we get

$$W_m^2(z) \leq 2\left[\left((1+\varepsilon)\sqrt{\Phi} + \sqrt{2\rho}\right)^2 D_m + 2\left(\|s\| \vee 1\right)^2 \Phi z^2 \left(\kappa(\varepsilon) + \frac{1}{\rho}\right)^2\right]$$

and therefore integrating with respect to z, we derive from (7.23) that

$$\frac{\varepsilon}{2+\varepsilon}\mathbb{E}_s\left[\|\tilde{s} - s\|^2\right] \leq \left(1 + \frac{2}{\varepsilon}\right)\|s - s_m\|^2 + \frac{(1+\varepsilon)^2 W_m^2(z)}{\varepsilon n} + \text{pen}(m)$$

with

$$W_m^2(z) \leq 3\left[(1+\varepsilon)^2 V_m + 2\rho^2 D_m + 4\Sigma\left(\|s\| \vee 1\right)^2 \Phi\left(\kappa(\varepsilon) + \frac{1}{\rho}\right)^2\right].$$

Finally

$$\Sigma = \sum_{m' \in \mathcal{M}} \exp\left[-\rho^2\left(\sqrt{\Phi}\left(\|s\| \vee 1\right)\right)^{-1}\sqrt{D_{m'}}\right]$$

$$\leq \Gamma \sum_{k=0}^{\infty} k^R \exp\left[-\rho^2\left(\sqrt{\Phi}\left(\|s\| \vee 1\right)\right)^{-1}\sqrt{k}\right],$$

hence, setting $a = \rho^2\left(\sqrt{\Phi}\left(\|s\| \vee 1\right)\right)^{-1}$

$$\Sigma \leq \Gamma \int_0^{\infty} (1+u)^R \exp\left[-a\sqrt{u}\right] du \leq \frac{\Gamma}{a^2} \int_0^{\infty} \left(1 + \frac{v}{a^2}\right)^R \exp\left[-\sqrt{v}\right] dv$$

$$\leq \frac{\Gamma}{(a \wedge 1)^{2+2R}} \int_0^{\infty} (1+v)^R \exp\left[-\sqrt{v}\right] dv$$

and the result follows. ∎

For every $m \in \mathcal{M}$, let

$$\Phi_m(x) = \sup\left\{t^2(x) : t \in S_m, \|t\|_2 = 1\right\} \quad \text{for all } x \in X$$

and consider the following condition

$$\|\Phi_m\|_\infty \leq \Phi D_m \quad \text{for all } m \in \mathcal{M}. \tag{7.24}$$

We shall see in Section 7.5.2 explicit examples of nested collections of models to which Theorem 7.5 applies.

Data-Driven Penalty Functions

We provide below a version of the above theorem which involves a data-driven penalty and includes the cross-validation procedure.

Theorem 7.6 *Under the same assumptions and notations as in Theorem 7.5, let for every $m \in \mathcal{M}$,*

$$\widehat{V}_m = P_n\left(\Phi_m\right) .$$

Let ε and η be positive real numbers. Take the penalty function pen $: \mathcal{M} \to \mathbb{R}_+$ *as either*

$$\mathrm{pen}\left(m\right) = \frac{\left(1+\varepsilon\right)^6}{n}\left(\sqrt{\widehat{V}_m} + \eta\sqrt{D_m}\right)^2 \quad \text{for all } m \in \mathcal{M} \qquad (7.25)$$

or

$$\mathrm{pen}\left(m\right) = \frac{\left(1+\varepsilon\right)^6 \widehat{V}_m}{n} \quad \text{for all } m \in \mathcal{M}. \qquad (7.26)$$

Let $\widetilde{s}$ be the penalized LSE associated with the collection of models $\{S_m\}_{m \in \mathcal{M}}$ and the penalty function pen. *If the penalty function is chosen according to 7.25, then*

$$\mathbb{E}_s\left[\|\widetilde{s} - s\|^2\right] \le C\left(\varepsilon, \eta, \Phi\right) \inf_{m \in \mathcal{M}}\left[d^2\left(s, S_m\right) + \frac{D_m}{n}\right]$$

$$+ \frac{C'\left(\varepsilon, \Phi, \eta, \Gamma\right)\left(1 + \|s\|^{4+2R}\right)}{n},$$

while if the penalty function is chosen according to 7.26, whenever

$$a\left(s\right) = \inf_{m \in \mathcal{M}} \frac{\int_{\mathbb{X}} s\left(x\right)\Phi_m\left(x\right)d\mu\left(x\right)}{D_m} > 0$$

the following bound is available

$$\mathbb{E}_s\left[\|\widetilde{s} - s\|^2\right] \le C\left(\varepsilon, \Phi\right)\inf_{m \in \mathcal{M}}\left[d^2\left(s, S_m\right) + \frac{D_m}{n}\right]$$

$$+ \frac{C'\left(\varepsilon, \Phi, \Gamma\right)}{n}\left[\left(\|s\|\, a^{-1}\left(s\right)\right)^{4+2R} + a^{-1}\left(s\right)\right].$$

Proof. We begin the proof exactly as that of Theorem 7.5. So that, given some positive real numbers ε' and z and defining ρ as the solution of the equation

$$2\rho + \kappa\left(\varepsilon\right)\rho^2 = \frac{\varepsilon'}{\sqrt{2}},$$

we have

$$\sqrt{n}\chi_n\left(m, \widehat{m}\right) \le W_m\left(z\right) + \left(1+\varepsilon\right)\sqrt{V_{\widehat{m}}} + \varepsilon'\sqrt{D_{\widehat{m}}} \qquad (7.27)$$

except on a set of probability less than Σe^{-z}, where

$$W_m\left(z\right) = \left(1+\varepsilon\right)\sqrt{\Phi D_m} + \sqrt{2}\left[\rho\sqrt{D_m} + \left(\|s\| \vee 1\right)\sqrt{\Phi}z\left(\kappa\left(\varepsilon\right) + \frac{1}{\rho}\right)\right]$$

and

$$\Sigma = \sum_{m' \in \mathcal{M}} \exp\left[-\rho^2 \left(\sqrt{\Phi}(\|s\| \vee 1)\right)^{-1} \sqrt{D_{m'}}\right].$$

Finishing the proof in the case of a random choice for the penalty function requires a little more efforts than in the case of a deterministic choice as in Theorem 7.5. What we have to show is that $\widehat{V}_{m'}$ is a good estimator of $V_{m'}$ uniformly over the possible values of m' in $\mathcal{M}$ in order to substitute $\widehat{V}_{\widehat{m}}$ to $V_{\widehat{m}}$ in (7.27). Towards this aim we note that for any $m' \in \mathcal{M}$, since $\widehat{V}_{m'} - V_{m'} = \nu_n[\Phi_{m'}]$, we can derive from Bernstein's inequality (2.23) that for any positive number $y_{m'}$

$$\mathbb{P}\left[\left|\widehat{V}_{m'} - V_{m'}\right| \geq \sqrt{\frac{2V_{m'}\|\Phi_{m'}\|_\infty y_{m'}}{n}} + \frac{\|\Phi_{m'}\|_\infty}{3n}y_{m'}\right]$$
$$\leq 2e^{-y_{m'}}.$$

Now by assumption $\|\Phi_{m'}\|_\infty \leq \Phi n$, hence, summing these probability bounds with respect to m' in $\mathcal{M}$ we derive that, except on a set of probability less than $2\sum_{m' \in \mathcal{M}} e^{-y_{m'}}$, for all $m' \in \mathcal{M}$

$$\left|\widehat{V}_{m'} - V_{m'}\right| \geq \sqrt{2V_{m'}\Phi y_{m'}} + \frac{\Phi}{3}y_{m'} \tag{7.28}$$

which in particular implies that

$$-V_{\widehat{m}} + \sqrt{2V_{\widehat{m}}\Phi y_{\widehat{m}}} + \frac{\Phi}{3}y_{\widehat{m}} + \widehat{V}_{\widehat{m}} \geq 0$$

thus

$$\sqrt{V_{\widehat{m}}} \leq \sqrt{\frac{\Phi y_{\widehat{m}}}{2}} + \sqrt{\frac{5\Phi y_{\widehat{m}}}{6} + \widehat{V}_{\widehat{m}}} \leq \left(\sqrt{\frac{1}{2}} + \sqrt{\frac{5}{6}}\right)\sqrt{\Phi y_{\widehat{m}}} + \sqrt{\widehat{V}_{\widehat{m}}}$$

and using $\left(\sqrt{1/2} + \sqrt{5/6}\right)^2 < 8/3$

$$\sqrt{V_{\widehat{m}}} \leq \sqrt{\frac{8}{3}\Phi y_{\widehat{m}}} + \sqrt{\widehat{V}_{\widehat{m}}}.$$

Setting for all $m' \in \mathcal{M}$

$$y_{m'} = \frac{3\varepsilon'^2}{8\Phi}D_{m'} + z$$

and defining

$$\Sigma' = \sum_{m' \in \mathcal{M}} \exp\left(-\frac{3\varepsilon'^2}{8\Phi}D_{m'}\right)$$

we derive that, except on a set of probability less than $2\Sigma'e^{-z}$

$$\sqrt{V_{\widehat{m}}} \leq \varepsilon'\sqrt{D_{\widehat{m}}} + \sqrt{\widehat{V}_{\widehat{m}}} + \sqrt{\frac{8}{3}\Phi z}. \qquad (7.29)$$

We now choose ε' differently, according to whether the penalty function is given by (7.25) or (7.26). In the first case we choose $\varepsilon' = \eta\,(1+\varepsilon)^2 / (2+\varepsilon)$ while in the latter case we take ε' as the solution of the equation

$$\left(1 + \varepsilon + \frac{\varepsilon'}{\sqrt{a\,(s)}}\right)\left(1 - \frac{\varepsilon'}{\sqrt{a\,(s)}}\right)^{-1} = (1+\varepsilon)^2. \qquad (7.30)$$

If the penalty function is defined by (7.25), combining (7.29) with (7.27), we derive that

$$\sqrt{n}\chi_n\,(m, \widehat{m}) \leq W_m\,(z) + (1+\varepsilon)\sqrt{\frac{8}{3}\Phi z} + (1+\varepsilon)\sqrt{\widehat{V}_{\widehat{m}}} + \varepsilon'\,(2+\varepsilon)\sqrt{D_{\widehat{m}}}$$

$$\leq W_m\,(z) + (1+\varepsilon)\sqrt{\frac{8}{3}\Phi z} + (1+\varepsilon)^{-1}\sqrt{\mathrm{pen}\,(\widehat{m})},$$

except on a set of probability less than $(2\Sigma' + \Sigma)\,e^{-z}$. If the penalty function is given by (7.26), we derive from (7.29) that

$$\sqrt{V_{\widehat{m}}} \leq \left[1 - \frac{\varepsilon'}{\sqrt{a\,(s)}}\right]^{-1}\left[\sqrt{\widehat{V}_{\widehat{m}}} + \sqrt{\frac{8}{3}\Phi z}\right],$$

which, combined with (7.27) yields via (7.30)

$$\sqrt{n}\chi_n\,(m, \widehat{m}) \leq W_m\,(z) + \left(1 + \varepsilon + \frac{\varepsilon'}{\sqrt{a\,(s)}}\right)\sqrt{V_{\widehat{m}}}$$

$$\leq W_m\,(z) + (1+\varepsilon)^2\sqrt{\frac{8}{3}\Phi z} + (1+\varepsilon)^2\sqrt{\widehat{V}_{\widehat{m}}}.$$

So that whatever the choice of the penalty function we have

$$\sqrt{n}\chi_n\,(m, \widehat{m}) \leq W'_m\,(z) + (1+\varepsilon)^{-1}\sqrt{\mathrm{pen}\,(\widehat{m})},$$

where

$$W'_m\,(z) = W_m\,(z) + (1+\varepsilon)^2\sqrt{\frac{8}{3}\Phi z}.$$

Hence by (7.6), we have except on a set of probability less than $(2\Sigma' + \Sigma)\,e^{-z}$

$$\left[\frac{n}{(1+\varepsilon)^2}\right]\chi_n^2\,(m, \widehat{m}) \leq \frac{W'^2_m\,(z)}{\varepsilon\,(1+\varepsilon)} + (1+\varepsilon)^{-1}\,\mathrm{pen}\,(\widehat{m})$$

which in turn implies

$$(1 + \varepsilon) \chi_n^2 (m, \widehat{m}) \leq \frac{(1 + \varepsilon)^2 W_m'^2 (z)}{\varepsilon n} + \text{pen} (\widehat{m}).$$

So that we get by (7.7)

$$\frac{\varepsilon}{2 + \varepsilon} \|\widetilde{s} - s\|^2 \leq \left(1 + \frac{2}{\varepsilon}\right) \|s - s_m\|^2 + \frac{(1 + \varepsilon)^2 W_m'^2 (z)}{\varepsilon n} + \text{pen} (m) \quad (7.31)$$

except on a set of probability less than $(2\Sigma' + \Sigma) e^{-z}$. Since the penalty function is given either by (7.26) or by (7.25) and in this case

$$\text{pen} (m) \leq 2 \frac{(1 + \varepsilon)^6}{n} \left(\widehat{V}_m + \eta^2 D_m\right),$$

it remains in any case to control $\widehat{V}_m$. But from (7.28) we derive that on the same set of probability where (7.31) holds, one also has

$$\widehat{V}_m \leq V_m + \sqrt{2 V_m \Phi y_m} + \frac{\Phi}{3} y_m \leq 2 V_m + \Phi y_m$$
$$\leq \left(2\Phi + \frac{3}{8} \varepsilon'^2\right) D_m + \Phi z.$$

Plugging this evaluation into (7.31) allows to finish the proof in the same way as that of Theorem 7.5 by integration with respect to z. $\blacksquare$

Some applications of this theorem to adaptive estimation on ellipsoids will be given in Section 7.5.

7.2.3 Model Subset Selection Within a Localized Basis

We assume that we are given a finite orthonormal system $\{\varphi_\lambda\}_{\lambda \in \Lambda_n}$ in $\mathbb{L}_2 (\mu)$. Given a family of subsets $\mathcal{M}$ of Λ_n, we are interested in the behavior of the penalized LSE associated with the collection of models $\{S_m\}_{m \in \mathcal{M}}$, where for every $m \in \mathcal{M}$, S_m is spanned by $\{\varphi_\lambda\}_{\lambda \in m}$.

Theorem 7.7 *Let $X_1, \ldots, X_n$ be a sample of the distribution $P = (s_0 + s) \mu$ where s_0 is a given function in $\mathbb{L}_2 (\mu)$ and $s \in \mathbb{L}_2 (\mu)$ with $\int s \, s_0 \, d\mu = 0$. Assume that the orthonormal system $\{\varphi_\lambda\}_{\lambda \in \Lambda_n}$ is orthogonal to s_0 and has the following property*

$$\left\| \sum_{\lambda \in \Lambda} a_\lambda \varphi_\lambda \right\|_\infty \leq B' \sqrt{|\Lambda_n|} \sup_{\lambda \in \Lambda} |a_\lambda| \quad \textit{for any } a \in \mathbb{R}^\Lambda$$

where B' does not depend on n. Let $\mathcal{M}$ be a collection of subsets of Λ_n and consider some family of nonnegative weights $\{L_m\}_{m \in \mathcal{M}}$ with such that

$$\sum_{m \in \mathcal{M}} \exp (-L_m |m|) \leq \Sigma,$$

where Σ does not depend on n. Let $\varepsilon > 0$ and $M > 0$ be given. Assume that $|\Lambda_n| \le (M \wedge 1) n (\ln (n))^{-2}$. For every $m \in \mathcal{M}$, let

$$\widehat{V}_m = P_n \left(\sum_{\lambda \in m} \varphi_\lambda^2 \right)$$

and S_m be the linear span of $\{\varphi_\lambda\}_{\lambda \in m}$. Choose the penalty function pen $:$ $\mathcal{M} \to \mathbb{R}_+$ as either

$$\mathrm{pen}\,(m) = \frac{M (1 + \varepsilon)^4 |m|}{n} \left(1 + \sqrt{2L_m} \right)^2 \quad \textit{for all } m \in \mathcal{M} \qquad (7.32)$$

or

$$\mathrm{pen}\,(m) = \frac{(1 + \varepsilon)^5}{n} \left(\sqrt{\widehat{V}_m} + \sqrt{2ML_m\,|m|} \right)^2 \quad \textit{for all } m \in \mathcal{M} \qquad (7.33)$$

and in that case suppose furthermore that

$$B'^2 |\Lambda_n| \le \frac{3}{4} M \left(\sqrt{1 + \varepsilon} - 1 \right)^2 n.$$

Let $\widetilde{s}$ be the penalized LSE associated with the collection of models $\{S_m\}_{m \in \mathcal{M}}$ and the penalty function pen. Then, provided that

$$\int_{\mathbb{X}} s\,(x) \left[\sum_{\lambda \in \Lambda_n} a_\lambda \varphi_\lambda\,(x) \right]^2 d\mu\,(x) \le M \sum_{\lambda \in \Lambda_n} a_\lambda^2$$

for all $a \in \mathbb{R}^{\Lambda_n}$ (which in particular holds whenever $\|s\|_\infty \le M$), the following risk bound holds

$$\mathbb{E}_s \left[\|\widetilde{s} - s\|^2 \right] \le C\,(\varepsilon) \inf_{m \in \mathcal{M}} \left[d^2\,(s, S_m) + \frac{M\,|m|}{n} (1 + L_m) \right]$$
$$+ C\,(\varepsilon) \frac{(1 + M\Sigma)}{n}$$

where $C\,(\varepsilon)$ depends only on ε and B'.

Proof. Let us define for any subset m of Λ_n,

$$\chi_n\,(m) = \left[\sum_{\lambda \in m} \nu_n^2\,(\varphi_\lambda) \right]^{1/2}.$$

We recall that inequality (7.7) can be written as

$$\frac{\varepsilon}{2 + \varepsilon} \|\widetilde{s} - s\|^2 \le \left(1 + \frac{2}{\varepsilon} \right) \|s - s_m\|^2 + (1 + \varepsilon) \chi_n^2\,(m \cup \widehat{m})$$
$$- \mathrm{pen}\,(\widehat{m}) + \mathrm{pen}\,(m). \qquad (7.34)$$

Let $z > 0$ to be chosen later. We can apply Theorem 7.4 to control $\chi_n (m \cup m')$ for all values of m' in $\mathcal{M}$ simultaneously. We note that for any $a \in \mathbb{R}^{\Lambda_n}$

$$\mathrm{Var} \left(\sum_{\lambda \in \Lambda_n} a_\lambda \varphi_\lambda (X_1) \right) \leq \int_{\mathbb{X}} \left[\sum_{\lambda \in \Lambda_n} a_\lambda \varphi_\lambda (x) \right]^2 s(x) \, d\mu(x)$$

and therefore, whenever $\sum_{\lambda \in \Lambda_n} a_\lambda^2 = 1$

$$\mathrm{Var} \left(\sum_{\lambda \in \Lambda_n} a_\lambda \varphi_\lambda (X_1) \right) \leq M.$$

Moreover, provided that $\sup_{\lambda \in \Lambda_n} |a_\lambda| = 1$, we have by assumption that

$$\left\| \sum_{\lambda \in \Lambda_n} a_\lambda \varphi_\lambda \right\|_\infty \leq B' \sqrt{|\Lambda_n|},$$

hence, it comes from Theorem 7.4 with $B = B' \sqrt{|\Lambda_n|}$ that there exists an event Ω_n (depending on ε') such that

$$\mathbb{P} [\Omega_n^c] \leq 2 |\Lambda_n| \exp \left(-\eta(\varepsilon) \frac{nM}{B'^2 |\Lambda_n|} \right)$$

and, for any subset m' of Λ_n

$$\mathbb{P} \left[\sqrt{n} \chi_n (m \cup m') \, \mathbb{1}_{\Omega_n} \geq (1 + \varepsilon) \left(\sqrt{V_{m \cup m'}} + \sqrt{2M (L_{m'} |m'| + z)} \right) \right]$$
$$\leq e^{-L_{m'} |m'| - z},$$

where
$$V_{m \cup m'} = \mathbb{E} [\Phi_{m \cup m'} (X_1)] \leq V_m + V_{m'} \leq M |m| + V_{m'}.$$

We begin by controlling the quadratic risk of the estimator $\tilde{s}$ on the set Ω_n. Summing up these probability bounds with respect to $m' \in \mathcal{M}$, we derive that, except on a set of probability less than Σe^{-z}

$$\left[\frac{\sqrt{n}}{1 + \varepsilon} \right] \chi_n (m \cup \hat{m}) \, \mathbb{1}_{\Omega_n} \leq \sqrt{M |m|} + \sqrt{V_{\hat{m}}} + \sqrt{2M \left(L_{\hat{m}} |\hat{m}| + z \right)}. \quad (7.35)$$

At this stage we have to follow two different strategies according to the choice of the penalty function. If the penalty function is chosen according to (7.32) then we can easily compare the square of this upper bound to $\mathrm{pen}(\hat{m})$ by simply using the inequality $V_{\hat{m}} \leq M |\hat{m}|$. Indeed by (7.6), inequality (7.35) yields

$$\left[\frac{n}{(1 + \varepsilon)^2} \right] \chi_n^2 (m \cup \hat{m}) \, \mathbb{1}_{\Omega_n} \leq 2 (1 + \varepsilon) \left[\frac{M (|m| + 2z)}{\varepsilon} \right]$$
$$+ (1 + \varepsilon) \left[\sqrt{V_{\hat{m}}} + \sqrt{2M L_{\hat{m}} |\hat{m}|} \right]^2.$$

Taking into account that $V_{\widehat{m}} \le M|\widehat{m}|$, we see that, except on a set with probability less than Σe^{-z} one has

$$\chi_n^2 (m \cup \widehat{m}) \, 1\!\!1_{\Omega_n} \le \frac{(1+\varepsilon)^3}{n} \left[\frac{2M (|m| + 2z)}{\varepsilon} \right] + \frac{\mathrm{pen}\,(\widehat{m})}{1+\varepsilon}. \tag{7.36}$$

Obtaining an analogous bound in the case of the random choice for the penalty function (7.33) requires a little more efforts. Essentially what we have to show is that $\widehat{V}_{m'}$ is a good estimator of $V_{m'}$ uniformly over the possible values of m' in $\mathcal{M}$ in order to substitute $\widehat{V}_{\widehat{m}}$ to $V_{\widehat{m}}$ in (7.35). Towards this aim we note that for any subset m' of Λ, since $\widehat{V}_{m'} - V_{m'} = \nu_n [\Phi_{m'}]$, we can derive from Bernstein's inequality (2.23) that

$$\mathbb{P} \left[\left| \widehat{V}_{m'} - V_{m'} \right| \ge \sqrt{\frac{2V_{m'} \|\Phi_{m'}\|_\infty x_{m'}}{n}} + \frac{\|\Phi_{m'}\|_\infty}{3n} x_{m'} \right]$$

$$\le 2 e^{-L_{m'} |m'| - z}.$$

where $x_{m'} = L_{m'} |m'| + z$. Now by assumption $\|\Phi_{m'}\|_\infty \le B'^2 |\Lambda_n| \le \theta(\varepsilon) n$ with

$$\theta(\varepsilon) = \frac{3}{4} M \left(\sqrt{1+\varepsilon} - 1 \right)^2. \tag{7.37}$$

Hence, summing these probability bounds with respect to m' in $\mathcal{M}$ we derive that, except on a set of probability less than $2\Sigma e^{-z}$, for all $m' \in \mathcal{M}$

$$\left| \widehat{V}_{m'} - V_{m'} \right| \ge \sqrt{2V_{m'} \theta(\varepsilon) x_{m'}} + \frac{\theta(\varepsilon)}{3} x_{m'} \tag{7.38}$$

which in particular implies that

$$-V_{\widehat{m}} + \sqrt{2V_{\widehat{m}} \theta(\varepsilon) x_{\widehat{m}}} + \frac{\theta(\varepsilon)}{3} x_{\widehat{m}} + \widehat{V}_{\widehat{m}} \ge 0$$

thus

$$\sqrt{V_{\widehat{m}}} \le \sqrt{\frac{\theta(\varepsilon) x_{\widehat{m}}}{2}} + \sqrt{\frac{5\theta(\varepsilon) x_{\widehat{m}}}{6} + \widehat{V}_{\widehat{m}}} \le \left(\sqrt{\frac{1}{2}} + \sqrt{\frac{5}{6}} \right) \sqrt{\theta(\varepsilon) x_{\widehat{m}}} + \sqrt{\widehat{V}_{\widehat{m}}}$$

and using (7.37) together with $\left(\sqrt{1/2} + \sqrt{5/6} \right)^2 < 8/3$

$$\sqrt{V_{\widehat{m}}} \le \left(\sqrt{1+\varepsilon} - 1 \right) \sqrt{2M x_{\widehat{m}}} + \sqrt{\widehat{V}_{\widehat{m}}}.$$

Plugging this inequality in (7.35), we derive that, except on a set of probability less than $3\Sigma e^{-z}$, the following inequality holds

$$\left[\frac{\sqrt{n}}{(1+\varepsilon)^{3/2}} \right] \chi_n (m \cup \widehat{m}) \, 1\!\!1_{\Omega_n} \le \sqrt{M|m|} + \sqrt{\widehat{V}_{\widehat{m}}} + \sqrt{2M \left(L_{\widehat{m}} |\widehat{m}| + z \right)}.$$

Hence, by using (7.6) again

$$\left[\frac{n}{(1+\varepsilon)^3}\right] \chi_n^2 (m \cup \widehat{m}) \, \mathbb{1}_{\Omega_n} \leq 2 (1+\varepsilon) \left[\frac{M (|m| + 2z)}{\varepsilon}\right]$$

$$+ (1+\varepsilon) \left[\sqrt{\widehat{V}_{\widehat{m}}} + \sqrt{2ML_{\widehat{m}} |\widehat{m}|}\right]^2 .$$

Therefore, setting $\alpha = 4$ if the penalty function is chosen according to (7.32) and $\alpha = 5$ if it is chosen according to (7.33) we can summarize this inequality together with (7.36) by the following statement. Except on a set of probability less than $3\Sigma e^{-z}$ we have

$$(1+\varepsilon) \chi_n^2 (m \cup \widehat{m}) \, \mathbb{1}_{\Omega_n} \leq \frac{(1+\varepsilon)^\alpha}{n} \left[\frac{2M (|m| + 2z)}{\varepsilon}\right] + \text{pen} \, (\widehat{m}) .$$

Coming back to (7.34) this implies that, except on a set with probability less than $3\Sigma e^{-z}$

$$\frac{\varepsilon}{2+\varepsilon} \|\widetilde{s} - s\|^2 \, \mathbb{1}_{\Omega_n} \leq \left(1 + \frac{2}{\varepsilon}\right) \|s - s_m\|^2 + \frac{(1+\varepsilon)^5}{n} \left[\frac{2M (|m| + 2z)}{\varepsilon}\right]$$

$$+ \text{pen} \, (m) . \tag{7.39}$$

It remains to bound $\text{pen} \, (m)$. We use inequality (7.6) again and get either

$$\text{pen} \, (m) \leq \frac{3 (1+\varepsilon)^4 M |m|}{n} (1 + L_m)$$

if (7.32) obtains, or

$$\text{pen} \, (m) \leq \frac{3 (1+\varepsilon)^5}{n} \left(\widehat{V}_m + M |m| L_m\right)$$

if (7.33) obtains. Now we know from (7.38), (7.6) and (7.37) that

$$\widehat{V}_m \leq 2V_m + \frac{5}{6} \theta \, (\varepsilon) \, x_m \leq 2M |m| + \frac{5}{32} \varepsilon^2 M (|m| L_m + z) .$$

This implies that, for an adequate choice of $C' \, (\varepsilon)$ we have whatever the choice of the penalty function

$$\text{pen} \, (m) \leq \frac{C' \, (\varepsilon) M}{n} \left[(1 + L_m) |m| + z\right] .$$

Plugging this bound in (7.39), we derive that, except on a set with probability less than $3\Sigma e^{-z}$ and possibly enlarging the value of $C' \, (\varepsilon)$

$$\frac{\varepsilon}{2+\varepsilon} \|\widetilde{s} - s\|^2 \, \mathbb{1}_{\Omega_n} \leq \left(1 + \frac{2}{\varepsilon}\right) \|s - s_m\|^2$$

$$+ \frac{2C' \, (\varepsilon) M}{n} \left[(1 + L_m) |m| + z\right] .$$

Integrating the latter inequality with respect to z yields

$$\frac{\varepsilon}{2+\varepsilon}\mathbb{E}\left[\|\widetilde{s}-s\|^2\,\mathbb{1}_{\Omega_n}\right] \leq \left(1+\frac{2}{\varepsilon}\right)\|s-s_m\|^2$$
$$+\frac{2C'(\varepsilon)M}{n}\left[(1+L_m)\,|m|+3\Sigma\right].$$

It remains to control the quadratic risk of $\widetilde{s}$ on Ω_n^c. First of all we notice that Ω_n^c has a small probability. Indeed we have by assumption on the cardinality of Λ_n

$$p_n = \mathbb{P}\left[\Omega_n^c\right] \leq 2\,|\Lambda_n|\exp\left(-\eta\,(\varepsilon)\,\frac{nM}{B'^2\,|\Lambda_n|}\right)$$

and therefore

$$p_n \leq 2n\exp\left(-\eta\,(\varepsilon)\,\frac{n\,(M\vee 1)\ln^2(n)}{B'^2}\right) \leq \frac{\kappa'(\varepsilon)}{n^2\,(M\vee 1)^2} \tag{7.40}$$

where $\kappa'(\varepsilon)$ is an adequate function of ε and B'. Since p_n is small, we can use a crude bound for $\|\widetilde{s}-s\|^2\,\mathbb{1}_{\Omega_n^c}$. Indeed, by Pythagore's Theorem we have

$$\|\widetilde{s}-s\|^2 = \left\|\widetilde{s}-s_{\widehat{m}}\right\|^2 + \left\|s_{\widehat{m}}-s\right\|^2 \leq \left\|\widetilde{s}-s_{\widehat{m}}\right\|^2 + \|s\|^2$$
$$\leq \sum_{\lambda\in\widehat{m}}\nu_n^2\,(\varphi_\lambda) + \|s\|^2 \leq \chi_n^2\,(\Lambda_n) + \|s\|^2.$$

Hence, by Cauchy–Schwarz inequality

$$\mathbb{E}\left[\|\widetilde{s}-s\|^2\,\mathbb{1}_{\Omega_n^c}\right] \leq p_n\,\|s\|^2 + \sqrt{p_n\mathbb{E}\left[\chi_n^4\,(\Lambda_n)\right]}. \tag{7.41}$$

All we have to do now is to bound the moments of $\chi_n\,(\Lambda_n)$. This can be obtained easily by application of Proposition 7.3 with

$$V_{\Lambda_n} \leq M\,|\Lambda_n| \leq Mn, \quad M_{\Lambda_n} \leq M \quad \text{and} \quad \|\Phi_{\Lambda_n}\|_\infty \leq B'^2\,|\Lambda_n| \leq B'^2 n.$$

Indeed, we derive from inequality (7.9) the existence of some constant C' depending only on B' such that for all positive x

$$\mathbb{P}\left[\chi_n\,(\Lambda_n) \geq C'\left(\sqrt{M}\vee 1\right)(1+x)\right] \leq e^{-x}$$

which by integration with respect to x yields

$$\mathbb{E}\left[\chi_n^4\,(\Lambda_n)\right] \leq \kappa'\,(M\vee 1)^2.$$

Combining this inequality with 7.41 and 7.40 we get

$$\mathbb{E}\left[\|\widetilde{s}-s\|^2\,\mathbb{1}_{\Omega_n^c}\right] \leq \frac{\kappa'(\varepsilon)}{n} + \frac{\sqrt{\kappa'\kappa'(\varepsilon)}}{n}$$

and the result follows. ∎

It is important to notice that the estimation procedure described in Theorem 7.7 is feasible provided that for instance we know an upper bound M for $\|s\|_\infty$, since M enters the penalty function. Actually, a little less is needed since M can be taken as an upper bound for the supremum of $\int_{\mathbb{X}} s(x) t^2(x) d\mu(x)$ when t varies in S_{Λ_n} with $\|t\| = 1$. In particular, if S_{Λ_n} is a space of piecewise polynomials of degree r on a given partition, M can be taken as an upper bound for $\|\Pi_{2r}(s)\|_\infty$, where Π_{2r} for the orthogonal projection operator onto the space of piecewise polynomials with degree $2r$ on the same partition. Generally speaking quantities like $\|s\|_\infty$ or $\|\Pi_{2r}(s)\|_\infty$ are unknown and one has to estimate it (or rather to overestimate it) by some statistics $\widehat{M}$.

7.3 Selecting the Best Histogram via Penalized Maximum Likelihood Estimation

We shall present in this section some of the results concerning the old standing problem of selecting "the best partition" when constructing some histogram, obtained by Castellan in [34]. We consider here the density framework where one observes n independent and identically distributed random variables with common density s with respect to the Lebesgue measure on $[0, 1]$. Let $\mathcal{M}$ be some finite (but possibly depending on n) collection of partitions of $[0, 1]$ into intervals. We could work with multivariate histograms as well (as in [34]) but since it does not causes any really new conceptual difficulty, we prefer to stay in the one dimensional framework for the sake of simplicity. For any partition m, we consider the corresponding histogram estimator $\widehat{s}_m$ defined by

$$\widehat{s}_m = \sum_{I \in m} (n\mu(I))^{-1} \left[\sum_{i=1}^{n} \mathbb{1}_I(\xi_i) \right] \mathbb{1}_I,$$

where μ denotes the Lebesgue measure on $[0, 1]$, and the purpose is to select "the best one." Recall that the histogram estimator on some partition m is known to be the MLE on the model S_m of densities which are piecewise constants on the corresponding partition m and therefore falls into our analysis. Then the natural loss function to be considered is the Kullback–Leibler loss and in order to understand the construction of Akaike's criterion, it is essential to describe the behavior of an oracle and therefore to analyze the Kullback–Leibler risk. First it easy to see that the Kullback–Leibler projection s_m of s on the histogram model S_m (i.e., the minimizer of $t \to \mathbf{K}(s, t)$ on S_m) is simply given by the orthogonal projection of s on the linear space of piecewise constant functions on the partition m and that the following Pythagore's type decomposition holds

$$\mathbf{K}(s, \widehat{s}_m) = \mathbf{K}(s, s_m) + \mathbf{K}(s_m, \widehat{s}_m). \tag{7.42}$$

Hence the oracle should minimize $\mathbf{K}(s, s_m) + \mathbb{E}[\mathbf{K}(s_m, \widehat{s}_m)]$ or equivalently, since $s - s_m$ is orthogonal to $\ln(s_m)$,

$$\mathbf{K}(s, s_m) + \mathbb{E}[\mathbf{K}(s_m, \widehat{s}_m)] - \int s \ln(s) = - \int s_m \ln(s_m) + \mathbb{E}[\mathbf{K}(s_m, \widehat{s}_m)].$$

Since $\int s_m \ln s_m$ depends on s, it has to be estimated. One could think of $\int \widehat{s}_m \ln \widehat{s}_m$ as being a good candidate for this purpose but since $\mathbb{E}[\widehat{s}_m] = s_m$, the following identity holds

$$\mathbb{E}\left[\int \widehat{s}_m \ln(\widehat{s}_m)\right] = \mathbb{E}[\mathbf{K}(\widehat{s}_m, s_m)] + \int s_m \ln(s_m),$$

which shows that it is necessary to remove the bias of $\int \widehat{s}_m \ln \widehat{s}_m$ if one wants to use it as an estimator of $\int s_m \ln s_m$. In order to summarize the previous analysis of the oracle procedure (with respect to the Kullback–Leibler loss), let us set

$$R_m = \int \widehat{s}_m \ln(\widehat{s}_m) - \mathbb{E}\left[\int \widehat{s}_m \ln(\widehat{s}_m)\right].$$

Then the oracle minimizes

$$- \int \widehat{s}_m \ln(\widehat{s}_m) + \mathbb{E}[\mathbf{K}(s_m, \widehat{s}_m)] + \mathbb{E}[\mathbf{K}(\widehat{s}_m, s_m)] + R_m. \qquad (7.43)$$

The idea underlying Akaike's criterion relies on two heuristics:

- neglecting the remainder term R_m (which is centered at its expectation),
- replacing $\mathbb{E}[\mathbf{K}(s_m, \widehat{s}_m)] + \mathbb{E}[\mathbf{K}(\widehat{s}_m, s_m)]$ by its asymptotic equivalent when n goes to infinity which is equal to D_m/n, where $1 + D_m$ denotes the number of pieces of the partition m (see [34] for a proof of this result).

Making these two approximations leads to Akaike's method which amounts to replace (7.43) by

$$- \int \widehat{s}_m \ln(\widehat{s}_m) + \frac{D_m}{n} \qquad (7.44)$$

and proposes to select a partition $\widehat{m}$ minimizing Akaike's criterion (7.44). An elementary computation shows that

$$P_n(-\ln(\widehat{s}_m)) = - \int \widehat{s}_m \ln(\widehat{s}_m).$$

If we denote by γ_n the maximum likelihood criterion, i.e., $\gamma_n(t) = P_n(-\ln(t))$, we derive that Akaike's criterion can be written as

$$\gamma_n(\widehat{s}_m) + \frac{D_m}{n},$$

and is indeed a penalized criterion of type (8.4) with pen$(m) = D_m/n$. It will be one of the main issues of Section 3 to discuss whether this heuristic approach can bevalidated or not but we can right now try to guess why concentration inequalities will be useful and in what circumstances Akaike's criterion should be corrected. Indeed, we have seen that Akaike's heuristics rely on the fact that some quantities R_m stay close to their expectations (they are actually all centered at 0). Moreover, this should hold with a certain uniformity over the list of partitions $\mathcal{M}$. This means that if the collection of partitions is not too rich, we can hope that the R_m's will be concentrated enough around their expectations to warrant that Akaike's heuristics works, while if the collection is too rich, concentration inequalities will turn to be an essential tool to understand how one should correct (substantially) Akaike's criterion. Our purpose is to study the selection procedure which consists in retaining a partition $\widehat{m}$ minimizing the *penalized log-likelihood criterion*

$$\gamma_n\left(\hat{s}_m\right) + \operatorname{pen}\left(m\right),$$

over $m \in \mathcal{M}$. We recall that Akaike's criterion corresponds to the choice $\operatorname{pen}\left(m\right) = D_m/n$, where $D_m + 1$ denotes the number of pieces of m and that Castellan's results presented below will allow to correct this criterion. Let us first explain the connection between the study of the *penalized MLE* $\tilde{s} = \hat{s}_{\widehat{m}}$ and the problem considered just above of controlling some chi-square statistics. The key is to control the Kullback–Leibler loss between s and $\tilde{s}$ in the following way

$$\mathbf{K}\left(s, \tilde{s}\right) \leq \mathbf{K}\left(s, s_m\right) + \nu_n\left(\ln \tilde{s} - \ln s_m\right) + \operatorname{pen}\left(m\right) - \operatorname{pen}\left(\widehat{m}\right), \qquad (7.45)$$

for every $m \in \mathcal{M}$, where

$$s_m = \sum_{I \in m} \frac{P\left(I\right)}{\mu\left(I\right)} \mathbb{1}_I.$$

Now the main task is to bound $\nu_n\left(\ln \tilde{s} - \ln s_m\right)$ as sharply as possible in order to determine what is the minimal value of $\operatorname{pen}\left(\widehat{m}\right)$ which is allowed for deriving a risk bound from (7.45). This will result from a uniform control of $\nu_n\left(\ln \hat{s}_{m'} - \ln s_m\right)$ with respect to $m' \in \mathcal{M}$. We write

$$\nu_n\left(\ln \frac{\hat{s}_{m'}}{s_m}\right) = \nu_n\left(\ln \frac{\hat{s}_{m'}}{s_{m'}}\right) + \nu_n\left(\ln \frac{s_{m'}}{s}\right) + \nu_n\left(\ln \frac{s}{s_m}\right)$$

and notice that the first term is the most delicate to handle since it involves the action of the empirical process on the estimator $\hat{s}_{m'}$ which is of course a random variable. This is precisely the control of this term which leads to the introduction of chi-square statistics. Indeed, setting

$$V^2\left(f, g\right) = \int s\left(\ln\left(\frac{f}{g}\right)\right)^2 d\mu,$$

for every densities f and g such that $\ln\left(f/g\right) \in \mathbb{L}_2\left(P\right)$, one derives that

$$\nu_n\left(\ln \frac{\hat{s}_{m'}}{s_{m'}}\right) \leq \sup_{t \in S_{m'}} \left|\nu_n\left(\frac{\ln t - \ln s_{m'}}{V\left(t, s_{m'}\right)}\right)\right| V\left(\hat{s}_{m'}, s_{m'}\right)$$

and if we set $\varphi_I = P(I)^{-1/2}\, \mathbb{1}_I$ for all $I \in m'$ then,

$$\sup_{t \in S_{m'}} \left| \nu_n \left(\frac{\ln t - \ln s_{m'}}{V(t, s_{m'})} \right) \right| = \sup_{a \in \mathbb{R}^{m'}, |a|_2 = 1} \left| \sum_{I \in m'} a_I \nu_n (\varphi_I) \right|$$

$$= \left[\sum_{I \in m'} \nu_n^2 (\varphi_I) \right]^{1/2} = \frac{1}{\sqrt{n}} \chi_n (m').$$

Hence (7.45) becomes

$$\mathbf{K}(s, \tilde{s}) \leq \mathbf{K}(s, s_m) + \mathrm{pen}(m) + \nu_n \left(\ln \frac{s}{s_m} \right)$$

$$+ n^{-1/2} V \left(\hat{s}_{\widehat{m}}, s_{\widehat{m}} \right) \chi_n (\widehat{m})$$

$$+ \nu_n \left(\ln \frac{s_{\widehat{m}}}{s} \right) - \mathrm{pen}(\hat{m}). \tag{7.46}$$

At this stage it becomes clear that what we need is a uniform control of $\chi_n(m')$ over $m' \in \mathcal{M}$. The key idea for improving on (5.54) is to upper bound $\chi_n(m')$ only on some part of the probability space where $P_n(\varphi_I)$ remains close to $P(\varphi_I)$ for every $I \in m'$.

7.3.1 Some Deepest Analysis of Chi-Square Statistics

This idea, introduced in the previous section in the context of subset selection within a conveniently localized basis can be fruitfully applied here. More precisely, Castellan proves in [34] the following inequality.

Proposition 7.8 *Let m be some partition of $[0,1]$ with $D_m + 1$ pieces and $\chi_n^2(m)$ be the chi-square statistics given by (5.53). Then for any positive real numbers ε and x,*

$$\mathbb{P}\left[\chi_n(m) \mathbb{1}_{\Omega_m(\varepsilon)} \geq (1 + \varepsilon) \left(\sqrt{D_m} + \sqrt{2x} \right) \right] \leq \exp(-x) \tag{7.47}$$

where $\Omega_m(\varepsilon) = \{ |P_n(I) - P(I)| \leq 2\varepsilon P(I)/\kappa(\varepsilon), \text{ for every } I \in m \}$ and $\kappa(\varepsilon) = 2\left(\varepsilon^{-1} + 1/3 \right)$.

Proof. Let $\theta = 2\varepsilon/\kappa(\varepsilon)$ and z be some positive number to be chosen later. Setting $\varphi_I = P(I)^{-1/2}\, \mathbb{1}_I$ for every $I \in m$ and denoting by $\mathcal{S}_m$ the unit sphere in $\mathbb{R}^m$ as before, we have on the one hand

$$\Omega_m(\varepsilon) = \left\{ |\nu_n(\varphi_I)| \leq \theta \sqrt{P(I)}, \text{ for every } I \in m \right\}$$

and on the other hand, by Cauchy–Schwarz inequality

$$n^{-1/2} \chi_n(m) = \left[\sum_{I \in m} \nu_n^2 (\varphi_I) \right]^{1/2} \geq \left| \sum_{I \in m} a_I \nu_n (\varphi_I) \right| \text{ for all } a \in \mathcal{S}_m,$$

with equality when $a_I = \nu_n\left(\varphi_I\right)\left(n^{-1/2}\chi_n\left(m\right)\right)^{-1}$ for all $I \in m$. Hence, defining $\mathcal{A}_m$ to be the set of those elements $a \in \mathcal{S}_m$ satisfying

$$\sup_{I \in m}\left(\left|a_I\right|\left(P\left(I\right)\right)^{-1/2}\right) \leq \sqrt{n}\theta/z,$$

we have that on the event $\Omega_m\left(\varepsilon\right) \cap \{\chi_n\left(m\right) \geq z\}$

$$n^{-1/2}\chi_n\left(m\right) = \sup_{a \in \mathcal{A}_m}\left|\sum_{I \in m}a_I\nu_n\left(\varphi_I\right)\right| = \sup_{a \in \mathcal{A}_m}\left|\nu_n\left[\sum_{I \in m}a_I\varphi_I\right]\right|. \qquad (7.48)$$

Moreover the same identity holds when replacing $\mathcal{A}_m$ by some countable and dense subset $\mathcal{A}'_m$ of $\mathcal{A}_m$, so that applying (5.50) to the countable set of functions

$$\left\{\sum_{I \in m}a_I\varphi_I, a \in \mathcal{A}'_m\right\},$$

we derive that for every positive x

$$\mathbb{P}\left[\sup_{a \in \mathcal{A}_m}\left|\nu_n\left(\sum_{I \in m}a_I\varphi_I\right)\right| \geq \left(1+\varepsilon\right)E_m + \sigma_m\sqrt{\frac{2x}{n}} + \kappa\left(\varepsilon\right)\frac{b_m}{n}x\right] \leq e^{-x},$$

$$(7.49)$$

where

$$E_m = \mathbb{E}\left[\sup_{a \in \mathcal{A}_m}\left|\nu_n\left(\sum_{I \in m}a_I\varphi_I\right)\right|\right] \leq \frac{\mathbb{E}\left(\chi_n\left(m\right)\right)}{\sqrt{n}} \leq \sqrt{\frac{D_m}{n}},$$

$$\sigma_m^2 = \sup_{a \in \mathcal{A}_m}\mathrm{Var}\left(\sum_{I \in m}a_I\varphi_I\left(\xi_1\right)\right) \leq \sup_{a \in \mathcal{S}_m}\mathbb{E}\left[\sum_{I \in m}a_I\varphi_I\left(\xi_1\right)\right]^2 \leq 1,$$

and

$$b_m = \sup_{a \in \mathcal{A}_m}\left\|\sum_{I \in m}a_I\varphi_I\right\|_\infty = \sup_{a \in \mathcal{A}_m}\sup_{I \in m}\frac{\left|a_I\right|}{\sqrt{P\left(I\right)}} \leq \frac{\theta\sqrt{n}}{z}.$$

Hence we get by (7.48) and (7.49),

$$\mathbb{P}\left[\chi_n\left(m\right)\mathbb{1}_{\Omega_m\left(\varepsilon\right)\cap\{\chi_n\left(m\right)\geq z\}} \geq \left(1+\varepsilon\right)\sqrt{D_m} + \sqrt{2x} + \kappa\left(\varepsilon\right)\frac{\theta}{z}x\right] \leq e^{-x}.$$

If we now choose $z = \sqrt{2x}$ and take into account the definition of θ, we get

$$\mathbb{P}\left[\chi_n\left(m\right)\mathbb{1}_{\Omega_m\left(\varepsilon\right)} \geq \left(1+\varepsilon\right)\left(\sqrt{D_m} + \sqrt{2x}\right)\right] \leq e^{-x}. \quad \blacksquare$$

Remark. It is well known that given some partition m of $[0,1]$, when $n \to +\infty$, $\chi_n\left(m\right)$ converges in distribution to $\|Y\|$, where Y is a standard

Gaussian vector in $\mathbb{R}^{D_m}$. An easy exercise consists in deriving from (5.1) a tail bound for the chi-square distribution with D_m degrees of freedom. Indeed by Cauchy–Schwarz inequality $\mathbb{E}\left[\|Y\|\right] \leq \sqrt{D_m}$ and therefore

$$\mathbb{P}\left[\|Y\| \geq \sqrt{D_m} + \sqrt{2x}\right] \leq e^{-x}. \tag{7.50}$$

One can see that (7.47) is very close to (7.50).

We are now in a position to control uniformly a collection of square roots of chi-square statistics $\{\chi_n(m),\ m \in \mathcal{M}\}$, under the following mild restriction on the collection of partitions $\mathcal{M}$.

$(\mathbf{H_0})$: *Let N be some integer such that $N \leq n\,(\ln(n))^{-2}$ and m_N be a partition of $[0, 1]$ the elements of which are intervals with equal length $(N + 1)^{-1}$. We assume that every element of any partition m belonging to $\mathcal{M}$ is the union of pieces of m_N.*

Assume that $(\mathbf{H_0})$ holds. Given $\eta \in (0, 1)$, setting

$$\Omega(\eta) = \{|P_n(I) - P(I)| \leq \eta P(I),\ \text{for every } I \in m_N\} \tag{7.51}$$

one has

$$\Omega(\eta) \subset \bigcap_{m \in \mathcal{M}} \{|P_n(I) - P(I)| \leq \eta P(I),\ \text{for every } I \in m\}.$$

Therefore, given some arbitrary family of positive numbers $(y_m)_{m \in \mathcal{M}}$, provided that $\eta \leq 2\varepsilon/\kappa(\varepsilon)$, we derive from (7.47) that

$$\mathbb{P}\left[\bigcup_{m \in \mathcal{M}} \left\{\chi_n(m)\mathbb{1}_{\Omega(\eta)} \geq (1 + \varepsilon)\left(\sqrt{D_m} + \sqrt{2y_m}\right)\right\}\right] \leq \sum_{m \in \mathcal{M}} e^{-y_m}. \tag{7.52}$$

This inequality is the required tool to evaluate the penalty function and establish a risk bound for the corresponding penalized MLE.

7.3.2 A Model Selection Result

Another advantage brought by the restriction to $\Omega(\eta)$ when assuming that $(\mathbf{H_0})$ holds, is that for every $m \in \mathcal{M}$, the ratios $\widehat{s}_m/s_m$ remain bounded on this set, which implies that $V^2(\widehat{s}_m, s_m)$ is of the order of $\mathbf{K}(s_m, \hat{s}_m)$. More precisely, on the set $\Omega(\eta)$, one has $P_n(I) \geq (1 - \eta)P(I)$ for every $I \in \mathcal{M}$ and therefore

$$\widehat{s}_m \geq (1 - \eta)s_m,$$

which implies by Lemma 7.24 (see the Appendix) that

$$\mathbf{K}(s_m, \hat{s}_m) \geq \left(\frac{1 - \eta}{2}\right)\int s_m \ln^2\left(\frac{\hat{s}_m}{s_m}\right) d\mu.$$

Since $\ln^2(\hat{s}_m/s_m)$ is piecewise constant on the partition m

$$\int s_m \ln^2 \left(\frac{\hat{s}_m}{s_m} \right) d\mu = \int s \ln^2 \left(\frac{\hat{s}_m}{s_m} \right) d\mu = V^2 \left(\hat{s}_m, s_m \right),$$

and therefore, for every $m \in \mathcal{M}$

$$\mathbf{K} \left(s_m, \hat{s}_m \right) \mathbb{1}_{\Omega(\eta)} \geq \left(\frac{1 - \eta}{2} \right) V^2 \left(\hat{s}_m, s_m \right) \mathbb{1}_{\Omega(\eta)}. \qquad (7.53)$$

This allows to better understand the structure of the proof of Theorem 7.9 below. Indeed, provided that

$$\eta \leq \frac{\varepsilon}{1 + \varepsilon}, \qquad (7.54)$$

one derives from (7.46) and (7.53) that on the set $\Omega(\eta)$,

$$\mathbf{K} \left(s, \tilde{s} \right) \leq \mathbf{K} \left(s, s_m \right) + \text{pen} \left(m \right) + \nu_n \left(\ln \frac{s}{s_m} \right)$$
$$+ n^{-1/2} \sqrt{2 \left(1 + \varepsilon \right) \mathbf{K} \left(s_{\widehat{m}}, \hat{s}_{\widehat{m}} \right)} \chi_n \left(\widehat{m} \right) + \nu_n \left(\ln \frac{s_{\widehat{m}}}{s} \right)$$
$$- \text{pen} \left(\widehat{m} \right).$$

Now, by (7.42) $\mathbf{K} \left(s, \tilde{s} \right) = \mathbf{K} \left(s, s_{\widehat{m}} \right) + \mathbf{K} \left(s_{\widehat{m}}, \tilde{s} \right)$, hence, taking into account that

$$n^{-1/2} \sqrt{2 \left(1 + \varepsilon \right) \mathbf{K} \left(s_{\widehat{m}}, \hat{s}_{\widehat{m}} \right)} \chi_n \left(\widehat{m} \right) \leq \left(1 + \varepsilon \right)^{-1} \mathbf{K} \left(s_{\widehat{m}}, \hat{s}_{\widehat{m}} \right) + \frac{\chi_n^2 \left(\widehat{m} \right) \left(1 + \varepsilon \right)^2}{2n},$$

one derives that on the set $\Omega(\eta)$,

$$\mathbf{K} \left(s, s_{\widehat{m}} \right) + \frac{\varepsilon}{1 + \varepsilon} \mathbf{K} \left(s_{\widehat{m}}, \tilde{s} \right) \leq \mathbf{K} \left(s, s_m \right) + \text{pen} \left(m \right) + \nu_n \left(\ln \frac{s}{s_m} \right)$$
$$+ \frac{\chi_n^2 \left(\widehat{m} \right) \left(1 + \varepsilon \right)^2}{2n} + \nu_n \left(\ln \frac{s_{\widehat{m}}}{s} \right)$$
$$- \text{pen} \left(\widehat{m} \right). \qquad (7.55)$$

Neglecting the terms $\nu_n \left(\ln s / s_m \right)$ and $\nu_n \left(\ln s_{\widehat{m}} / s \right)$, we see that the penalty pen $\left(\widehat{m} \right)$ should be large enough to compensate $\chi_n^2 \left(\widehat{m} \right) \left(1 + \varepsilon \right)^2 / 2n$ with high probability. Since we have at our disposal the appropriate exponential bound to control chi-square statistics uniformly over the family of partitions $\mathcal{M}$, it remains to control

$$\nu_n \left(\ln \left(s / s_m \right) \right) + \nu_n \left(\ln \left(s_{\widehat{m}} / s \right) \right).$$

The trouble is that there is no way to warrant that the ratios $s_{\widehat{m}} / s$ remain bounded except by making some extra unpleasant preliminary assumption on s. This makes delicate the control of $\nu_n \left[\ln \left(s_{\widehat{m}} / s \right) \right]$ as a function of $\mathbf{K} \left(s, s_{\widehat{m}} \right)$

as one should expect. This is the reason why we shall rather pass to the control of Hellinger loss rather than Kullback–Leibler loss.

Let us recall that the Hellinger distance $\mathbf{h}(f,g)$ between two densities f and g on $[0,1]$ is defined by

$$\mathbf{h}^2(f,g) = \frac{1}{2} \int \left(\sqrt{f} - \sqrt{g} \right)^2.$$

It is known (see Lemma 7.23 below) that

$$2\mathbf{h}^2(f,g) \leq \mathbf{K}(f,g), \tag{7.56}$$

and that a converse inequality exists whenever $\|\ln(f/g)\|_\infty < \infty$. This in some sense confirms that it is slightly easier (although very close by essence) to control Hellinger risk as compared to Kullback–Leibler risk. The following result is due to Castellan (it is in fact a particular case of Theorem 3.2. in [34]).

Theorem 7.9 *Let $\xi_1, \ldots, \xi_n$ be some independent $[0,1]$-valued random variables with common distribution $P = s\mu$, where μ denotes the Lebesgue measure. Consider a finite family $\mathcal{M}$ of partitions of $[0,1]$ satisfying to assumption* $(\mathbf{H_0})$. *Let, for every partition m*

$$\hat{s}_m = \sum_{I \in m} \frac{P_n(I)}{\mu(I)} \mathbb{1}_I \quad and \quad s_m = \sum_{I \in m} \frac{P(I)}{\mu(I)} \mathbb{1}_I$$

be respectively the histogram estimator and the histogram projection of s, based on m. Consider some absolute constant Σ and some family of nonnegative weights $\{x_m\}_{m \in \mathcal{M}}$ such that

$$\sum_{m \in \mathcal{M}} e^{-x_m} \leq \Sigma. \tag{7.57}$$

Let $c_1 > 1/2$, $c_2 = 2\left(1 + c_1^{-1}\right)$ and consider some penalty function pen $: \mathcal{M} \to \mathbb{R}_+$ *such that*

$$\text{pen}(m) \geq \frac{c_1}{n} \left(\sqrt{D_m} + \sqrt{c_2 x_m} \right)^2, \quad for\ all\ m \in \mathcal{M},$$

where $D_m + 1$ denotes the number of elements of partition m. Let $\hat{m}$ minimizing the penalized log-likelihood criterion

$$-\int \hat{s}_m \ln(\hat{s}_m)\, d\mu + \text{pen}(m)$$

over $m \in \mathcal{M}$ and define the penalized MLE by $\tilde{s} = s_{\hat{m}}$. If for some positive real number ρ, $s \geq \rho$ almost everywhere and $\int s (\ln s)^2 d\mu \leq L < \infty$, then for some constant $C(c_1, \rho, L, \Sigma)$,

$$\mathbb{E}\left[\mathbf{h}^2\left(s,\tilde{s}\right)\right] \le \frac{(2c_1)^{1/5}}{(2c_1)^{1/5}-1} \inf_{m\in\mathcal{M}}\left\{\mathbf{K}\left(s,s_m\right)+\text{pen}\left(m\right)\right\}$$
$$+\frac{C\left(c_1,\rho,L,\Sigma\right)}{n}.$$

Proof. Let z be given, $\varepsilon > 0$ to be chosen later and

$$\eta = \frac{\varepsilon}{1+\varepsilon} \wedge \frac{2\varepsilon}{\kappa\left(\varepsilon\right)}.$$

Hellinger distance will appear in the above analysis of the Kullback–Leibler risk (see (7.55)) through the control of $\nu_n\left(\ln\left(s_{\widehat{m}}/s\right)\right)$ which can be performed via Proposition 7.27 of the Appendix. Indeed, except on a set with probability less that $\sum_{m'\in\mathcal{M}}\exp\left(-y_{m'}\right)$ one has for every $m'\in\mathcal{M}$

$$\nu_n\left(\ln\left(s_{m'}/s\right)\right) \le \mathbf{K}\left(s,s_{m'}\right) - 2\mathbf{h}^2\left(s,s_{m'}\right) + 2y_{m'}/n$$

and therefore a fortiori

$$\nu_n\left(\ln\left(s_{\widehat{m}}/s\right)\right) \le \mathbf{K}\left(s,s_{\widehat{m}}\right) - 2\mathbf{h}^2\left(s,s_{\widehat{m}}\right) + 2y_{\widehat{m}}/n.$$

Let $\Omega\left(\eta\right)$ be defined by (7.51). Setting for every $m\in\mathcal{M}$, $y_m = x_m + z$, since (7.54) holds because of our choice of η, it comes from the previous inequality and (7.55) that on the set $\Omega\left(\eta\right)$ and except on a set with probability less than Σe^{-z}, one has for every $m\in\mathcal{M}$

$$\mathbf{K}\left(s,s_{\widehat{m}}\right) + \frac{\varepsilon}{1+\varepsilon}\mathbf{K}\left(s_{\widehat{m}},\tilde{s}\right) \le \mathbf{K}\left(s,s_m\right) + \text{pen}\left(m\right) + \nu_n\left(\ln\left(\frac{s}{s_m}\right)\right)$$
$$+ \frac{\chi_n^2\left(\widehat{m}\right)\left(1+\varepsilon\right)^2}{2n} + \mathbf{K}\left(s,s_{\widehat{m}}\right)$$
$$- 2\mathbf{h}^2\left(s,s_{\widehat{m}}\right) + 2\frac{x_{\widehat{m}}+z}{n} - \text{pen}\left(\widehat{m}\right).$$

Equivalently

$$2\mathbf{h}^2\left(s,s_{\widehat{m}}\right) + \frac{\varepsilon}{1+\varepsilon}\mathbf{K}\left(s_{\widehat{m}},\tilde{s}\right) \le \mathbf{K}\left(s,s_m\right) + \text{pen}\left(m\right)$$
$$+ \nu_n\left(\ln\left(\frac{s}{s_m}\right)\right) + \frac{\left(1+\varepsilon\right)^2\chi_n^2\left(\widehat{m}\right)}{2n}$$
$$+ 2\frac{x_{\widehat{m}}+z}{n} - \text{pen}\left(\widehat{m}\right).$$

Now, by the triangle inequality,

$$\mathbf{h}^2\left(s,\tilde{s}\right) \le 2\mathbf{h}^2\left(s,s_{\widehat{m}}\right) + 2\mathbf{h}^2\left(s_{\widehat{m}},\tilde{s}\right).$$

Hence, using (7.56), we derive that on $\Omega\left(\eta\right)$ and except on a set with probability less than Σe^{-z}, the following inequality holds:

$$\frac{\varepsilon}{1+\varepsilon} \mathbf{h}^2 (s, \tilde{s}) \leq \mathbf{K} (s, s_m) + \text{pen} (m) + \nu_n \left(\ln \left(\frac{s}{s_m} \right) \right)$$
$$+ \frac{(1+\varepsilon)^2 \chi_n^2 (\widehat{m})}{2n} + 2 \frac{x_{\widehat{m}} + z}{n} - \text{pen} (\widehat{m}) . \tag{7.58}$$

Now we can use the above uniform control of chi-square statistics and derive from (7.52) that on the set $\Omega (\eta)$ and except on a set with probability less than Σe^{-z}

$$\chi_n^2 (\widehat{m}) \leq (1+\varepsilon)^2 \left(\sqrt{D_{\widehat{m}}} + \sqrt{2\kappa (x_{\widehat{m}} + z)} \right)^2$$
$$\leq (1+\varepsilon)^2 \left[(1+\varepsilon) \left(\sqrt{D_{\widehat{m}}} + \sqrt{2\kappa x_{\widehat{m}}} \right)^2 + 2\kappa z \left(1 + \varepsilon^{-1} \right) \right] .$$

Plugging this inequality in (7.58) implies that on the set $\Omega (\eta)$ and except on a set with probability less than $2\Sigma e^{-z}$,

$$\frac{\varepsilon}{1+\varepsilon} \mathbf{h}^2 (s, \tilde{s}) \leq \mathbf{K} (s, s_m) + \text{pen} (m) + \nu_n \left(\ln \left(\frac{s}{s_m} \right) \right)$$
$$+ \frac{(1+\varepsilon)^5}{2n} \left(\sqrt{D_{\widehat{m}}} + \sqrt{2\kappa x_{\widehat{m}}} \right)^2 + 2 \frac{x_{\widehat{m}}}{n} - \text{pen} (\widehat{m})$$
$$+ \frac{z}{n} \left(\kappa \varepsilon^{-1} (1+\varepsilon)^5 + 2 \right) .$$

Now we can notice that choosing ε adequately, i.e., such that $c_1 = (1+\varepsilon)^5 / 2$ ensures that

$$\frac{(1+\varepsilon)^5}{2n} \left(\sqrt{D_{\widehat{m}}} + \sqrt{2\kappa x_{\widehat{m}}} \right)^2 + 2 \frac{x_{\widehat{m}}}{n} - \text{pen} (\widehat{m}) \leq 0.$$

Hence, except on a set of probability less than $2\Sigma e^{-z}$, the following inequality is available:

$$\frac{\varepsilon}{1+\varepsilon} \mathbf{h}^2 (s, \tilde{s}) \, \mathbb{1}_{\Omega(\eta)} \leq \mathbf{K} (s, s_m) + \text{pen} (m) + \nu_n \left(\ln \left(\frac{s}{s_m} \right) \right) \mathbb{1}_{\Omega(\eta)}$$
$$+ \frac{z}{n} \left(\varepsilon^{-1} (1+\varepsilon)^5 + 2 \right) .$$

Integrating this inequality with respect to z leads to

$$\frac{\varepsilon}{1+\varepsilon} \mathbb{E} \left[\mathbf{h}^2 (s, \tilde{s}) \, \mathbb{1}_{\Omega(\eta)} \right] \leq \mathbf{K} (s, s_m) + \text{pen} (m)$$
$$+ \mathbb{E} \left[\nu_n \left(\ln \left(\frac{s}{s_m} \right) \right) \mathbb{1}_{\Omega(\eta)} \right]$$
$$+ \frac{2\Sigma}{n} \left(\varepsilon^{-1} (1+\varepsilon)^5 + 2 \right) .$$

Since $\nu_n \left(\ln \left(s / s_m \right) \right)$ is centered at expectation and the Hellinger distance is bounded by 1, it follows from the above inequality that

$$\frac{\varepsilon}{1 + \varepsilon} \mathbb{E} \left[\mathbf{h}^2 \left(s, \tilde{s} \right) \right] \leq \mathbf{K} \left(s, s_m \right) + \text{pen} \left(m \right) + \frac{2 \Sigma}{n} \left(\varepsilon^{-1} \left(1 + \varepsilon \right)^5 + 2 \right)$$
$$+ \mathbb{E} \left[\left(- \nu_n \left(\ln \left(\frac{s}{s_m} \right) \right) + 1 \right) \mathbb{1}_{\Omega^c (\eta)} \right]. \qquad (7.59)$$

It remains to bound the last term of the right-hand side of the above inequality. By Cauchy–Schwarz inequality

$$\mathbb{E} \left[- \nu_n \left(\ln \left(\frac{s}{s_m} \right) \right) \mathbb{1}_{\Omega^c (\eta)} \right] \leq \left(\frac{1}{n} \int s \left(\ln \left(\frac{s}{s_m} \right) \right)^2 d\mu \right)^{1/2} \left(\mathbb{P} \left[\Omega^c (\eta) \right] \right)^{1/2},$$

and

$$\int s \left(\ln \frac{s}{s_m} \right)^2 d\mu \leq 2 \left(\int s \left(\ln s \right)^2 d\mu + \int s \left(\ln s_m \right)^2 d\mu \right)$$
$$\leq 2 \left(\int s \left(\ln s \right)^2 d\mu + \left(\ln \left(\frac{1}{\rho} \vee n \right) \right)^2 \right),$$

since $\rho \leq s_m \leq n$. Moreover, setting $\delta = \inf_{I \in m_N} P \left(I \right)$ it follows from Bernstein's inequality that

$$\mathbb{P} \left[\Omega^c (\eta) \right] \leq 2 \left(N + 1 \right) \exp \left(- \frac{n \eta^2 \delta}{2 \left(1 + \eta / 3 \right)} \right)$$

yielding, because of the restriction $N + 1 \leq n \left(\ln \left(n \right) \right)^{-2}$ (see $(\mathbf{H_0})$),

$$\mathbb{P} \left[\Omega^c (\eta) \right] \leq 2 n \exp \left(- \frac{\eta^2 \rho \left(\ln \left(n \right) \right)^2}{2 \left(1 + \eta / 3 \right)} \right).$$

This shows that, as a function of n, $\mathbb{P} \left[\Omega^c (\eta) \right]$ tends to 0 faster than any power of n. Collecting the above inequalities and plugging them into (7.59) finishes the proof of the theorem. ∎

Theorem 7.9 suggests to take a penalty function of the form:

$$\text{pen} \left(m \right) = \frac{c_1}{n} \left(\sqrt{D_m} + \sqrt{c_2 x_m} \right)^2,$$

where the weights x_m satisfy (7.57) and, of course, the constant c_1 and c_2 are independent of the density s. The choice $c_1 > 1/2$ provides an upper bound for the Hellinger risk of the penalized MLE:

$$\mathbb{E} \left[\mathbf{h}^2 \left(s, \tilde{s} \right) \right] \leq C_1 \inf_{m \in \mathcal{M}} \left\{ \mathbf{K} \left(s, s_m \right) + \text{pen} \left(m \right) \right\} + \frac{C_2}{n} \qquad (7.60)$$

where the constant C_1 does not depend on s whereas the constant C_2 depends on s (via ρ and L) and on the family of models (via Σ). Furthermore, the constant C_1, which depends only on c_1, converges to infinity when c_1 tends to $1/2$. This suggests that on the one hand c_1 should be chosen substantially larger than $1/2$ and on the other hand that one could get into trouble when choosing $c_1 < 1/2$. Using further refinements of the above method, it is proved in [34] that the special choice $c_1 = 1$ optimizes the risk bound (7.60).

7.3.3 Choice of the Weights $\{x_m, m \in \mathcal{M}\}$

The penalty function depends on the family $\mathcal{M}$ through the choice of the weights x_m satisfying (7.57). A reasonable way of choosing those weights is to make them depend on m only through the dimension D_m. More precisely, we are interested in weights of the form

$$x_m = L(D_m) D_m.$$

With such a definition the number of histogram models S_m having the same dimension plays a fundamental role for bounding the series (7.57) and therefore to decide what value of $L(D)$ should be taken in order to get a reasonable value for Σ. Let us consider two extreme examples.

- **Case of regular histograms.** Let J be the largest integer such that 2^J is not larger than $n(\ln(n))^{-2}$. Let $\mathcal{M}_J^r$ be the collection of regular partitions with 2^j pieces with $j \leq J$. Then assumption $(\mathbf{H_0})$ is satisfied and since there is only one model per dimension, $L(D)$ can be taken as some arbitrary positive constant η and

$$\sum_{m \in \mathcal{M}_J^r} e^{-\eta D_m} \leq \sum_{j=0}^{\infty} e^{-\eta 2^j} \leq \eta^{-1}.$$

Consequently, all penalties of the form

$$\text{pen}(m) = c \frac{D_m}{n},$$

with $c > 1/2$ are allowed, including that of Akaike, namely $c = 1$. Since $\mathbf{K}(s, s_m) + D_m/2$ represents actually the order of the Kullback–Leibler risk of the histogram estimator $\hat{s}_m$ (see ([34])), the meaning of (7.60) is that, up to constant, $\tilde{s}$ behaves like an oracle. This is not exactly true in terms of the Kullback–Leibler loss since we have bounded the Hellinger risk instead of the Kullback–Leibler risk. However when the log-ratios $\ln(s/s_m)$ remain uniformly bounded, then the Kullback–Leibler bias $\mathbf{K}(s, s_m)$ is of the order of $\mathrm{h}^2(s, s_m)$ and (7.60) can be interpreted as an oracle inequality for the Hellinger loss. It should be noticed that the statement of the theorem provides some flexibility concerning the choice of the penalty function so that we could take as well

$$\text{pen}(m) = c\frac{D_m}{n} + c'\frac{D_m^{\alpha}}{n}$$

for some $\alpha \in (0,1)$. As already mentioned, the choice $c = 1$ can be shown to optimize the risk bound for the corresponding penalized MLE and the structure of the proof made in ([34]) tends to indicate that it would be desirable to choose a penalty function which is slightly heavier than what is proposed in Akaike's criterion. This is indeed confirmed by simulations in [26], the gain being especially spectacular for small or moderate values of the sample size n (we mean less than 200).

- **Case of irregular histograms.** We consider here the family $\mathcal{M}_N^{ir}$ of all partitions built from a single regular partition m_N with $N+1$ pieces where N is less than $n(\ln(n))^{-2}$. Then the cardinality of the family of partitions belonging to $\mathcal{M}_N^{ir}$ with a number of pieces equal to $D+1$ is bounded by $\binom{N}{D}$. Hence

$$\sum_{m \in \mathcal{M}_N^{ir}} e^{-x_m} \leq \sum_{D=1}^{N} \binom{N}{D} e^{-L(D)D} \leq \sum_{D \geq 1} \left(\frac{eN}{D}\right)^D e^{-L(D)D},$$

and the choice $L(D) = L + \ln(eN/D)$ implies that condition (7.57) holds with $\Sigma = (e^L - 1)^{-1}$. This leads to a penalty function of the form

$$\text{pen}(m) = c\frac{D_m}{n}\ln\left(\frac{N}{D_m}\right) + c'\frac{D_m}{n},$$

for large enough constants c and c'. The corresponding risk bound can be written as:

$$\mathbb{E}\left[\mathbf{h}^2(s,\tilde{s})\right] \leq C\left[\inf_{D \leq N} \inf_{\mathcal{M}_N^{ir}(D)}\left\{\mathbf{K}(s,s_m) + \frac{D}{n}\left(1 + \ln\left(\frac{N}{D}\right)\right)\right\}\right]$$

where $\mathcal{M}_N^{ir}(D)$ denotes the set of partitions m with dimension $D_m = D$. This means that, given some integer D, whenever s belongs to $\mathcal{S}_D = \cup_{m \in \mathcal{M}_N^{ir}(D)} S_m$, the Hellinger risk of s is bounded by $CD/n(1 + \ln(N/D))$. This shows that, because of the extra logarithmic factor, the penalized MLE fails to mimic the oracle in terms of Hellinger loss. One can wonder whether this is due to a weakness of the method or not. The necessity of this extra logarithmic factor is proved in [21] (see Proposition 2 therein) where the minimax risk over the set $\mathcal{S}_D$ is shown to be bounded from below by $D/n(1 + \ln(N/D))$, up to some constant. In this sense the above risk bound is optimal.

7.3.4 Lower Bound for the Penalty Function

One can also wonder whether the condition $c_1 > 1/2$ in Theorem 7.9 is necessary or not. We cannot answer this question in full generality. The following result shows that, when there are only a few models per dimension, taking $\text{pen}(m) = cD_m/n$ for some arbitrary constant $c < 1/2$ leads to a disaster in the sense that, if the true s is uniform, the penalized log-likelihood selection criterion will choose models of large dimension with high probability and the Hellinger risk will be bounded away from 0 when n goes to infinity. The proof of this result heavily relies on the inequalities for the right and also for the left tails of chi-square statistics. The proof being quite similar to that of Theorem 7.9, we skip it and refer the interested reader to [34] (and also to Chapter 4, where a similar result is proved in the Gaussian framework).

Theorem 7.10 *Let* $\xi_1, \ldots, \xi_n$ *be some independent* $[0, 1]$*-valued random variables with common distribution* $P = s\mu$ *with* $s = \mathbb{1}_{[0,1]}$. *Consider some finite family of partitions* $\mathcal{M}$ *such that for each integer* D, *there exists only one partition* m *such that* $D_m = D$. *Moreover, let us assume that* $\mu(I) \geq (\ln(n))^2/n$ *for every* $I \in m$ *and* $m \in \mathcal{M}$.

Assume that for some partition $m_N \in \mathcal{M}$ *with* $N + 1$ *pieces one has*

$$\text{pen}(m_N) = c\,\frac{N}{n}$$

with $c < 1/2$. *Let* $\widehat{m}$ *be the minimizer over* $\mathcal{M}$ *of the penalized log-likelihood criterion*

$$-\int \widehat{s}_m \ln(\widehat{s}_m) + \text{pen}(m).$$

Then, whatever the values of $\text{pen}(m)$ *for* $m \neq m_N$ *there exist positive numbers* N_0 *and* L, *depending only on* c, *such that, for all* $N \geq N_0$,

$$\mathbb{P}\left[D_{\widehat{m}} \geq \frac{1 - 4c^2}{4} N\right] \geq 1 - \beta(c),$$

where

$$\beta(c) = \Sigma(L) \exp\left[-\frac{L}{2}\left((1 - 4c^2)\,N\right)^{\frac{1}{2}}\right] + \frac{C(c)}{n} \quad \text{with } \Sigma(L) = \sum_{D \geq 1} e^{-L\sqrt{D}}.$$

Moreover, if $\tilde{s} = \widehat{s}_{\widehat{m}}$,

$$\mathbb{E}\left[\mathbf{K}(s, \tilde{s})\right] \geq \delta(c)\left[1 - \beta(c)\right]\frac{N}{n}$$

where $\delta(c) = (1 - 2c)(1 + 2c)^2/16$.

7.4 A General Model Selection Theorem for MLE

The previous approach for studying histogram selection heavily relies on the linearity of the centered log-likelihood process on histograms. This property carries over to exponential families of a linear finite dimensional space and

Castellan has indeed been able to extend the previous results to exponential families of piecewise polynomials (see [35]). Our purpose is now to consider more general models. The price to pay for generality is that the absolute constants involved in the penalty terms proposed in the general model selection theorem below will become unrealistic.

7.4.1 Local Entropy with Bracketing Conditions

An adequate general tool to bound the Hellinger risk of a MLE on a given model is entropy with bracketing. More precisely, if S is a set of probability densities with respect to some positive measure μ, we denote by $\sqrt{S}$ the set $\{\sqrt{t}, t \in S\}$ and consider $\sqrt{S}$ as a metric subspace of $\mathbb{L}_2(\mu)$. Denoting by $\mathbf{h}$ the Hellinger distance between probability densities, the relationship

$$\left\| \sqrt{t} - \sqrt{u} \right\|_2 = \sqrt{2}\mathbf{h}(t, u),$$

implies that $\left(\sqrt{S}, \|.\|_2 \right)$ is isometric to $(S, \sqrt{2}\mathbf{h})$. We denote by $H_{[.]} \left(., \sqrt{S} \right)$, the $\mathbb{L}_2(\mu)$ entropy with bracketing of $\sqrt{S}$. Recall that for every positive ε, $H_{[.]} \left(\varepsilon, \sqrt{S} \right)$ is the logarithm of the minimal number of intervals (or brackets) $[f_L, f_U]$ with extremities f_L, f_U belonging to $\mathbb{L}_2(\mu)$ such that $\|f_L - f_U\|_2 \leq \varepsilon$, which are necessary to cover $\sqrt{S}$. Of course since $0 \leq (f_U \vee 0) - (f_L \vee 0) \leq f_U - f_L$, possibly changing f_L and f_U into $f_L \vee 0$ and $f_U \vee 0$, we may always assume that the brackets involved in a covering of $\sqrt{S}$ are of the form $\left[\sqrt{t_L}, \sqrt{t_U} \right]$, where t_L and t_U belong to $\mathbb{L}_1(\mu)$. This is what we shall do in what follows. Given some collection of models $\{S_m\}_{m \in \mathcal{M}}$, we shall assume for each model S_m the square entropy with bracketing $\sqrt{H_{[.]} \left(\varepsilon, \sqrt{S_m} \right)}$ to be integrable at 0. We consider some function ϕ_m on $\mathbb{R}_+$ with the following properties

(i) ϕ_m is nondecreasing, $x \to \phi_m(x)/x$ is nonincreasing on $(0, +\infty)$ and for every $\sigma \in \mathbb{R}_+$ and every $u \in S_m$

$$\int_0^\sigma \sqrt{H_{[.]} \left(x, \sqrt{S_m(u, \sigma)} \right)} \, dx \leq \phi_m(\sigma),$$

where $S_m(u, \sigma) = \{t \in S_m : \left\| \sqrt{t} - \sqrt{u} \right\|_2 \leq \sigma\}$.

Of course, we may always take ϕ_m as the concave function

$$\sigma \to \int_0^\sigma \sqrt{H_{[.]} \left(\varepsilon, \sqrt{S_m} \right)} \, d\varepsilon,$$

but we shall see that when S_m is defined by a finite number of parameters, it is better to consider this more local version of the integrated square entropy in order to avoid undesirable logarithmic factors. In order to avoid measurability problems we shall consider the following separability condition on the models.

(M) There exists some countable subset S'_m of S_m and a set $\mathcal{X}'$ with $\mu(\mathcal{X} \setminus \mathcal{X}') = 0$ such that for every $t \in S_m$, there exists some sequence $(t_k)_{k \geq 1}$ of elements of S'_m such that for every $x \in \mathcal{X}'$, $\ln(t_k(x))$ tends to $\ln(t(x))$ as k tends to infinity.

Theorem 7.11 *Let $X_1, \ldots, X_n$ be i.i.d. random variables with unknown density s with respect to some positive measure μ. Let $\{S_m\}_{m \in \mathcal{M}}$ be some at most countable collection of models, where for each $m \in \mathcal{M}$, the elements of S_m are assumed to be probability densities with respect to μ and S_m fulfills (**M**). We consider a corresponding collection of ρ-MLEs $(\widehat{s}_m)_{m \in \mathcal{M}}$ which means that for every $m \in \mathcal{M}$*

$$P_n\left(-\ln(\widehat{s}_m)\right) \leq \inf_{t \in S_m} P_n\left(-\ln(t)\right) + \rho.$$

Let $\{x_m\}_{m \in \mathcal{M}}$ be some family of nonnegative numbers such that

$$\sum_{m \in \mathcal{M}} e^{-x_m} = \Sigma < \infty,$$

and for every $m \in \mathcal{M}$ considering ϕ_m with property (i) define σ_m as the unique positive solution of the equation

$$\phi_m(\sigma) = \sqrt{n}\sigma^2.$$

Let pen *: $\mathcal{M} \to \mathbb{R}_+$ and consider the penalized log-likelihood criterion*

$$\mathrm{crit}(m) = P_n\left(-\ln(\widehat{s}_m)\right) + \mathrm{pen}(m).$$

Then, there exists some absolute constants κ and C such that whenever

$$\mathrm{pen}(m) \geq \kappa\left(\sigma_m^2 + \frac{x_m}{n}\right) \text{ for every } m \in \mathcal{M}, \tag{7.61}$$

some random variable $\widehat{m}$ minimizing crit *over $\mathcal{M}$ does exist and moreover, whatever the density s*

$$\mathbb{E}_s\left[\mathbf{h}^2\left(s, \widehat{s}_{\widehat{m}}\right)\right] \leq C\left(\inf_{m \in \mathcal{M}}\left(\mathbf{K}(s, S_m) + \mathrm{pen}(m)\right) + \rho + \frac{\Sigma}{n}\right), \tag{7.62}$$

where, for every $m \in \mathcal{M}$, $\mathbf{K}(s, S_m) = \inf_{t \in S_m} \mathbf{K}(s, t)$.

Proof. For the sake of simplicity, we shall assume that $\rho = 0$. For every $m \in \mathcal{M}$, there exists some point $\overline{s}_m \in S_m$ such that $2\mathbf{K}(s, S_m) \geq \mathbf{K}(s, \overline{s}_m)$ and some point $s_m \in S_m$ such that $\mathbf{h}^2(s, s_m) \leq 2\inf_{t \in S_m} \mathbf{h}^2(s, t)$. Let us consider the family of functions

$$g_m = -\frac{1}{2}\ln\left(\frac{\overline{s}_m}{s}\right), \; f_m = -\ln\left(\frac{s + s_m}{2s}\right),$$

$$\widehat{g}_m = -\frac{1}{2}\ln\left(\frac{\widehat{s}_m}{s}\right) \text{ and } \widehat{f}_m = -\ln\left(\frac{s + \widehat{s}_m}{2s}\right), \; m \in \mathcal{M}.$$

We now fix some $m \in \mathcal{M}$ such that $\mathbf{K}(s, \bar{s}_m) < \infty$ and define $\mathcal{M}' = \{m' \in \mathcal{M}, \text{ crit}(m') \leq \text{crit}(m)\}$. By definition, for every $m' \in \mathcal{M}'$

$$P_n(\widehat{g}_{m'}) + \text{pen}(m') \leq P_n(\widehat{g}_m) + \text{pen}(m) \leq P_n(g_m) + \text{pen}(m).$$

Hence, since by concavity of the logarithm $\widehat{f}_{m'} \leq \widehat{g}_{m'}$ one has for every $m' \in \mathcal{M}'$

$$P\left(\widehat{f}_{m'}\right) = -\nu_n\left(\widehat{f}_{m'}\right) + P_n\left(\widehat{f}_{m'}\right)$$

$$\leq P_n(g_m) + \text{pen}(m) - \nu_n\left(\widehat{f}_{m'}\right) - \text{pen}(m')$$

and therefore

$$P\left(\widehat{f}_{m'}\right) + U_m \leq P(g_m) + \text{pen}(m) - \nu_n\left(\widehat{f}_{m'}\right) - \text{pen}(m'),$$

where $U_m = -\nu_n(g_m)$. Taking into account the definitions of f_m and g_m above this inequality also implies that for every $m' \in \mathcal{M}'$

$$\mathbf{K}\left(s, \frac{s + \widehat{s}_{m'}}{2}\right) + U_m \leq 2\mathbf{K}(s, S_m) + \text{pen}(m) - \nu_n\left(\widehat{f}_{m'}\right) - \text{pen}(m'). \quad (7.63)$$

Our purpose is now to control $-\nu_n\left(\widehat{f}_{m'}\right)$. We shall perform this control by first deriving from the maximal inequalities of the previous chapter some exponential tail bounds for $-\nu_n\left(\widehat{f}_{m'}\right)$ and then sum up these bounds over all the possible values $m' \in \mathcal{M}$.

Control of $-\nu_n\left(\widehat{f}_{m'}\right)$

We first consider for every positive σ

$$W_{m'}(\sigma) = \sup_{t \in S_{m'}, \left\|\sqrt{t} - \sqrt{s_{m'}}\right\|_2 \leq \sigma} \sqrt{n} \nu_n\left(\ln\left(\frac{s + t}{s + s_{m'}}\right)\right)$$

and intend to use Theorem 6.8. By Lemma 7.26, we know that if t is such that $\left\|\sqrt{t} - \sqrt{s_{m'}}\right\|_2 \leq \sigma$, then (since $\left\|\sqrt{t} - \sqrt{s_{m'}}\right\|_2 \leq \sqrt{2}$ anyway)

$$P\left(\left|\ln\left(\frac{s + t}{s + s_{m'}}\right)\right|\right)^k \leq \frac{2^{k-2}k!}{2}\left(\frac{9\left(\sigma^2 \wedge 2\right)}{8}\right).$$

Moreover if $\sqrt{t}$ belongs to some bracket $[\sqrt{t_L}, \sqrt{t_U}]$ with $\left\|\sqrt{t_L} - \sqrt{t_U}\right\|_2 \leq \delta$, then

$$\ln\left(\frac{s + t_L}{s + s_{m'}}\right) \leq \ln\left(\frac{s + t}{s + s_{m'}}\right) \leq \ln\left(\frac{s + t_U}{s + s_{m'}}\right)$$

with by Lemma 7.26

$$P\left(\ln\left(\frac{s + t_U}{s + t_L}\right)\right)^k \leq \frac{2^{k-2}k!}{2}\left(\frac{9\delta^2}{8}\right).$$

We are therefore in position to apply Theorem 6.8 which implies via condition (i) that for every measurable set A with $\mathbb{P}[A] > 0$

$$\mathbb{E}^A\left[W_{m'}(\sigma)\right] \leq \frac{81}{2\sqrt{2}}\phi_{m'}(\sigma) + \frac{7}{\sqrt{n}}H_{[.]}\left(\sigma, \sqrt{S_{m'}(s_{m'}, \sigma)}\right)$$
$$+ \frac{21}{2}\sigma\sqrt{\ln\left(\frac{1}{\mathbb{P}[A]}\right)} + \frac{4}{\sqrt{n}}\ln\left(\frac{1}{\mathbb{P}[A]}\right). \tag{7.64}$$

Now since $\delta \to H_{[.]}\left(\delta, \sqrt{S_{m'}(s_{m'}, \sigma)}\right)$ and $\delta \to \delta^{-1}\phi_{m'}(\delta)$ are nonincreasing we derive from the definition of $\sigma_{m'}$ that for $\sigma \geq \sigma_{m'}$

$$H_{[.]}\left(\sigma, \sqrt{S_{m'}(s_{m'}, \sigma)}\right) \leq \sigma^{-2}\phi_{m'}^2(\sigma) \leq \frac{\sigma_{m'}}{\sigma}\sqrt{n}\phi_{m'}(\sigma) \leq \sqrt{n}\phi_{m'}(\sigma).$$

Plugging this inequality in (7.64), we derive that for every σ such that $\sigma_{m'} \leq \sigma$ one has for some absolute constant κ'

$$\mathbb{E}^A\left[W_{m'}(\sigma)\right] \leq \kappa'\left(\phi_{m'}(\sigma) + \sigma\sqrt{\ln\left(\frac{1}{\mathbb{P}[A]}\right)} + \frac{1}{\sqrt{n}}\ln\left(\frac{1}{\mathbb{P}[A]}\right)\right).$$

Provided that $y_{m'} \geq \sigma_{m'}$, Lemma 4.23 yields

$$\frac{y_{m'}}{4\kappa'}\mathbb{E}^A\left[\sup_{t \in S_{m'}} \nu_n\left(\frac{\ln(s+t) - \ln(s + s_{m'})}{y_{m'}^2 + \left\|\sqrt{s_{m'}} - \sqrt{t}\right\|^2}\right)\right] \leq \frac{\phi_{m'}(y_{m'})}{\sqrt{n}y_{m'}} + A_{m'},$$

where

$$A_{m'} = \frac{1}{\sqrt{n}}\sqrt{\ln\left(\frac{1}{\mathbb{P}[A]}\right)} + \frac{1}{y_{m'}n}\ln\left(\frac{1}{\mathbb{P}[A]}\right).$$

Using again the monotonicity of $\delta \to \delta^{-1}\phi_{m'}(\delta)$ we get by definition of $\sigma_{m'}$

$$\phi_{m'}(y_{m'}) \leq \frac{y_{m'}}{\sigma_{m'}}\phi_{m'}(\sigma_{m'}) \leq y_{m'}\sqrt{n}\sigma_{m'}$$

and therefore

$$\frac{y_{m'}}{4\kappa'}\mathbb{E}^A\left[\sup_{t \in S_{m'}} \nu_n\left(\frac{\ln(s+t) - \ln(s + s_{m'})}{y_{m'}^2 + \left\|\sqrt{s_{m'}} - \sqrt{t}\right\|^2}\right)\right] \leq \sigma_{m'} + A_{m'}.$$

Now by definition of $s_{m'}$ we have for every $t \in S_{m'}$

$$\left\|\sqrt{s_{m'}} - \sqrt{s}\right\|^2 \leq 2\left\|\sqrt{t} - \sqrt{s}\right\|^2 \tag{7.65}$$

and also

$$\left\|\sqrt{s_{m'}} - \sqrt{t}\right\|^2 \leq \left(\left\|\sqrt{s_{m'}} - \sqrt{s}\right\| + \left\|\sqrt{t} - \sqrt{s}\right\|\right)^2 \leq 6\left\|\sqrt{t} - \sqrt{s}\right\|^2$$

hence

$$\frac{y_{m'}}{4\kappa'} \mathbb{E}^A \left[\sup_{t \in S_{m'}} \nu_n \left(\frac{\ln(s+t) - \ln(s+s_{m'})}{y_{m'}^2 + 6 \left\| \sqrt{s} - \sqrt{t} \right\|^2} \right) \right] \le \sigma_{m'} + A_{m'}$$

which implies that

$$\frac{y_{m'}}{4\kappa'} \mathbb{E}^A \left[\nu_n \left(\frac{f_{m'} - \widehat{f}_{m'}}{y_{m'}^2 + 6 \left\| \sqrt{s} - \sqrt{\widehat{s}_{m'}} \right\|^2} \right) \right] \le \sigma_{m'} + A_{m'}.$$

Using Lemma 2.4, we can derive an exponential bounds from this inequality. Namely, for every positive x the following inequality holds except on a set with probability less than $e^{-x_{m'} - x}$

$$\nu_n \left(\frac{f_{m'} - \widehat{f}_{m'}}{y_{m'}^2 + 6 \left\| \sqrt{s} - \sqrt{\widehat{s}_{m'}} \right\|^2} \right) \le \frac{4\kappa'}{y_{m'}} \left(\sigma_{m'} + \sqrt{\frac{x_{m'} + x}{n}} + \frac{x_{m'} + x}{n y_{m'}} \right). \tag{7.66}$$

It remains to bound $-\nu_n(f_{m'})$. We simply combine Lemma 7.26 with Bernstein's inequality (2.20) and derive that except on a set with probability less than $e^{-x_{m'} - x}$

$$-\nu_n(f_{m'}) \le \frac{1}{\sqrt{n}} \left(\frac{3}{2} \left\| \sqrt{s} - \sqrt{s_{m'}} \right\| \sqrt{x_{m'} + x} + \frac{2(x_{m'} + x)}{\sqrt{n}} \right).$$

We derive from this inequality that

$$\mathbb{P} \left[\frac{-\nu_n(f_{m'})}{y_{m'}^2 + \left\| \sqrt{s} - \sqrt{s_{m'}} \right\|^2} \ge \frac{3}{4 y_{m'}} \sqrt{\frac{x_{m'} + x}{n}} + \frac{2(x_{m'} + x)}{n y_{m'}^2} \right] \le e^{-x_{m'} - x}$$

which in turn via (7.66) and (7.65) implies that, except on a set with probability less than $2 \exp(-x_{m'} - x)$, the following inequality holds for every positive x, every $y_{m'} \ge \sigma_{m'}$ and some absolute constant κ''

$$\frac{-\nu_n \left(\widehat{f}_{m'} \right)}{y_{m'}^2 + \left\| \sqrt{s} - \sqrt{\widehat{s}_{m'}} \right\|^2} \le \frac{\kappa''}{y_{m'}} \left(\sigma_{m'} + \sqrt{\frac{x_{m'} + x}{n}} + \frac{x_{m'} + x}{n y_{m'}} \right) \tag{7.67}$$

End of the proof

It remains to choose adequately $y_{m'}$ and sum up the tail bounds (7.67) over the possible values of m'. Defining $y_{m'}$ as

$$y_{m'} = \theta^{-1} \sqrt{\sigma_{m'}^2 + \frac{(x_{m'} + x)}{n}}$$

where θ denotes some constant to be chosen later, we derive from (7.67) that, except on a set with probability less than $2\Sigma \exp(-x)$ the following inequality is valid for all $m' \in \mathcal{M}$ simultaneously

$$\frac{-\nu_n\left(\widehat{f}_{m'}\right)}{y_{m'}^2 + \left\|\sqrt{s} - \sqrt{\widehat{s}_{m'}}\right\|^2} \le \kappa''\left(2\theta + \theta^2\right)$$

Now setting $\alpha = \ln(2) - (1/2)$ it comes from Lemma 7.23 and (7.63) that for every $m' \in \mathcal{M}'$

$$\alpha\left\|\sqrt{s} - \sqrt{\widehat{s}_{m'}}\right\|^2 + U_m \le 2\mathbf{K}(s, S_m) + \mathrm{pen}(m) - \nu_n\left(\widehat{f}_{m'}\right) - \mathrm{pen}(m').$$

Hence, choosing θ in such a way that $\kappa''\left(2\theta + \theta^2\right) = \alpha/2$, except on a set with probability less than $2\Sigma \exp(-x)$, we have for every $m' \in \mathcal{M}'$

$$\frac{\alpha}{2}\left\|\sqrt{s} - \sqrt{\widehat{s}_{m'}}\right\|^2 + U_m \le 2\mathbf{K}(s, S_m) + \mathrm{pen}(m) + \frac{\alpha}{2}y_{m'}^2 - \mathrm{pen}(m').$$

Finally, if we define $\kappa = \alpha\theta^{-2}$, then by (7.61) $\left(\alpha y_{m'}^2/2\right) - \left(\mathrm{pen}(m')/2\right) \le \kappa x/(2n)$ and therefore, except on a set with probability less than $2\Sigma \exp(-x)$, the following inequality holds for every $m' \in \mathcal{M}'$

$$\frac{\alpha}{2}\left\|\sqrt{s} - \sqrt{\widehat{s}_{m'}}\right\|^2 + U_m + \frac{\mathrm{pen}(m')}{2} \le 2\mathbf{K}(s, S_m) + \mathrm{pen}(m) + \frac{\kappa x}{2n}. \quad (7.68)$$

We can use this bound in two different ways. First, since U_m is integrable, we derive from (7.68) that $M = \sup_{m' \in \mathcal{M}'} \mathrm{pen}(m')$ is almost surely finite, so that since by (4.73), $\kappa x_{m'}/n \le M$ for every $m' \in \mathcal{M}'$, the following bound is valid

$$\Sigma \ge \sum_{m' \in \mathcal{M}'} \exp(-x_{m'}) \ge |\mathcal{M}'|\exp\left(-\frac{Mn}{\kappa}\right)$$

and therefore $\mathcal{M}'$ is almost surely a finite set. This proves of course that some minimizer $\widehat{m}$ of crit over $\mathcal{M}'$ and hence over $\mathcal{M}$ does exist. For such a minimizer, (7.68) implies that

$$\frac{\alpha}{2}\left\|\sqrt{s} - \sqrt{\widehat{s}_{\widehat{m}}}\right\|^2 + U_m \le 2\mathbf{K}(s, S_m) + \mathrm{pen}(m) + \frac{\kappa x}{2n}.$$

The proof can be easily completed by integrating this tail bound, noticing that U_m is centered at expectation. ∎

The oracle type inequality (7.62) makes some unpleasant bias term appear since one would expect Hellinger loss rather than Kullback–Leibler loss. The following result will be convenient to overcome this difficulty.

Lemma 7.12 *Let s be some probability density and $f \in \mathbb{L}_2(\mu)$, such that $\sqrt{s} \le f$. Let t be the density $t = f^2/\|f\|^2$, then*

$$\left\|\sqrt{s} - \sqrt{t}\right\| \le \left\|\sqrt{s} - f\right\| \quad (7.69)$$

and

$$1 \wedge \mathbf{K}(s, t) \le 3\left\|\sqrt{s} - f\right\|^2. \quad (7.70)$$

Proof. The first task is to relate $\left\| \sqrt{s} - \sqrt{t} \right\|$ to $\left\| \sqrt{s} - f \right\|$. Let $\lambda \geq 1$, then

$$
\left\| \sqrt{s} - \lambda \sqrt{t} \right\|^2 = \left\| \left(\sqrt{s} - \sqrt{t} \right) - (\lambda - 1) \sqrt{t} \right\|^2
$$

$$
= \left\| \sqrt{s} - \sqrt{t} \right\|^2 + (\lambda - 1)^2 - 2 (\lambda - 1) \int \sqrt{t} \left(\sqrt{s} - \sqrt{t} \right) d\mu
$$

$$
= \left\| \sqrt{s} - \sqrt{t} \right\|^2 + (\lambda - 1)^2 + (\lambda - 1) \left\| \sqrt{s} - \sqrt{t} \right\|^2
$$

$$
\geq \left\| \sqrt{s} - \sqrt{t} \right\|^2 .
$$

Applying this inequality with $\lambda = \|f\|$ provides the desired comparison i.e., (7.69). Combining this inequality with (7.104) we derive that

$$
\mathbf{K}(s, t) \leq (2 + \ln (\|f\|)) \left\| \sqrt{s} - f \right\|^2 .
$$

Setting $\varepsilon = \left\| \sqrt{s} - f \right\|$, we get via the triangle inequality

$$
\mathbf{K}(s, t) \leq (2 + \ln (1 + \varepsilon)) \varepsilon^2 .
$$

If $\varepsilon \leq 0.6$ then $2 + \ln (1 + \varepsilon) \leq 3$ and (7.70) derives from the previous inequality, while if $\varepsilon > 0.6$ $3\varepsilon^2 > 1$ and (7.70) remains valid. ∎

This Lemma will turn to be useful in the sequel since it allows to construct some good approximating density $f^2 / \|f\|^2$ of a given density s in Kullback–Leibler loss if one starts from some upper approximation f of the square root of s in $\mathbb{L}_2$ loss. If this approximation happens to belong to some model S_m, this provides an upper bound for the bias term $\mathbf{K}(s, S_m)$.

7.4.2 Finite Dimensional Models

Our purpose is now to provide some illustration of the previous general theorem. All along this section, we assume that μ is a probability measure. Following the lines of [12], we focus on the situation where $\sqrt{S_m}$ is a subset of some finite dimensional space $\overline{S}_m$ of $\mathbb{L}_2$. In this case, one can hope to compute the quantity σ_m^2 involved in the definition of the penalty restriction (7.61) as a function of the dimension D_m of $\overline{S}_m$. Unfortunately, the dimension itself is not enough to compute the local entropy with bracketing of the linear space $\overline{S}_m$. Essentially, this computation requires to say something on the $\mathbb{L}_\infty$-structure of $\overline{S}_m$. As in [12], we define some index which will turn to be extremely convenient for this purpose. Given some D-dimensional linear subspace $\overline{S}$ of $\mathbb{L}_\infty$, for every orthonormal basis $\varphi = \{\varphi_\lambda\}_{\lambda \in \Lambda}$ of $\overline{S}$ we define

$$
\overline{r}(\varphi) = \frac{1}{\sqrt{D}} \sup_{\beta \neq 0} \frac{\left\| \sum_{\lambda \in \Lambda} \beta_\lambda \varphi_\lambda \right\|_\infty}{|\beta|_\infty}
$$

where $|\beta|_\infty = \sup_{\lambda \in \Lambda} |\beta_\lambda|$ and next the $\mathbb{L}_\infty$-*index* of $\overline{S}$

$$\overline{r} = \inf_\varphi \overline{r}(\varphi),$$

where the infimum is taken over the set of all orthonormal basis of $\overline{S}$. Note that since μ is a probability measure $\left\| \sum_{\lambda \in \Lambda} \beta_\lambda \varphi_\lambda \right\|_\infty \geq \left\| \sum_{\lambda \in \Lambda} \beta_\lambda \varphi_\lambda \right\|_2$ for every basis $\{\varphi_\lambda\}_{\lambda \in \Lambda}$ and therefore $\overline{r} \geq 1$. This index is easy to handle for linear spaces of interest. A quite typical example is the following.

Lemma 7.13 *Let μ be the Lebesgue measure on $[0,1]$, $\mathcal{P}$ be some finite partition of $[0,1]$, the pieces of which are intervals and r be some integer. Let $\overline{S}_{\mathcal{P},r}$ be the space of piecewise polynomials on $\mathcal{P}$ with degree less than or equal to r. Then, denoting by $|\mathcal{P}|$ the number of pieces of $\mathcal{P}$, one has*

$$\overline{r} \leq \frac{2r+1}{\sqrt{|\mathcal{P}| \inf_{I \in \mathcal{P}} \mu(I)}}.$$

Proof. Let $\{Q_j\}_{j \geq 0}$ be the orthogonal basis of Legendre polynomials in $\mathbb{L}_2([-1,1], dx)$, then the following properties hold for all $j \geq 0$ (see [125], pp. 302–305 for details):

$$\|Q_j\|_\infty \leq 1 \quad \text{and} \quad \int_{-1}^1 Q_j^2(x)\, dx = \frac{2}{2j+1}.$$

Let I be some interval belonging to $\mathcal{P}$ and denote by a and b its extremities, $a < b$. We define for every $x \in [0,1]$ and $j \geq 0$

$$\varphi_{I,j}(x) = \sqrt{\frac{2j+1}{b-a}} Q_j \left(\frac{2x-a-b}{b-a} \right) \mathbb{1}_I(x).$$

Obviously, $\{\varphi_{I,j}\}_{I \in \mathcal{P}, 0 \leq j \leq r}$ is an orthonormal basis of $\overline{S}_{\mathcal{P},r}$ and for every piecewise polynomial

$$Q = \sum_{I \in \mathcal{P}} \sum_{j=0}^r \beta_{I,j} \varphi_{I,j}$$

one has

$$\|Q\|_\infty = \sup_{I \in \mathcal{P}} \left\| \sum_{j=0}^r \beta_{I,j} \varphi_{I,j} \right\|_\infty \leq |\beta|_\infty \sum_{j=0}^r \|\varphi_{I,j}\|_\infty$$

$$\leq \frac{|\beta|_\infty}{\sqrt{\inf_{I \in \mathcal{P}} \mu(I)}} \sum_{j=0}^r \sqrt{2j+1}$$

and since $\overline{S}$ has dimension $|\mathcal{P}|(r+1)$, we derive by definition of $\overline{r}$ that

$$\overline{r} \leq \frac{\sum_{j=0}^r \sqrt{2j+1}}{\sqrt{(r+1)|\mathcal{P}| \inf_{I \in \mathcal{P}} \mu(I)}}.$$

Since $\sum_{j=0}^r \sqrt{2j+1} \leq (r+1)\sqrt{2r+1}$, the result follows easily. ∎

In particular, whenever $\overline{S}$ is the space of piecewise constant functions on a regular partition with D pieces of $[0,1]$, Lemma 7.13 ensures that $\overline{r} = 1$. More generally, we derive from Lemma 7.13 that whenever $\mathcal{P}$ is a regular, if $\overline{S}$ is the space of piecewise polynomials on $\mathcal{P}$ with degree less or equal to r, the index $\overline{r}$ is bounded by $2r + 1$.

Another example of interest is wavelet expansions as explained in [12]. We contents ourselves here with Haar expansions just as an illustration. We set for every integer $j \geq 0$

$$\Lambda\left(j\right) = \left\{\left(j,k\right); 1 \leq k \leq 2^j\right\}.$$

Let $\varphi = \mathbf{1}_{[0,1/2]} - \mathbf{1}_{(1/2,1]}$ and for every integers $j \geq 0$, $1 \leq k \leq 2^j$

$$\varphi_{j,k}\left(x\right) = 2^{j/2} \, \varphi\left(2^j x - k + 1\right) \text{ for all } x \in [0,1].$$

If we consider for every integer m, the linear span $\overline{S}_m$ of $\left\{\varphi_\lambda, \lambda \in \bigcup_{j=0}^m \Lambda\left(j\right)\right\}$, it is possible to bound the $\mathbb{L}_\infty$-index $\overline{r}_m$ of $\overline{S}_m$. Indeed for every $j \geq 0$

$$\left\|\sum_{k=1}^{2^j} \beta_{j,k}\varphi_{j,k}\right\|_\infty \leq 2^{j/2} \sup_k |\beta_{j,k}|$$

and therefore

$$\overline{r}_m \leq \left[\sum_{j=0}^m 2^{j/2}\right] \Big/ \sqrt{\left[\sum_{j=0}^m 2^j\right]} < 1 + \sqrt{2}. \tag{7.71}$$

The interesting feature here (as in the case of regular polynomials) is that the $\mathbb{L}_\infty$-index $\overline{r}_m$ of $\overline{S}_m$ is bounded independently of m. This property is preserved when one deals with wavelet expansions (the upper bound depending on the father wavelet ψ which is considered).

It remains to connect the local entropy with bracketing of a linear finite dimensional subspace $\overline{S}$ of $\mathbb{L}_\infty$ with its dimension D and its $\mathbb{L}_\infty$-index $\overline{r}$. This will be a trivial consequence of the following result.

Lemma 7.14 *Let $\overline{S}$ be some D-dimensional subspace of $\mathbb{L}_\infty$ with $\mathbb{L}_\infty$-index $\overline{r}$. Let $u \in \overline{S}$ and $\sigma > 0$ be given. Denote by $\mathcal{B}_2\left(u,\sigma\right)$ the closed $\mathbb{L}_2$ ball of $\overline{S}$ centered at u with radius σ. For every $\delta \in (0,\sigma]$, let $N'_\infty\left(\delta, \mathcal{B}_2\left(u,\sigma\right)\right)$ denote the minimal number of closed $\mathbb{L}_\infty$-balls with radius δ which are necessary to cover $\mathcal{B}_2\left(u,\sigma\right)$. Then for some absolute constant κ_∞ ($\kappa_\infty = \sqrt{3\pi e/2}$ works)*

$$N'_\infty\left(\delta, \mathcal{B}_2\left(u,\sigma\right)\right) \leq \left(\frac{\kappa_\infty \overline{r} \sigma}{\delta}\right)^D.$$

Proof. Without loss of generality we may assume that $u = 0$ and that $\bar{r} = \bar{r}(\varphi)$, for some orthonormal basis $\varphi = \{\varphi_j\}_{1 \le j \le D}$ of $\bar{S}$. Using the natural isometry between the Euclidean space $\mathbb{R}^D$ and $\bar{S}$ corresponding to the basis φ, one defines the countable set T as the image of the lattice $T = \left[\left(2\delta/\bar{r}\sqrt{D}\right)\mathbb{Z}\right]^D$ i.e.,

$$\overline{T} = \left\{\sum_{j=1}^{D} \beta_j \varphi_j : \beta \in T\right\}.$$

Considering the partition of $\mathbb{R}^D$ into cubes of vertices with length $2\delta/\bar{r}\sqrt{D}$ centered on the points of T, we retain the set of cubes T^σ which intersect the Euclidean ball centered at 0 with radius σ and denote by $\overline{T}^\sigma$ the corresponding set in $\bar{S}$. We define the mapping $\Pi^\sigma : \mathbb{R}^D \to T^\sigma$ such that $\Pi^\sigma(\beta)$ and β belong to the same cube. Then for every $\beta \in \mathbb{R}^D$,

$$|\beta - \Pi^\sigma(\beta)|_\infty \le \frac{\delta}{\bar{r}\sqrt{D}}$$

and therefore, for every $t = \sum_{j=1}^{D} \beta_j \varphi_j \in \mathcal{B}_2(0, \sigma)$, there exists some point $\overline{\Pi}^\sigma(t) = \sum_{j=1}^{D} \Pi_j^\sigma(\beta) \varphi_j \in \overline{T}^\sigma$ such that

$$\left\|t - \overline{\Pi}^\sigma(t)\right\|_\infty \le \bar{r}\sqrt{D}\,|\beta - \Pi^\sigma(\beta)|_\infty \le \delta,$$

which means that the closed $\mathbb{L}_\infty$-balls with radius δ are covering $\mathcal{B}_2(0, \sigma)$. Since $\left|\overline{T}^\sigma\right| = |T^\sigma|$, we have to control $|T^\sigma|$. Towards this aim, we notice that if β belongs to some cube intersecting $\mathcal{B}_2(0, \sigma)$, then for some point $\beta^\sigma \in \mathcal{B}_2(0, \sigma)$, one has

$$|\beta - \beta^\sigma|_\infty \le \frac{2\delta}{\bar{r}\sqrt{D}}$$

which implies that

$$|\beta - \beta^\sigma|_2 \le \frac{2\delta}{\bar{r}}$$

and therefore β belongs to the Euclidean ball centered at 0 with radius $\sigma + (2\delta/\bar{r})$. Hence the disjoint cubes of vertices with length $2\delta/\bar{r}\sqrt{D}$ centered on the points of T^σ are packed into this euclidean ball which yields

$$|T^\sigma|\left(\frac{2\delta}{\bar{r}\sqrt{D}}\right)^D \le \left(\sigma + \left(\frac{2\delta}{\bar{r}}\right)\right)^D V_D,$$

where V_D denotes the volume of the Euclidean unit ball of $\mathbb{R}^D$. So since $\sigma\bar{r}/\delta \ge 1$

$$|T^\sigma| \le \left(\left(\frac{\sigma\bar{r}}{2\delta}\right) + 1\right)^D D^{D/2} V_D \le \left(\frac{3\sigma\bar{r}}{2\delta}\right)^D D^{D/2} V_D \qquad (7.72)$$

and it remains to bound V_D. It is well known that $V_D = \pi^{D/2} \left(\Gamma \left(1 + D/2 \right) \right)^{-1}$. We intend to prove by induction that

$$V_D \leq \left(\frac{2\pi e}{D} \right)^{D/2}. \tag{7.73}$$

Since $V_1 = 2$ and $V_2 = \sqrt{\pi}$, this bound holds true for $D = 1$ and $D = 2$. Now

$$V_{D+2} = \frac{2\pi}{2D+1} V_D,$$

so, assuming that (7.73) holds true, since $\ln\left(1 + x\right) \leq x$ we derive that

$$V_{D+2} \left(\left(\frac{2\pi e}{D+2} \right)^{-1-D/2} \right) \leq \frac{1}{e} \left(1 + \frac{2}{D} \right)^{D/2} \leq 1$$

which means that (7.73) is valid when we substitute $D + 2$ to D. Hence we have proved (7.73) by induction. Combining (7.73) with (7.72) leads to the result. ∎

Keeping the notations of the previous Lemma, if we consider some covering of a ball $\mathcal{B}_2\left(u, \sigma\right)$ in $\overline{S}$ by some finite family of $\mathbb{L}_\infty$-balls with radius $\delta/2$, denoting by T the family of centers of these balls, for every $t \in \mathcal{B}_2\left(u, \sigma\right)$, there exists some point $\Pi\left(t\right)$ belonging to T, such that

$$\|t - \Pi\left(t\right)\|_\infty \leq \delta/2$$

or equivalently

$$\Pi\left(t\right) - \delta/2 \leq t \leq \Pi\left(t\right) + \delta/2.$$

Hence, the family brackets $\{[t - \delta/2, t + \delta/2], t \in T\}$ is a covering of $\mathcal{B}_2\left(u, \sigma\right)$. Since μ is a probability measure, the $\mathbb{L}_2$-diameter of each of these brackets is less or equal to δ and we derive from Lemma 7.14 that

$$H_{[\cdot]}\left(\delta, \mathcal{B}_2\left(u, \sigma\right)\right) \leq \ln N'_\infty\left(\delta/2, \mathcal{B}_2\left(u, \sigma\right)\right) \leq D \ln \left(\frac{2\kappa_\infty \overline{r} \sigma}{\delta} \right).$$

Coming back to the situation where $\sqrt{S_m}$ is a subset of some D_m-dimensional $\overline{S}_m$ subspace of $\mathbb{L}_\infty$, denoting by $\overline{r}_m$ the $\mathbb{L}_\infty$-index of $\overline{S}_m$, since $\sqrt{a+b} \leq \sqrt{a} + \sqrt{b}$, we derive from this inequality that

$$\int_0^\sigma \sqrt{D_m^{-1} H_{[\cdot]}\left(x, \sqrt{S_m\left(u, \sigma\right)}\right)} dx \leq \int_0^\sigma \sqrt{\ln \left(\frac{2\kappa_\infty \overline{r}_m \sigma}{x} \right)} dx$$

$$\leq \sigma \sqrt{\ln\left(\overline{r}_m\right)} + \int_0^\sigma \sqrt{\ln \left(\frac{2\kappa_\infty \sigma}{x} \right)} dx.$$

Setting $x = y\sigma$ in the integral above yields

$$\int_0^\sigma \sqrt{D_m^{-1} H_{[.]}\left(x, \sqrt{S_m\left(u,\sigma\right)}\right)} dx \leq \sigma \left(\sqrt{\ln\left(\overline{r}_m\right)} + \int_0^1 \sqrt{\ln\left(\frac{2\kappa_\infty}{y}\right)} dy\right).$$

Hence, for some absolute constant κ'_∞, the function ϕ_m defined by

$$\phi_m\left(\sigma\right) = \sqrt{D_m}\sigma\left(\sqrt{\ln\left(\overline{r}_m\right)} + \kappa'_\infty\right)$$

satisfies to assumption (i). Since the solution σ_m of the equation $\phi_m\left(\sigma\right) = \sqrt{n}\sigma^2$ satisfies to

$$\sigma_m^2 = \frac{D_m}{n}\left(\sqrt{\ln\left(\overline{r}_m\right)} + \kappa'_\infty\right)^2 \leq \frac{2D_m}{n}\left(\ln\left(\overline{r}_m\right) + \kappa'_\infty\right),$$

applying Theorem 7.11, we obtain the following interesting result (which is exactly Theorem 2 in [12]).

Corollary 7.15 Let $X_1, \ldots, X_n$ be i.i.d. random variables with unknown density s with respect to some positive measure μ. Let $\left(\overline{S}_m\right)_{m \in \mathcal{M}}$ be some at most countable collection of finite dimensional linear subspaces of $\mathbb{L}_\infty$. For any $m \in \mathcal{M}$ we denote respectively by D_m and $\overline{r}_m$ the dimension and the $\mathbb{L}_\infty$-index of $\overline{S}_m$. Consider for every $m \in \mathcal{M}$ the set S_m of probability densities t (with respect to μ) such that $\sqrt{t} \in \overline{S}_m$. Let $\{x_m\}_{m \in \mathcal{M}}$ be some family of nonnegative numbers such that

$$\sum_{m \in \mathcal{M}} e^{-x_m} = \Sigma < \infty.$$

Let pen $: \mathcal{M} \to \mathbb{R}_+$ be such that

$$\text{pen}\left(m\right) \geq \frac{\kappa_1}{n}\left(D_m\left(\ln\left(1 + \overline{r}_m\right)\right) + x_m\right), \tag{7.74}$$

where κ_1 is a suitable numerical constant and let $\widetilde{s}$ be the corresponding penalized MLE which is a minimizer with respect to $m \in \mathcal{M}$ and $t \in S_m$ of

$$\left(P_n\left(-\ln\left(t\right)\right) + \text{pen}\left(m\right)\right) \text{ if } t \in S_m.$$

Then, for some absolute constant C_1, whatever the density s

$$\mathbb{E}_s\left[\mathbf{h}^2\left(s, \widetilde{s}\right)\right] \leq C_1\left(\inf_{m \in \mathcal{M}}\left(\mathbf{K}\left(s, S_m\right) + \text{pen}\left(m\right)\right) + \frac{\Sigma}{n}\right). \tag{7.75}$$

Since the $\mathbb{L}_\infty$-index of the space of piecewise polynomials for instance can be easily bounded (see Lemma 7.13 above), Corollary 7.15 can be applied to a variety of problems as shown in [12]. Coming back to the problem of selecting histograms on $[0, 1]$ based on a partition with end points on the regular grid

$\{j/N, 0 \leq j \leq N\}$, we derive from Lemma 7.13 that the $\mathbb{L}_\infty$-index $\overline{r}_m$ of the space $\overline{S}_m$ of piecewise constant functions on such a partition m satisfies to

$$\overline{r}_m \leq \sqrt{\frac{N}{D_m}},$$

where D_m denotes the number of pieces of m (which is of course also the dimension of $\overline{S}_m$). Since the considerations on the weights $\{x_m\}_{m \in \mathcal{M}}$ are exactly the same here as in Section 7.3.3, we readily see that (up to numerical constants) the penalties that we derive from Corollary 7.15 are similar to those of Section 7.3.3 i.e., for some suitable constant K

$$\text{pen}(m) = \frac{KD_m}{n}\left(1 + \ln\left(\frac{N}{D_m}\right)\right)$$

for the problem of selecting irregular partitions and

$$\text{pen}(m) = \frac{KD_m}{n}$$

for the problem of selecting a regular partition. Of course the very general entropy with bracketing arguments that we used to derive Theorem 7.11 and therefore Corollary 7.15 are not sharp enough to recover the very precise results concerning the numerical constants obtained in Theorem 7.9 via the concentration inequalities for chi-square type statistics.

7.5 Adaptive Estimation in the Minimax Sense

Exactly as in the Gaussian case, it is possible to study the adaptive properties in the minimax sense of penalized LSE or MLE in a huge variety of density estimation problems as illustrated in [20] and [12]. We present below a list of examples that we hope to be significant illustrations of the general idea that we have already developed in the Gaussian case: the link between model selection and adaptive estimation is made through approximation theory. This means that one of the main gain that one gets when working with lists of models which may depend on the sample size n is that it offers the possibility to use models because of their known approximation qualities with respect to target classes of densities.

7.5.1 Lower Bounds for the Minimax Risk

As in the Gaussian case we need some benchmarks to understand whether our estimators are approximately minimax or not on a variety of target parameter spaces.

Hölder Classes

Given some positive real numbers α and R, let us consider the largest integer r smaller than α and the Hölder class $\mathcal{H}(\alpha, R)$ of functions f on $[0, 1]$ such that f if r-times differentiable with

$$\left| f^{(r)}(x) - f^{(r)}(y) \right| \leq R |x - y|^{\alpha - r}.$$

Our purpose is to build a minimax lower bound for the squared Hellinger risk on the class $\mathcal{S}(\alpha, R)$ of densities s such that $\sqrt{s} \in \mathcal{H}(\alpha, R)$.

Proposition 7.16 *Suppose that one observes independent random variables* $X_1, \ldots, X_n$ *with common density s with respect to the Lebesgue measure on* $[0, 1]$. *For every $\alpha > 0$, there exists some positive constant κ_α such that whatever the estimator $\tilde{s}$ of s the following lower bound is valid for all $R \geq 1/\sqrt{n}$*

$$\sup_{s \in \mathcal{S}(\alpha, R)} \mathbb{E}_s \left[h^2(s, \tilde{s}) \right] \geq \kappa_\alpha \left[\left(R^{2/(2\alpha+1)} n^{-2\alpha/(2\alpha+1)} \right) \wedge 1 \right].$$

Proof. Let us take some infinitely differentiable function $\varphi : \mathbb{R} \to \mathbb{R}$ with compact support included in $(1/4, 3/4)$ such that

$$\int \varphi(x)\, dx = 0 \text{ and } \int \varphi^2(x)\, dx = 1.$$

We set $C = \max_{0 \leq k \leq r+1} \left\| \varphi^{(k)} \right\|_\infty > 1$. Given some positive integer D to be chosen later, we define for every positive integer $j \leq D$ and every $x \in [0, 1]$

$$\varphi_j(x) = \frac{R}{8C} D^{-\alpha} \varphi(Dx - j + 1).$$

Note that for every j, φ_j is supported by $((j-1)/D, j/D)$ so that the functions $\{\varphi_j, 1 \leq j \leq D\}$ have disjoint supports (and are therefore a fortiori orthogonal). For every $\theta \in \{0, 1\}^D$ we introduce

$$f_\theta = 1 + \sum_{j=1}^{D} (2\theta_j - 1)\varphi_j.$$

Then on the one hand

$$\|f_\theta\|_2^2 = 1 + \frac{R^2 D^{-2\alpha}}{64 C^2} = \rho^2$$

and on the other hand, whenever $RD^{-\alpha} \leq 2$

$$\left\| \sum_{j=1}^{D} (2\theta_j - 1)\varphi_j \right\|_\infty \leq \frac{RD^{-\alpha}}{8} \leq 1/4,$$

so that $3/4 \leq f_\theta \leq 5/4$. In particular this means that f_θ is positive and if we define the probability density $s_\theta = f_\theta^2/\rho^2$, we have $\sqrt{s_\theta} = f_\theta/\rho$. Since $\rho > 1$, to check that $s_\theta \in \mathcal{S}(\alpha, R)$ it is enough to prove that $f_\theta \in \mathcal{H}(\alpha, R)$. Noticing that if $x \in ((j-1)/D, j/D)$ and $y \in ((j'-1)/D, j'/D)$ one has

$$\left| f_\theta^{(r)}(x) - f_\theta^{(r)}(y) \right| = \left| (2\theta_j - 1)\varphi_j^{(r)}(x) - (2\theta_{j'} - 1)\varphi_{j'}^{(r)}(y) \right|.$$

Now two situations occur. Assuming first that $|x - y| < 1/(4D)$, one has either $\left| f_\theta^{(r)}(x) - f_\theta^{(r)}(y) \right| = 0$ if $j \neq j'$ (by construction of the φ_k's) or if $j = j'$ by the mean value theorem

$$\left| f_\theta^{(r)}(x) - f_\theta^{(r)}(y) \right| \leq |x - y|^{\alpha - r}(4D)^{-1-r+\alpha}\left\| \varphi_j^{(r+1)} \right\|_\infty$$
$$\leq \frac{R}{8}|x - y|^{\alpha - r}.$$

Assuming now that $|x - y| \geq 1/(4D)$ we simply write

$$\left| f_\theta^{(r)}(x) - f_\theta^{(r)}(y) \right| = \left\| \varphi_j^{(r)} \right\|_\infty + \left\| \varphi_{j'}^{(r)} \right\|_\infty \leq \frac{R}{4}D^{r-\alpha}$$
$$\leq R|x - y|^{\alpha - r}.$$

This proves that f_θ indeed belongs to $\mathcal{H}(\alpha, R)$ whatever $\theta \in \{0, 1\}^D$. To apply the strategy for deriving lower bounds from Birgé's lemma that we have already experimented in the Gaussian framework, we need to evaluate for every $\theta, \theta' \in \{0, 1\}^D$, the Kullback–Leibler loss $\mathbf{K}(s_\theta, s_{\theta'})$ and the Hellinger loss $\mathbf{h}^2(s_\theta, s_{\theta'})$. We derive from (7.104) in the Appendix that

$$\mathbf{K}(s_\theta, s_{\theta'}) \leq \frac{2}{\rho^2}(1 + \ln(\|f_\theta/f_{\theta'}\|_\infty))\|f_\theta - f_{\theta'}\|^2 \leq 3\|f_\theta - f_{\theta'}\|^2$$
$$\leq \frac{3R^2 D^{-2\alpha}}{16C^2}.$$

Moreover,

$$\mathbf{h}^2(s_\theta, s_{\theta'}) = \frac{1}{2\rho^2}\|f_\theta - f_{\theta'}\|^2 = \frac{R^2 D^{-2\alpha-1}}{32\rho^2 C^2}\sum_{j=1}^{D}\mathbb{1}_{\theta_j \neq \theta'_j}$$

so that, restricting ourselves to the subset Θ of $\{0, 1\}^D$ coming from Lemma 4.7 and applying Corollary 2.19 we derive that for any estimator $\widetilde{s}$ one has

$$\sup_{\theta \in \Theta} \mathbb{E}_{s_\theta}\left[\mathbf{h}^2(s_\theta, \widetilde{s})\right] \geq 2^{-9}R^2 D^{-2\alpha}(1 - \kappa)\rho^{-2}C^{-2} \qquad (7.76)$$

provided that

$$\max_{\theta, \theta'}\left[n\mathbf{K}(s_\theta, s_{\theta'})\right] \leq \kappa D/8,$$

where κ denotes the absolute constant of Lemma 2.17. The restriction above on the maximal Kullback–Leibler mutual information is a fortiori satisfied if

$$\frac{3R^2 D^{-2\alpha}}{16C^2} \leq \frac{\kappa D}{8n} \tag{7.77}$$

or equivalently $D^{2\alpha+1} \geq 1.5nR^2/(\kappa C^2)$. We know that $\kappa \geq 1/2$, so that choosing D as

$$D = \min\left\{k \geq 1 : k^{2\alpha+1} \geq 3nR^2\right\}$$

warrants that (7.77) is fulfilled. Assuming first that $R \leq n^\alpha$, we see that our choice of D fulfills the constraint $RD^{-\alpha} \leq 1 < 2$. On the other hand $\rho^2 \leq 17/16$ and $nR^2 \geq 1$ warrants that $D \geq 2$ so that by definition of D

$$D \leq 2\left(3nR^2\right)^{1/(2\alpha+1)}.$$

Plugging these inequalities in (7.76) leads to

$$\sup_{\theta \in \Theta} \mathbb{E}_{s_\theta}\left[\mathbf{h}^2\left(s_\theta, \widetilde{s}\right)\right] \geq \kappa_\alpha R^{2/(2\alpha+1)} n^{-2\alpha/(2\alpha+1)}$$

by setting

$$\kappa_\alpha = \frac{2^{-5-2\alpha}\left(1-\kappa\right)}{51C^2}.$$

On the contrary, if $R > n^\alpha$, the maximum risk on $\mathcal{S}(\alpha, R)$ is always not smaller than the maximum risk on the smaller class $\mathcal{S}(\alpha, n^\alpha)$ for which the previous proof works, leading to the desired result. ∎

We turn now to ellipsoids.

Ellipsoids

For densities this structure is somehow less natural and easy to deal with as compared to the Gaussian case, the reason being that the positivity restriction that must be fulfilled by a density can lead to a parameter space which can be substantially smaller than the whole ellipsoid itself. For this reason lower bounds on ellipsoids are more delicate to establish in the density case than in the Gaussian case and require to make some assumptions on the underlying orthonormal basis. Consider some countable family of functions $\{\varphi_\lambda\}_{\lambda \in \Lambda}$ such that $\{\mathbb{1}\} \cup \{\varphi_\lambda\}_{\lambda \in \Lambda}$ is an orthonormal basis of $\mathbb{L}_2(\mu)$. For any function $t \in \mathbb{L}_2(\mu)$ and any $\lambda \in \Lambda$, we denote by $\beta_\lambda(t)$ the coefficient of t in the direction of φ_λ, $\beta_\lambda(t) = \int t\varphi_\lambda d\mu$. We provide a hierarchical structure to this basis by taking some partition of Λ

$$\Lambda = \bigcup_{j \in \mathbb{N}} \Lambda(j)$$

where, for each integer j, $\Lambda(j)$ is a finite set. Setting for any integer j, $D_j = \sum_{i=0}^{j} |\Lambda(i)|$, we define for any positive numbers α and R, the set

$$\bar{\mathcal{E}}_2\left(\alpha, R\right) = \left\{ s \in \mathbb{L}_2\left(\mu\right) : \sum_{j=0}^{\infty} D_j^{2\alpha} \sum_{\lambda \in \Lambda(j)} \beta_\lambda^2\left(s\right) \leq R^2 \text{ and } 1 + s \geq 0 \right\},$$

(7.78)

which is a subset of an ellipsoid of $\mathbb{L}_2\left(\mu\right)$. The conditions that we have to impose on the basis in order to build our lower bounds are as follows:

- there exists some positive constant Φ such that for all integer m, the function $\Phi_m = \sum_{j=0}^{m} \sum_{\lambda \in \Lambda(j)} \varphi_\lambda^2$ satisfies to

$$\|\Phi_m\|_\infty \leq \Phi D_m,$$

(7.79)

- there exists some positive constant C such that

$$D_{m+1} \leq C D_m, \text{ for all integer } m.$$

(7.80)

Note that these conditions are satisfied for all the orthonormal basis that we know to play a role in approximation theory (see [12] for several examples including eigenfunctions of the Laplacian operator on a compact Riemannian manifold). Two typical examples are the trigonometric and the Haar systems for which $\Phi = 1$ and $C = 3$ as easily seen from their description that we recall hereunder.

- μ is the uniform distribution on the torus $[0, 2\pi]$ and for each integer j $\Lambda\left(j\right) = \{2j, 2j + 1\}$ with $\varphi_{2j}\left(x\right) = \sqrt{2}\cos\left((j + 1)x\right)$, $\varphi_{2j+1}\left(x\right) = \sqrt{2}\sin\left((j + 1)x\right)$ for all $x \in [0, 2\pi]$.
- μ is the uniform distribution on $[0, 1]$. Moreover let $\varphi = 1\!\!1_{[0,1/2)} - 1\!\!1_{(1/2,1]}$ and for each integer j, $\Lambda\left(j\right) = \{(j, k) \mid 1 \leq k \leq 2^j\}$ with $\varphi_{j,k}\left(x\right) = 2^{j/2}\varphi\left(2^j x - k + 1\right)$ for all $(j, k) \in \Lambda\left(j\right)$ and $x \in [0, 1]$.

We shall also use the following stronger version of (7.79):

- there exists some positive constant Φ such that for all integer m and every $\beta \in \mathbb{R}^{\Lambda_m}$

$$\left\| \sum_{\lambda \in \Lambda_m} \beta_\lambda \varphi_\lambda \right\|_\infty \leq \left(\sup_{\lambda \in \Lambda_m} |\beta_\lambda| \right) \sqrt{\Phi D_m}$$

(7.81)

This condition is directly connected to the properties of the $\mathbb{L}_\infty$-index of the linear spans of $\{\varphi_\lambda, \lambda \in \Lambda_m\}$ and typically holds for a wavelet basis. For instance, we derive from (7.71) that it holds for the Haar basis with $\Phi = \left(1 + \sqrt{2}\right)^2$. We can now state our lower bound (which is a simplified version of Proposition 2 in [12]).

Proposition 7.17 Let $X_1, \ldots, X_n$ be i.i.d. observations with density $1 + s$ with respect to the probability measure μ. Assume that the orthonormal system $\{\varphi_\lambda, \lambda \in \Lambda\}$ satisfies to conditions (7.80) and (7.79). Then, there exists some *numerical constant κ_1 such that, whenever $nR^2 \geq D_0^{2\alpha+1}$, for every estimator $\widetilde{s}$ of s, one has*

$$\sup_{s \in \overline{\mathcal{E}}_2(\alpha, R)} \mathbb{E}_s \|s - \widetilde{s}\|^2 \geq \frac{\kappa_1}{C\varPhi} R^{2/(2\alpha+1)} n^{-2\alpha/(2\alpha+1)}$$

provided that

$$R^2 \leq n^{\alpha-1/2}. \tag{7.82}$$

Moreover if the stronger condition (7.81) holds, then (7.82) can be replaced by the milder restriction $R \leq n^\alpha$.

Proof. The proof is very similar to that of Proposition 7.16. Since $nR^2 \geq D_0^{2\alpha+1}$, the set $\left\{j \geq 0 : D_j \leq \left(nR^2\right)^{1/(2\alpha+1)}\right\}$ is nonvoid. We define

$$m = \sup\left\{j \geq 0 : D_j \leq \left(nR^2\right)^{1/(2\alpha+1)}\right\}$$

and write $D = D_m$ for short. It comes from (7.80) and the very definition of D that

$$\frac{nR^2}{C^{2\alpha+1}} \leq D^{2\alpha+1} \leq nR^2. \tag{7.83}$$

We introduce for every $\theta \in \{0,1\}^{\Lambda_m}$

$$s_\theta = \rho \sum_{\lambda \in \Lambda_m} \theta_\lambda \varphi_\lambda \text{ with } \rho = \frac{1}{4\sqrt{n\varPhi}}.$$

Condition (7.79) implies by Cauchy -Schwartz inequality that

$$\|s_\theta\|_\infty \leq \rho\sqrt{\varPhi}D \leq \frac{D}{4\sqrt{n}}$$

and therefore by (7.83) and (7.82) $\|s_\theta\|_\infty \leq 1/4$. Similarly if the stronger condition (7.81) holds, then

$$\|s_\theta\|_\infty \leq \rho\sqrt{\varPhi D} \leq \frac{1}{4}\sqrt{\frac{D}{n}}$$

so that combining (7.83) with the condition $R \leq n^\alpha$ we still get $\|s_\theta\|_\infty \leq 1/4$. Moreover for every $\theta, \theta' \in \{0,1\}^{\Lambda_m}$ one has

$$\|s_\theta - s_{\theta'}\|^2 = \rho^2 \sum_{\lambda \in \Lambda_m} (\theta_\lambda - \theta'_\lambda)^2. \tag{7.84}$$

Since $3/4 \leq 1 + s_\theta \leq 5/4$ for every $\theta \in \{0,1\}^{\Lambda_m}$, Kullback–Leibler, Hellinger and square losses are of the same order when restricted to the family of densities $\left\{1 + s_\theta; \theta \in \{0,1\}^{\Lambda_m}\right\}$. Indeed

$$\mathbf{h}^2\left(1+s_\theta, 1+s_{\theta'}\right) = \frac{1}{2}\left\|\frac{s_\theta - s_{\theta'}}{\sqrt{1+s_\theta} + \sqrt{1+s_{\theta'}}}\right\|^2 \leq \frac{1}{6}\left\|s_\theta - s_{\theta'}\right\|^2$$

which implies via (7.104) and (7.84) that

$$\mathbf{K}\left(1+s_\theta, 1+s_{\theta'}\right) \leq \frac{1}{3}\left(2 + \ln\left(\frac{5}{3}\right)\right)\left\|s_\theta - s_{\theta'}\right\|^2 \leq \left\|s_\theta - s_{\theta'}\right\|^2$$

$$\leq \rho^2 D.$$

Restricting to the subset Θ of $\{0,1\}^{\Lambda_m}$ coming from Lemma 4.7 and applying Corollary 2.19 we derive that for any estimator $\tilde{s}$ one has

$$\sup_{\theta \in \Theta} \mathbb{E}_{s_\theta}\left[\left\|s_\theta - s_{\theta'}\right\|^2\right] \geq \frac{1}{16}\rho^2 D\left(1 - \kappa\right) \tag{7.85}$$

provided that

$$\max_{\theta,\theta'}\left[n\mathbf{K}\left(1+s_\theta, 1+s_{\theta'}\right)\right] \leq \kappa D/8,$$

where κ denotes the absolute constant of Lemma 2.17. This restriction on the Kullback–Leibler mutual information is a fortiori satisfied if

$$\rho^2 D \leq \frac{\kappa D}{8n}$$

which is indeed fulfilled because of our choice of ρ (we recall again that $\kappa \geq 1/2$). Using (7.83), we derive from (7.85) that

$$\sup_{\theta \in \Theta} \mathbb{E}_{s_\theta}\left[\left\|s_\theta - s_{\theta'}\right\|^2\right] \geq \frac{2^{-8}\left(1 - \kappa\right)}{n\Phi C}\left(nR^2\right)^{1/(2\alpha+1)}$$

achieving the proof of our lower bound. ∎

The lower bounds that we have proved under assumption (7.79) are relevant in the Hilbert–Schmidt case, i.e., when $\alpha > 1/2$. Assumption (7.81) allows to relax this restriction on α. At the price of additional technicalities, it is also possible to show that for Fourier ellipsoids the restriction on α is also useless although this basis obviously does not satisfy to (7.81). The interested reader will find details on this result in [12].

Lower Bounds under Metric Entropy Conditions

Our aim is to prove an analogue in the density estimation framework of Proposition 4.13. The difficulty is that Hellinger and Kullback–Leibler loss are not necessarily of the same order. Hence metric entropy conditions alone do not lead to lower bounds. Interestingly, entropy with bracketing which was at the heart of our analysis of MLEs is also an adequate tool to derive minimax lower bounds in the density estimation framework. We may consider $\mathbb{L}_1$ or Hellinger metric entropy with bracketing conditions.

Lemma 7.18 *Let S be some set of probability densities. Let $\delta \in (0,1)$ and $C > 1$. Assume that there exists some covering of $\sqrt{S}$ by brackets $\left[\sqrt{\tau^-}; \sqrt{\tau^+}\right]$, $\tau \in \mathcal{T}_\delta$, with diameter less than or equal to $C\delta$ in $\mathbb{L}_2(\mu)$. Let $\mathcal{R}_\delta$ be some δ-net of S with respect to Hellinger distance. Then, there exists some subset S_δ of $\mathcal{R}_\delta$, some $\tau \in \mathcal{T}_\delta$ and some $\theta \in (0,1)$ such that, setting either*

$$s_\theta = (1-\theta)\, s + \theta \frac{\tau^+}{\left\|\sqrt{\tau^+}\right\|^2} \text{ for every } s \in S_\delta \tag{i}$$

or

$$s_\theta = \frac{\left((1-\theta)\sqrt{s} + \theta\sqrt{\tau^+}\right)^2}{\left\|(1-\theta)\sqrt{s} + \theta\sqrt{\tau^+}\right\|^2} \text{ for every } s \in S_\delta, \tag{ii}$$

the following properties hold:

$$\ln|S_\delta| \geq \ln|\mathcal{R}_\delta| - \ln|\mathcal{T}_\delta|$$

and for every $s \neq t \in S_\delta$

$$\mathbf{K}(s_\theta, t_\theta) \leq 4\,(2 + \ln(C))\, C^2 \delta^2 \text{ and } \mathbf{h}(s_\theta, t_\theta) \geq \frac{\delta}{2}. \tag{7.86}$$

Proof. There exists some bracket $[\tau^-, \tau^+]$ which contains at least $|\mathcal{R}_\delta|/|\mathcal{T}_\delta|$ points of $\mathcal{R}_\delta$. Hence, the set $S_\delta = [\tau^-, \tau^+] \bigcap \mathcal{R}_\delta$ satisfies to

$$\ln|S_\delta| \geq \ln|\mathcal{R}_\delta| - \ln|\mathcal{T}_\delta|.$$

Let $\theta \in (0,1)$ to be chosen later and $s, t \in S_\delta$ with $s \neq t$. Assume first that the family of densities $\{u_\theta,\ u \in S_\delta\}$ is defined by (i). By Lemma 7.25, we note that since $s, t \in [\tau^-, \tau^+]$

$$\mathbf{h}^2(s_\theta, t_\theta) \leq (1-\theta)\,\mathbf{h}^2(s,t) \leq \frac{(1-\theta)}{2}\left\|\sqrt{\tau^+} - \sqrt{\tau^-}\right\|^2,$$

so that on the one hand

$$\mathbf{h}^2(s_\theta, t_\theta) \leq \frac{(1-\theta)\,C^2\delta^2}{2}. \tag{7.87}$$

while on the other hand, we derive from (7.69) that

$$\mathbf{h}^2(s, s_\theta) \leq \theta \mathbf{h}^2\left(s, \frac{\tau^+}{\left\|\sqrt{\tau^+}\right\|^2}\right) \leq \frac{\theta}{2}\left\|\sqrt{s} - \sqrt{\tau^+}\right\|^2 \leq \frac{\theta C^2\delta^2}{2}. \tag{7.88}$$

In order to upper bound $\mathbf{K}(s_\theta, t_\theta)$, we notice that

$$1 \leq \left\| \sqrt{\tau^+} \right\| \leq 1 + C\delta, \tag{7.89}$$

which implies that $s_\theta \leq \tau^+$ and $t_\theta \geq \theta \left(1 + C\delta\right)^{-2} \tau^+$. Hence

$$\left\| \frac{s_\theta}{t_\theta} \right\|_\infty \leq \frac{(1 + C\delta)^2}{\theta} < \frac{(1 + C)^2}{\theta}$$

and we derive from (7.104) and (7.87) that

$$\mathbf{K}(s_\theta, t_\theta) \leq \left(2 + \ln \left(\frac{(1+C)^2}{\theta} \right) \right) C^2 \delta^2.$$

Now by the triangle inequality

$$\mathbf{h}(s_\theta, t_\theta) \geq \mathbf{h}(s, t) - \mathbf{h}(s, s_\theta) - \mathbf{h}(t, t_\theta).$$

But (7.88) implies that

$$\mathbf{h}(s, s_\theta) + \mathbf{h}(t, t_\theta) \leq \sqrt{2\theta} C\delta,$$

hence, choosing $\theta = 1/\left(8C^2\right)$ warrants that

$$\mathbf{h}(s, t_\theta) \geq \delta - \delta/2 \geq \delta/2.$$

Moreover our choice for θ leads to

$$\mathbf{K}(s_\theta, t_\theta) \leq \left(2 + \ln \left(8C^2 (1 + C)^2 \right) \right) C^2 \delta^2$$
$$\leq 4 \left(2 + \ln(C) \right) C^2 \delta^2$$

and therefore property (7.86) is fulfilled. Assume now that the family of densities $\{u_\theta, \, u \in S_\delta\}$ is defined by (ii). In this case, setting

$$f_\theta = (1 - \theta) \sqrt{s} + \theta \sqrt{\tau^+} \quad \text{and} \quad g_\theta = (1 - \theta) \sqrt{t} + \theta \sqrt{\tau^+},$$

one has $\sqrt{s_\theta} = f_\theta \|f_\theta\|^{-1}$ and $\sqrt{t_\theta} = g_\theta \|g_\theta\|^{-1}$. We note that since $\sqrt{s} \leq f_\theta$, $\|f_\theta\| \geq 1$ and one derives from (7.89) that

$$\left\| \sqrt{s_\theta} - f_\theta \right\| = \|f_\theta\| - 1 \leq \theta C\delta. \tag{7.90}$$

Similarly $\left\| \sqrt{t_\theta} - g_\theta \right\| \leq \theta C\delta$, so that the triangle inequality yields

$$\left\| \sqrt{s_\theta} - \sqrt{t_\theta} \right\| \leq 2\theta C\delta + \|f_\theta - g_\theta\| \leq 2\theta C\delta + (1 - \theta) \left\| \sqrt{s} - \sqrt{t} \right\|$$
$$\leq (1 + \theta) C\delta$$

and

$$\left\|\sqrt{s_\theta} - \sqrt{s}\right\| \le \theta C\delta + \left\|f_\theta - \sqrt{s}\right\| \le \theta C\delta + \theta\left\|\sqrt{\tau^+} - \sqrt{s}\right\|$$
$$\le 2\theta C\delta.$$

By the triangle inequality again, we derive from the latter inequality that

$$\mathbf{h}(s_\theta, t_\theta) \ge \mathbf{h}(s,t) - \mathbf{h}(s, s_\theta) - \mathbf{h}(t, t_\theta) \ge \mathbf{h}(s,t) - 2\sqrt{2}\theta C\delta$$

and therefore, choosing $\theta = \left(4\sqrt{2}C\right)^{-1}$ yields $\mathbf{h}(s_\theta, t_\theta) \ge \delta/2$. Finally, since by (7.90)

$$\|g_\theta\|^2 (1 + \theta C\delta)^{-2} \le 1 \le \|f_\theta\|^2$$

$f_\theta^2 \le \tau^+ \le \theta^{-2}g_\theta^2$ implies that

$$\left\|\frac{s_\theta}{t_\theta}\right\|_\infty \le \frac{(1 + \theta C\delta)^2}{\theta^2} \le \left(\theta^{-1} + C\right)^2$$

and we derive from (7.104) that

$$\mathbf{K}(s_\theta, t_\theta) \le 2\left(1 + \ln\left(\theta^{-1} + C\right)\right)\left\|\sqrt{s_\theta} - \sqrt{t_\theta}\right\|^2$$
$$\le 2\left(1 + \ln\left(\theta^{-1} + C\right)\right)(1 + \theta)^2 C^2\delta^2.$$

Plugging the value of θ in the latter inequality, some elementary computations allow to conclude. ∎

It is now quite easy to derive a minimax lower bound from the preceding Lemma. The key is to warrant that the net $\{s_\theta, s \in S_\delta\}$ which is constructed above does belong to S. This is indeed true in various situations. In Proposition 7.19 below we investigate the simplest possibility which is to consider $\mathbb{L}_\infty$ brackets for covering either S or $\sqrt{S}$ and make some convexity type assumptions on S.

Proposition 7.19 *Let $\overline{S}$ be some linear subspace of $\mathbb{L}_\infty$ equipped with some seminorm $\mathbf{n}$ such that $1 \in \overline{S}$ and $\mathbf{n}(1) = 0$. For every positive R, let $\mathcal{B}_\mathbf{n}(R) = \{t \in \overline{S}, \mathbf{n}(t) \le R\}$. Given $c > 0$, we denote by $\mathcal{S}_c$ the set of densities s such that $s \ge c$. Consider S_R to be the set of densities such that: either i) $S_R = \mathcal{S}_c \bigcap \mathcal{B}_\mathbf{n}(R)$ or ii) $\sqrt{S_R} = \sqrt{\mathcal{S}} \bigcap \mathcal{B}_\mathbf{n}(R)$. Assume S_R to be totally bounded in $\mathbb{L}_\infty$ and denote respectively by $H(., S_R)$ and $H_\infty(., S_R)$, the metric entropy of S_R with respect to Hellinger distance and the $\mathbb{L}_\infty$-distance. Assume that for some positive constants α, C_1 and C_2 the following inequalities hold for every positive $\delta \le R \wedge 1/2$,*

$$H(\delta, S_R) \ge C_1 \left(\frac{R}{\delta}\right)^{1/\alpha}$$

and either

$$H_\infty (\delta, S_R) \leq C_2 \left(\frac{R}{\delta}\right)^{1/\alpha} \quad \textit{in case i)}$$

or

$$H_\infty \left(\delta, \sqrt{S_R}\right) \leq C_2 \left(\frac{R}{\delta}\right)^{1/\alpha} \quad \textit{in case ii)}.$$

Then, there exists some positive constant κ_1 (depending on α, C_1, C_2 and also on c in case i)) such that for every estimator $\widetilde{s}$, provided that $R \geq 1/\sqrt{n}$ the following lower bound is valid

$$\sup_{s \in S_R} \mathbb{E}_s \left[\mathbf{h}^2 (s, \widetilde{s})\right] \geq \kappa_1 \left(\left(R^{2/(2\alpha+1)} n^{-2\alpha/(2\alpha+1)}\right) \wedge 1\right).$$

Proof. Let $\delta > 0$ and $C > 1$ to be chosen later. In case i), we consider some $C\sqrt{c}\delta$-net $\mathcal{T}_\delta$ of S_R (with respect to the $\mathbb{L}_\infty$-distance) with log-cardinality $H_\infty (\delta, S_R)$ and set for every $\tau \in \mathcal{T}_\delta$

$$\tau^- = \left(\tau - C\sqrt{c}\delta\right) \vee c \text{ and } \tau^+ = \tau + C\sqrt{c}\delta.$$

Then $\tau^- \leq \tau^+$ and

$$\left\|\sqrt{\tau^+} - \sqrt{\tau^-}\right\|^2 = \left\|\frac{(\tau^+ - \tau^-)}{\left(\sqrt{\tau^+} + \sqrt{\tau^-}\right)}\right\|^2 \leq C^2 \delta^2$$

which means that $\left\{\left[\sqrt{\tau^-}, \sqrt{\tau^+}\right], \tau \in \mathcal{T}_\delta\right\}$ is a collection of brackets with diameter less than or equal to $C\delta$ covering $\sqrt{S_R}$. Let us consider the $\delta/2$-net $\{s_\theta, s \in S_\delta\}$ of S coming from Lemma 7.18, starting from a δ-net $\mathcal{R}_\delta$ of S_R with respect to Hellinger distance with log-cardinality $H (\delta, S_R)$. Then for every $s \in S_\delta$,

$$\mathbf{n} (s_\theta) = \mathbf{n}\left(\theta s + \frac{(1 - \theta)}{1 + C\sqrt{c}\delta}\tau\right) \leq \theta \mathbf{n} (s) + \frac{(1 - \theta)}{1 + C\sqrt{c}\delta}\mathbf{n} (\tau) \leq R$$

which means that the density s_θ does belong to S_R. We derive from Corollary 2.19 that for any estimator $\widetilde{s}$ one has

$$16 \sup_{s \in S} \mathbb{E}_s \left[\mathbf{h}^2 (s, \widetilde{s})\right] \geq \delta^2 (1 - \kappa)$$

provided that

$$4 (2 + \ln (C)) C^2 \delta^2 n \leq \kappa \left(H (\delta, S_R) - H_\infty \left(C\sqrt{c}\delta, S_R\right)\right).$$

Choosing $C = c^{-1/2} (2C_2/C_1)^\alpha$ warrants that

$$H (\delta, S_R) - H_\infty \left(C\sqrt{c}\delta, S_R\right) \geq (C_1/2) \left(\frac{R}{\delta}\right)^{1/\alpha}$$

for every $\delta \leq (R \wedge (1/2)) / (C\sqrt{c})$. It suffices to take

$$\delta = \left(\frac{1}{2C\sqrt{c}} \wedge \left(\frac{C_1\kappa}{8C^2(2+\ln(C))}\right)^{\alpha/(2\alpha+1)}\right)\left(1 \wedge \left(R^{1/(2\alpha+1)}n^{-\alpha/(2\alpha+1)}\right)\right)$$

to achieve the proof in case **i)**.

In case **ii)**, we consider some $C\delta/2$-net $\mathcal{T}_\delta$ of $\sqrt{S_R}$ (with respect to the $\mathbb{L}_\infty$-distance) with log-cardinality $H_\infty\left(C\delta/2, \sqrt{S_R}\right)$ and set for every $\tau \in \mathcal{T}_\delta$

$$\sqrt{\tau^-} = \left(\sqrt{\tau} - \frac{C\delta}{2}\right)_+ \quad \text{and} \quad \sqrt{\tau^+} = \sqrt{\tau} + \frac{C\delta}{2}.$$

Then $\left\{\left[\sqrt{\tau^-}, \sqrt{\tau^+}\right], \tau \in \mathcal{T}_\delta\right\}$ is a collection of brackets with diameter less than or equal to $C\delta$ covering $\sqrt{S_R}$. Let $\delta/2$-net $\{s_\theta, s \in S_\delta\}$ of S coming from Lemma 7.18, starting from a δ-net $\mathcal{R}_\delta$ of S_R with respect to Hellinger distance with log-cardinality $H(\delta, S_R)$. Then for every $s \in S_\delta$, $\sqrt{s_\theta}$ belongs to $\overline{S}$ with $\left\|(1-\theta)\sqrt{s} + \theta\sqrt{\tau^+}\right\| \geq 1$ and

$$\mathbf{n}\left(\sqrt{s_\theta}\right) = \mathbf{n}\left(\frac{\theta\sqrt{s} + (1-\theta)\sqrt{\tau}}{\left\|(1-\theta)\sqrt{s} + \theta\sqrt{\tau^+}\right\|}\right) \leq \theta\mathbf{n}\left(\sqrt{s}\right) + (1-\theta)\mathbf{n}\left(\sqrt{\tau}\right) \leq R$$

which means that the density s_θ does belong to S_R. As in case **i)**, we derive from Corollary 2.19 that for any estimator $\tilde{s}$ one has

$$16 \sup_{s \in S} \mathbb{E}_s\left[\mathbf{h}^2(s, \tilde{s})\right] \geq \delta^2(1-\kappa)$$

provided that

$$4(2+\ln(C))C^2\delta^2 n \leq \kappa\left(H(\delta, S_R) - H_\infty\left(C\delta, \sqrt{S_R}\right)\right).$$

Choosing $C = (2C_2/C_1)^\alpha$ warrants that

$$H(\delta, S_R) - H_\infty\left(C\delta, \sqrt{S_R}\right) \geq (C_1/2)\left(\frac{R}{\delta}\right)^{1/\alpha}$$

for every $\delta \leq (R \wedge (1/2))/C$. It suffices to take

$$\delta = \left(\frac{1}{2C} \wedge \left(\frac{C_1\kappa}{8C^2(2+\ln(C))}\right)^{\alpha/(2\alpha+1)}\right)\left(1 \wedge \left(R^{1/(2\alpha+1)}n^{-\alpha/(2\alpha+1)}\right)\right)$$

to achieve the proof of the proposition. ∎

Assuming that a density s belongs to a ball centered at zero with radius R with respect to some seminorm is a common formulation for a smoothness assumption. This is true for Hölder, Sobolev or Besov classes of smooth functions. Assuming that $\sqrt{s}$ instead of s belongs to such a ball is a less commonly used assumption. It turns out that this way of formulating a smoothness condition is interesting for building a neat minimax lower bound since it allows to avoid the unpleasant condition that s remains bounded away from zero.

7.5.2 Adaptive Properties of Penalized LSE

We provide some illustrations of the adaptive properties of penalized LSE. It is clear that the list of examples is not all exhaustive. For example we shall not explain how to use the special strategies for Besov bodies introduced in the Gaussian framework in order to remove the extra logarithmic factors appearing in the risk bounds for Besov bodies which are proved below but this is indeed possible as shown in [20].

Adaptive Estimation on Ellipsoids

We consider the same framework as for our minimax lower bounds, i.e., we take some countable family of functions $\{\varphi_\lambda\}_{\lambda \in \Lambda}$ such that $\{1\!\!1\} \cup \{\varphi_\lambda\}_{\lambda \in \Lambda}$ which is an orthonormal basis of $\mathbb{L}_2(\mu)$. We consider some countable partition of Λ

$$\Lambda = \bigcup_{j \in \mathbb{N}} \Lambda(j)$$

where, for each integer j, $\Lambda(j)$ is a finite set. We set for any integer m, $D_m = \sum_{j=0}^m |\Lambda(j)|$ and consider for any positive numbers α and R the subset $\overline{\mathcal{E}}_2(\alpha, R)$ defined by (7.78). We have in view to show that under some mild condition on the basis, it is possible to define a penalized LSE which achieves the minimax risk up to some universal constant, on many sets $\mathcal{E}_2(\alpha, R)$ at the same time. The conditions that we have to impose on the basis in order to apply our theorems are the same as for the lower bounds i.e., (7.79) and (7.80).

We have to define a proper collection of models. Given some integer m, we denote by Λ_m the set $\cup_{j \leq m} \Lambda(j)$ and by S_m the linear span of $\{\varphi_\lambda, \lambda \in \Lambda_m\}$. Then we consider the collection of models $\{S_m\}_{m \in \mathcal{M}}$, where $\mathcal{M}$ is the set of integers m such that $D_m = |\Lambda_m| \leq n$. This collection is nested so that condition (7.15) holds with $\Gamma = 1$ and $R = 0$. Therefore we are in position to apply Theorem 7.5 with the deterministic penalty function given by (7.17) or Theorem 7.6 with the random choice for the penalty function given by (7.25). In both cases we get

$$\mathbb{E}_s\left[\|\widetilde{s} - s\|^2\right] \leq C_1(\varepsilon, \Phi) \inf_{m \in \mathcal{M}}\left[d^2(s, S_m) + \frac{D_m}{n}\right]$$
$$+ \frac{C_2(\varepsilon, \Phi)\left(1 + \|s\|^4\right)}{n}. \tag{7.91}$$

Now, whenever s belongs to $\mathcal{E}_2(\alpha, R)$ we have by monotonicity

$$d^2(s, S_m) = \sum_{j > m} \sum_{\lambda \in \Lambda(j)} \beta_\lambda^2(s) \leq D_{m+1}^{-2\alpha} \sum_{j > m} D_j^{2\alpha} \sum_{\lambda \in \Lambda(j)} \beta_\lambda^2(s)$$
$$\leq R^2 D_{m+1}^{-2\alpha}. \tag{7.92}$$

Let $m(n) = \inf \{m \geq 0 : D_m/n \geq R^2 D_{m+1}^{-2\alpha}\}$, then, provided that $D_{m(n)} \leq n$, one has

$$I = \inf_{m \in \mathcal{M}} \left[d^2(s, S_m) + \frac{D_m}{n} \right] \leq 2 \frac{D_{m(n)}}{n}.$$

If $m(n) = 0$, then $I \leq 2D_0/n$. Otherwise

$$nR^2 \geq D_{m(n)-1} D_{m(n)}^{2\alpha} \geq C^{-1} D_{m(n)}^{1+2\alpha}$$

and therefore

$$D_{m(n)} \leq C^{\frac{1}{1+2\alpha}} R^{\frac{2}{1+2\alpha}} n^{\frac{1}{1+2\alpha}}.$$

Hence, assuming that $CR^2 \leq n^{2\alpha}$ warrants that $D_{m(n)} \leq n$ and we get

$$I \leq 2 \left[\frac{D_0}{n} \vee \left(C^{\frac{1}{1+2\alpha}} R^{\frac{2}{1+2\alpha}} n^{-\frac{2\alpha}{1+2\alpha}} \right) \right] \leq 2 \left[\frac{D_0}{n} + CR^{\frac{2}{1+2\alpha}} n^{-\frac{2\alpha}{1+2\alpha}} \right].$$

It remains to bound $\|s\|$ when s belongs to $\mathcal{E}_2(\alpha, R)$. For this purpose, we note that by Jensen's inequality one has for any $\lambda \in \Lambda$

$$\beta_\lambda^2(s) = \left(\int (1+s) \varphi_\lambda d\mu \right)^2 \leq \int (1+s) \varphi_\lambda^2 d\mu,$$

which yields for any integer m

$$\sum_{j \leq m} \sum_{\lambda \in \Lambda(j)} \beta_\lambda^2(s) \leq \int (1+s) \Phi_m d\mu \leq \Phi D_m.$$

Combining this inequality with (7.92) we derive that

$$\|s\|^2 \leq \inf_{m \in \mathbb{N}} \left[\Phi D_m + R^2 D_{m+1}^{-2\alpha} \right].$$

Setting $J = \inf \{m : R^2 D_{m+1}^{-2\alpha} \leq \Phi D_m\}$ and arguing exactly as for the majorization of I we get

$$\|s\|^2 \leq 2\Phi D_J \leq 2 \left[\Phi D_0 \vee \left(C^{\frac{1}{1+2\alpha}} R^{\frac{2}{1+2\alpha}} \Phi^{\frac{2\alpha}{1+2\alpha}} \right) \right] \leq 2C\Phi \left(D_0 \vee R^{\frac{2}{1+2\alpha}} \right),$$

which in turn implies if $R^2 \leq n$

$$\frac{\|s\|^4}{n} \leq 4C^2 \Phi^2 \left(\frac{D_0^2}{n} + R^{\frac{2}{1+2\alpha}} n^{-\frac{2\alpha}{1+2\alpha}} \right).$$

Collecting the above evaluations and coming back to (7.91), we finally obtain

$$\sup_{s \in \mathcal{E}_2(\alpha, R)} \mathbb{E}_s \left[\|\tilde{s} - s\|^2 \right] \leq C_3(\varepsilon, \Phi, C) R^{\frac{2}{1+2\alpha}} n^{-\frac{2\alpha}{1+2\alpha}},$$

provided that $D_0^{4\alpha+2}/n \leq R^2 \leq C^{-1} n^{1 \wedge 2\alpha}$. By using the lower bound established in the previous section, we know that at least for $\alpha > 1/2$ (or for Haar ellipsoids for instance when $\alpha \leq 1/2$), this upper bound is optimal (up to constants).

Adaptive Estimation on ℓ_p-Bodies

Consider some countable orthonormal basis of $\mathbb{L}_2(\mu)$, $\{\varphi_\lambda\}_{\lambda \in \Lambda_0 \cup \Lambda}$, where Λ_0 is a finite set and Λ is totally ordered in such a way that setting for all $\lambda \in \Lambda$, $\Lambda_\lambda = \{\lambda' \in \Lambda : \lambda' \le \lambda\}$ and $r(\lambda) = |\Lambda_\lambda|$, $\lambda \to r(\lambda)$ is an increasing mapping from Λ onto $\mathbb{N}^*$. For any function $t \in \mathbb{L}_2(\mu)$ and any $\lambda \in \Lambda_0 \cup \Lambda$, we denote by $\beta_\lambda(t)$ the coefficient of t in the direction of φ_λ, i.e., $\beta_\lambda(t) = \int t\varphi_\lambda d\mu$. We define for any positive numbers p and M and any nonincreasing sequence $(c_k)_{k \in \mathbb{N}^*}$ tending to 0 at infinity, the set

$$\mathcal{E}_p(c, M) = \left\{ s \in \mathbb{L}_2(\mu) \cap \mathcal{S} : \sum_{\lambda \in \Lambda} \left| \beta_\lambda(s) / c_{r(\lambda)} \right|^p \le 1, \ \|s\|_\infty \le M \right\},$$

which is a subset of an l_p-body of $\mathbb{L}_2(\mu)$. We have in view to show that under some mild condition on the basis, it is possible to define a penalized LSE which achieves the minimax risk up to some slowly varying function of n on many sets $\mathcal{E}_p(c, M)$ at the same time. The condition that we have to impose on the basis in order to apply our theorem is as follows: there exists some positive constant B such that for all $\lambda \in \Lambda$

$$\left\| \sum_{\lambda' \in \Lambda_\lambda} a_{\lambda'} \varphi_{\lambda'} \right\|_\infty \le B\sqrt{|\Lambda_\lambda|} \sup_{\lambda' \in \Lambda_\lambda} |a_{\lambda'}| \quad \text{for any } a \in \mathbb{R}^{\Lambda_\lambda}. \tag{7.93}$$

This condition is typically satisfied for a wavelet basis which is ordered according to the lexicographical ordering. We have now to define a proper collection of models. We first define N to be the largest element of Λ such that $r(N) + |\Lambda_0| \le n(\ln n)^{-2}$ and consider $\mathcal{M} = \{\Lambda_0 \cup m' : m' \subset \Lambda_N\}$. We are in position to apply Theorem 7.7 with for instance the deterministic penalty function given by (7.32) and for each $m \in \mathcal{M}$, $L_m = \ln n$ which warrants that

$$\sum_{m \in \mathcal{M}} \exp(-L_m |m|) \le \sum_{D \le n} \binom{n}{D} \exp(-D \ln n) \le \sum_{D=1}^{\infty} \left(\frac{e}{D} \right)^D = \Sigma.$$

We know that this choice of collection of models and of the penalty term means that the resulting penalized LSE is a hard thresholding estimator. Its performance can be analyzed thanks to the risk bound

$$\mathbb{E}_s \left[\|\tilde{s} - s\|^2 \right] \le C(\varepsilon) \inf_{m \in \mathcal{M}} \left[d^2(s, S_m) + \frac{M|m| \ln n}{n} \right]$$
$$+ C(\varepsilon) \frac{(1 + M\Sigma)}{n},$$

which yields by Pythagore's Theorem

$$\mathbb{E}_s \left[\|\tilde{s} - s\|^2 \right] \le C(\varepsilon) \inf_{m' \subset \Lambda} \left[d^2(s, S_{\Lambda_0 \cup m'}) + d^2(s, S_{\Lambda_0 \cup \Lambda_N}) + \frac{M|m'| \ln n}{n} \right]$$
$$+ C(\varepsilon) \frac{(1 + M\Sigma + M|\Lambda_0| \ln n)}{n}.$$

We can also write this inequality as

$$\mathbb{E}_s \left[\|\tilde{s} - s\|^2 \right] \leq C\left(\varepsilon\right) \inf_{D \in \mathbb{N}^*} \left[\inf_{|m'| \leq 2D} d^2\left(s, S_{\Lambda_0 \cup m'}\right) + \frac{2MD \ln n}{n} \right]$$
$$+ C\left(\varepsilon\right) \left[d^2\left(s, S_{\Lambda_0 \cup \Lambda_N}\right) + \frac{\left(1 + M\Sigma + M \left|\Lambda_0\right| \ln n\right)}{n} \right].$$

Now, whenever s belongs to $\mathcal{E}_p\left(c, M\right)$ we have by monotonicity

$$d^2\left(s, S_{\Lambda_0 \cup \Lambda_N}\right) = \sum_{\lambda > N} \beta_\lambda^2\left(s\right) \leq \left[\sum_{\lambda > N} \left|\beta_\lambda\left(s\right)\right|^p \right]^{2/p} \leq c_{r(N)+1}^2.$$

It remains to bound, for any $D \in \mathbb{N}^*$, the bias term

$$b_D^2\left(s\right) = \inf_{|m'| \leq 2D} d^2\left(s, S_{\Lambda_0 \cup m'}\right).$$

This comes easily from approximation theory in sequence spaces. Indeed, we know that the nonlinear strategy consisting in keeping the $2D$ largest coefficients (in absolute value) $\beta_\lambda\left(s\right)$ for $\lambda \in \Lambda$ provides a set m' with cardinality $2D$ such that

$$\sum_{\lambda \notin m'} \beta_\lambda^2\left(s\right) \leq D^{1-2/p} \left[\sum_{r(\lambda) > D} \left|\beta_\lambda\left(s\right)\right|^p \right]^{2/p}.$$

Hence $b_D^2\left(s\right) \leq D^{1-2/p} c_D^2$ and we get

$$C^{-1}\left(\varepsilon\right) \sup_{s \in \mathcal{E}_p\left(c, M\right)} \mathbb{E}_s \left[\|\tilde{s} - s\|^2 \right] \leq 2 \inf_{D \in \mathbb{N}^*} \left[D^{1-2/p} c_D^2 + \frac{MD \ln n}{n} \right] + \left[c_{r(N)+1}^2 \right.$$
$$\left. + \frac{\left(1 + M\Sigma + M \left|\Lambda_0\right| \ln n\right)}{n} \right]. \qquad (7.94)$$

We see that whenever the sequence $\left(c_k\right)_k$ decreases rapidly enough, the order of the maximal risk of $\tilde{s}$ will be given by

$$\inf_{D \in \mathbb{N}^*} \left[D^{1-2/p} c_D^2 + n^{-1} MD \ln n \right].$$

As compared to the Gaussian case, the novelty here is the presence of M which controls the infinite norm of s and also the fact that the linear truncation term $c_{r(N)+1}^2$ can influence the order of the risk. If we consider the example of a Besov body for which $c_k = Rk^{1/p-1/2-\alpha}$, $k \in \mathbb{N}^*$, with $\alpha > 1/p - 1/2$, using the fact that $1 + r\left(N\right) \geq n \left(\ln n\right)^{-2} - \left|\Lambda_0\right|$ and arguing as in the treatment of ellipsoids above, we derive from (7.94) that, whenever $n \left(\ln n\right)^{-2} \geq 2 \left|\Lambda_0\right|$ and $n^{-1} \left(1 + M\right) \ln n \leq R^2$,

$$\sup_{s \in \mathcal{E}_p(c,M)} \mathbb{E}_s \left[\|\tilde{s} - s\|^2 \right] \leq C' \left(\varepsilon, |\Lambda_0| \right) R^{\frac{2}{1+2\alpha}} \left(\frac{n}{(1+M)\ln n} \right)^{-\frac{2\alpha}{1+2\alpha}}$$
$$+ C' \left(\varepsilon, |\Lambda_0| \right) R^2 \left(\frac{n}{\ln^2 n} \right)^{-2\alpha - 1 + 2/p}$$

and therefore

$$\sup_{s \in \mathcal{E}_p(c,M)} \mathbb{E}_s \left[\|\tilde{s} - s\|^2 \right] \leq C' \left(\varepsilon, |\Lambda_0| \right) R^{\frac{2}{1+2\alpha}} \left(\frac{n}{(1+M)\ln n} \right)^{-\frac{2\alpha}{1+2\alpha}}$$
$$+ C' \left(\varepsilon, |\Lambda_0| \right) R^2 n^{-2\alpha - 1 + 2/p} \left(\ln n \right)^{4\alpha}.$$

Combining this inequality with elementary computations, we derive that whenever

$$n^{-1} \ln n \leq \frac{R^2}{1+M} \leq n^{2\alpha + \left(1 + (2\alpha)^{-1}\right)(1 - 2/p)} \left(\ln n \right)^{-4\alpha - 1} \tag{7.95}$$

the maximum risk of the thresholding estimator is controlled by

$$\sup_{s \in \mathcal{E}_p(c,M)} \mathbb{E}_s \left[\|\tilde{s} - s\|^2 \right] \leq 2C' \left(\varepsilon, |\Lambda_0| \right) R^{\frac{2}{1+2\alpha}} \left(\frac{n}{(1+M)\ln n} \right)^{-\frac{2\alpha}{1+2\alpha}}.$$

Since $\mathcal{E}_p(c, M) \subset \mathcal{E}_2(c, M)$, we can conclude from the lower bounds for ellipsoids built from a conveniently localized basis that this is indeed the order of the minimax risk up to some constant depending only on $\varepsilon, |\Lambda_0|$ and M. It is of course desirable that the right-hand side in (7.95) tends to infinity as n goes to infinity (otherwise (7.95) really appears as a stringent condition on R). This is indeed the case if and only if $\alpha > \alpha_0$, where α_0 is the nonnegative root of the equation $2\alpha + \left(1 + (2\alpha)^{-1}\right)(1 - 2/p) = 0$, that is

$$\alpha_0 = \frac{\sqrt{2-p}}{4p} \left[\sqrt{2-p} + \sqrt{3+p} \right] \geq \frac{1}{p} - \frac{1}{2}.$$

It is easy to see that the condition $\alpha > \alpha_0$ is less stringent than the one assumed by Donoho, Johnstone, Kerkyacharian and Picard (see [54]) which is namely $p \geq 1$ and $\alpha > 1/p$. Whether our condition is optimal or not is an opened question.

7.5.3 Adaptive Properties of Penalized MLE

In the spirit of [12], many applications to adaptive estimation can be derived from the selection theorems for MLEs above. Again our purpose here is not to try to be exhaustive but rather to present some representative illustration of this idea. A major step in deriving adaptation results from Corollary 7.15 consists in controlling the bias term in (7.75). Using Lemma 7.12 it is possible to bound the Kullback–Leibler bias by the $\mathbb{L}_\infty$-bias.

Proposition 7.20 *Let $\overline{S}$ be some linear subspace of $\mathbb{L}_\infty$ such that $1 \in \overline{S}$ and $\sqrt{S}$ be the set of nonnegative elements of $\mathbb{L}_2$-norm equal to 1 in $\overline{S}$. Setting $S = \left\{ f^2, f \in \sqrt{S} \right\}$, one has*

$$1 \wedge \mathbf{K}(s, S) \leq 12 \inf_{f \in \overline{S}} \left\| \sqrt{s} - f \right\|_\infty^2 .$$

Proof. The proof is a straightforward consequence of Lemma 7.12. Indeed given $f \in \overline{S}$, setting $\varepsilon = \left\| \sqrt{s} - f \right\|_\infty$ we define $f^+ = f + \varepsilon$. Then $f^+ \geq \sqrt{s}$ with $\| f^+ - \sqrt{s} \| \leq \| f^+ - \sqrt{s} \|_\infty \leq 2\varepsilon$ and we derive from (7.70) that

$$1 \wedge \mathbf{K}(s, t) \leq 12\varepsilon^2 .$$

Since by assumption $f^+ \in \overline{S}$, so that $t = f^{+2} / \| f^+ \|^2$ belongs to S, this achieves the proof of the result. ∎

We are now in position to study a significant example.

Adaptive Estimation on Hölder Classes

We intend to use Corollary 7.15 for selecting a piecewise polynomial with degree r on a regular partition of $[0, 1]$ with N pieces, r and N being unknown. Because of the well know approximation properties of piecewise polynomials, we shall show that the corresponding penalized MLE is adaptive in the minimax sense when $\sqrt{s}$ belongs to some Hölder class $\mathcal{H}(\alpha, R)$ as defined in Section 7.5.1.

Proposition 7.21 *We consider $\mathcal{M} = \mathbb{N} \times \mathbb{N}^*$ and define for every $m = (r, \Delta) \in \mathcal{M}$, $\overline{S}_m$ to be the linear space of piecewise polynomials with maximal degree r on a regular partition of $[0, 1]$ with Δ pieces. Finally, for every $m \in \mathcal{M}$, we consider the model S_m of densities t such that $\sqrt{t} \in \overline{S}_m$. Let $\widetilde{s}$ be the penalized MLE defined by a penalty function*

$$\text{pen}(r, \Delta) = \frac{\kappa_1}{n} (r + 1) \Delta (2 + \ln(r + 1)) \tag{7.96}$$

where κ_1 is the absolute constant coming from Corollary 7.15. Then there exists a constant $C(\alpha)$ such for all density s with $\sqrt{s} \in \mathcal{H}(\alpha, R)$

$$\mathbb{E}_s \left[\mathbf{h}^2(s, \widetilde{s}) \right] \leq C(\alpha) R^{2/(2\alpha+1)} n^{-2\alpha/(2\alpha+1)}, \tag{7.97}$$

provided that $R \geq 1/\sqrt{n}$.

Proof. In order to apply Corollary 7.15 we consider the family of weights $\{x_m\}_{m \in \mathcal{M}}$, defined by

$$x_{r,\Delta} = r + \Delta \text{ for every } (r, \Delta) \in \mathbb{N} \times \mathbb{N}^*.$$

This choice of the weights leads to

$$\Sigma = \sum_{r\geq 0, \Delta\geq 1} \exp\left(-x_{r,\Delta}\right) = \left(\sum_{r\geq 0} \exp\left(-r\right)\right)\left(\sum_{\Delta\geq 1} \exp\left(-\Delta\right)\right)$$

$$= \frac{e}{(e-1)^2} < 1.$$

Since for every $m = (r, \Delta) \in \mathcal{M}$, one has $x_m = r + \Delta \leq (r+1)\Delta = D_m$ and $\bar{r}_m \leq 2r + 1$ (by Lemma 7.13), our choice (7.96) indeed satisfies to the restriction (7.74). Given s with $\sqrt{s} \in \mathcal{H}(\alpha, R)$, in order to bound the bias term $\mathbf{K}(s, S_m)$ we provide a control of the $\mathbb{L}_\infty$-distance between $f = \sqrt{s}$ and $\overline{S}_m$ and apply Proposition 7.20. Defining r as the largest integer smaller than α, it is known (see for instance [107]) that for a given interval I with length δ, there exists a polynomial Q_I with degree not greater than r such that

$$\|f|_I - Q_I\|_\infty \leq C'(r)\,\delta^r \omega(\delta)$$

where $C'(r)$ depends only on r and

$$\omega(\delta) = \sup_{x,y\in I} \left|f^{(r)}(x) - f^{(r)}(y)\right|.$$

This implies from the definition of $\mathcal{H}(\alpha, R)$ that, for $m = (r, \Delta)$, setting $f_m = \sum_{I\in\mathcal{P}_\Delta} Q_I \mathbb{1}_I$, where $\mathcal{P}_\Delta$ is a regular partition of $[0, 1]$ with Δ pieces, one has

$$\|f - f_m\|_\infty \leq C'(r)\,R\Delta^{-\alpha}$$

and therefore by Proposition 7.20

$$1 \wedge \mathbf{K}(s, S_m) \leq 12C'(r)\,R^2\Delta^{-2\alpha}.$$

Combining this bias term control with (7.75) yields

$$C_1^{-1}\mathbb{E}_s\left[\mathbf{h}^2(s, \tilde{s})\right]$$
$$\leq \inf_{\Delta\geq 1}\left\{12C'(r)\,R^2\Delta^{-2\alpha} + \frac{(1+\kappa_1)}{n}(r+1)\Delta\,(2+\ln(r+1))\right\}.$$

Recalling that $nR^2 \geq 1$, it remains to optimize this bound by choosing Δ as

$$\Delta = \min\left\{k : k \geq \left(nR^2\right)^{1/(2\alpha+1)}\right\},$$

which easily leads to (7.97). ∎

Recalling that $\mathcal{S}(\alpha, R)$ is the set of densities s such that $\sqrt{s} \in \mathcal{H}(\alpha, R)$, it comes from the minimax lower bound provided by Proposition 7.16 that $\tilde{s}$ is approximately minimax on each set $\mathcal{S}(\alpha, R)$ provided that $R \geq 1/\sqrt{n}$. These results can be generalized in different directions. For instance one can extend these results to multivariate functions with anisotropic smoothness (see [12]).

Selecting Bracketed Nets Towards Adaptation on Totally Bounded Sets

Like in the Gaussian case. It is possible use our general theorem for penalized MLE to select conveniently calibrated nets towards adaptation on (almost) arbitrary compact sets. Let us first notice that our general theorem for penalized MLE has the following straightforward consequence.

Corollary 7.22 *Let* $\{S_m\}_{m \in \mathcal{M}}$ *be some at most countable collection of models, where for each* $m \in \mathcal{M}$, S_m *is assumed to be a finite set of probability densities with respect to* μ. *We consider a corresponding collection of MLEs* $(\widehat{s}_m)_{m \in \mathcal{M}}$ *which means that for every* $m \in \mathcal{M}$

$$P_n\left(-\ln\left(\widehat{s}_m\right)\right) = \inf_{t \in S_m} P_n\left(-\ln\left(t\right)\right).$$

Let pen $: \mathcal{M} \to \mathbb{R}_+$ *and consider some random variable* $\widehat{m}$ *such that*

$$P_n\left(-\ln\left(\widehat{s}_{\widehat{m}}\right)\right) + \text{pen}\left(\widehat{m}\right) = \inf_{m \in \mathcal{M}}\left(P_n\left(-\ln\left(\widehat{s}_m\right)\right) + \text{pen}\left(m\right)\right).$$

Let $\{\Delta_m\}_{m \in \mathcal{M}}$, $\{x_m\}_{m \in \mathcal{M}}$ *be some family of nonnegative numbers such that* $\Delta_m \geq \ln\left(|S_m|\right)$ *for every* $m \in \mathcal{M}$ *and*

$$\sum_{m \in \mathcal{M}} e^{-x_m} = \Sigma < \infty.$$

Assume that

$$\text{pen}\left(m\right) = \frac{\kappa''\left(\Delta_m + x_m\right)}{n} \text{ for every } m \in \mathcal{M} \tag{7.98}$$

for some suitable numerical constant κ''. *Then, whatever the density* s

$$\mathbb{E}_s\left[\mathbf{h}^2\left(s, \widehat{s}_{\widehat{m}}\right)\right] \leq C''\left(\inf_{m \in \mathcal{M}}\left(\mathbf{K}\left(s, S_m\right) + \frac{\kappa\left(\Delta_m + x_m\right)}{n}\right) + \frac{\Sigma}{n}\right). \tag{7.99}$$

It is possible to mimic the approach developed in the Gaussian case in order to build adaptive estimators on a given collection of compact subsets of the set of densities, equipped with Hellinger distance. The main difference being that the metric structure with respect to Hellinger distance alone is not enough to deal with a discretized penalized MLE. In order to be able to control the bias term in (7.99) we need to use conveniently calibrated *bracketed* nets rather than simply nets as in the Gaussian case.

The general idea goes as follows. to use such a result in order to build some adaptive estimator on a collection of compact sets of $\mathbb{H}$ can be roughly described as follows. Let us consider some at most countable collection of subsets $(\mathcal{T}_m)_{m \in \mathcal{M}}$ of the set $\mathcal{S}$ of all probability densities. Assume that each set

$$\sqrt{\mathcal{T}_m} = \left\{ \sqrt{t}; t \in \mathcal{T}_m \right\}$$

has a finite entropy with bracketing $H_{[.]}\left(\delta, \sqrt{\mathcal{T}_m}\right)$ in $\mathbb{L}_2\left(\mu\right)$, for every $\delta > 0$. Given δ_m to be chosen later, one covers $\sqrt{\mathcal{T}_m}$ by at most $\exp\left(H_{[.]}\left(\delta_m, \sqrt{\mathcal{T}_m}\right)\right)$ brackets with diameter δ_m. In particular, denoting by $\sqrt{S_m^+}$ the set of upper extremities of those brackets, for all $m \in \mathcal{M}$ and all $s \in \mathcal{T}_m$, there exists some point $s_m^+ \in S_m^+$ such that

$$\sqrt{s} \leq \sqrt{s_m^+} \text{ and } \left\| \sqrt{s} - \sqrt{s_m^+} \right\| \leq \delta_m.$$

Setting $s_m = s_m^+ / \int s_m^+$, Lemma 7.12 ensures that

$$1 \wedge \mathbf{K}\left(s, s_m\right) \leq 3\delta_m^2.$$

For every $m \in \mathcal{M}$, if we take as a model

$$S_m = \left\{ t / \int t; t \in S_m^+ \right\},$$

then $\mathbf{K}\left(s, S_m\right) \leq 3\delta_m^2$ whenever $s \in \mathcal{T}_m$. We derive from Corollary 7.22 that, setting $\Delta_m = \left(n\delta_m^2\right) \vee H_{[.]}\left(\delta_m, \sqrt{\mathcal{T}_m}\right)$ and $x_m = \Delta_m$, provided that

$$\Sigma = \sum_{m \in \mathcal{M}} e^{-\Delta_m} < \infty,$$

the penalized MLE with penalty given by (7.98) satisfies for all $m \in \mathcal{M}$

$$\mathbb{E}_s\left[\mathbf{h}^2\left(s, \hat{s}_{\hat{m}}\right)\right] \leq \frac{C''}{n}\left(\Delta_m\left(3 + 2\kappa''\right) + \Sigma\right), \text{ for every } s \in \mathcal{T}_m. \qquad (7.100)$$

Choosing δ_m in such a way that $\varepsilon^2 H_{[.]}\left(\delta_m, \sqrt{\mathcal{T}_m}\right)$ is of order δ_m leads to a risk bound of order δ_m^2, (at least if Σ remains under control). In view of Proposition 7.19, we know that under some proper conditions δ_m^2 can be taken of the order of the minimax risk on $\mathcal{T}_m$, so that if Σ is kept under control, (7.100) implies that $\hat{s}_{\hat{m}}$ is approximately minimax on each parameter set $\mathcal{T}_m$. As in the Gaussian case, one can even hope adaptation on a wider continuously indexed collection of parameter spaces instead of the original discrete one. In analogy with the Gaussian case, let us illustrate these ideas by the following example.

As an exercise illustrating these general ideas, let us apply Corollary 4.24 to build an adaptive estimator in the minimax sense over some collection of compact subsets $\{\mathcal{S}_{\alpha,R}, \alpha \in \mathbb{N}^*, R > 0\}$ of $\mathcal{S}$ with the following structure:

$$\sqrt{\mathcal{S}_{\alpha,R}} = \sqrt{\mathcal{S}} \bigcap R\mathcal{H}_{\alpha,1},$$

where $\mathcal{H}_{\alpha,1}$ is star shaped at 0. Assume moreover that for some positive constant $C_2(\alpha)$

$$H_{[.]}\left(\delta, \sqrt{S_{\alpha,R}}\right) \leq C_2(\alpha) \left(\frac{R}{\delta}\right)^{1/\alpha}$$

for every $\delta \leq R \wedge 1/2$. In order to build an estimator which achieves (up to constants) the minimax risk over all the compact sets $S_{\alpha,R}$, we simply consider for every positive integers α and k a bracketed net $S_{\alpha,k}$ of $S_{\alpha,k/\sqrt{n}}$ constructed as explained above with $\delta_{\alpha,k} = k^{1/(2\alpha+1)}/\sqrt{n}$, so that

$$\ln|S_{\alpha,k}| \leq C_2(\alpha) \left(\frac{k}{\sqrt{n}\delta_{\alpha,k}}\right)^{1/\alpha} \leq C_2(\alpha) k^{2/(2\alpha+1)}$$

and

$$\mathbf{K}(s, S_{\alpha,k}) \leq 3\delta_{\alpha,k}^2, \text{ for all } s \in S_{\alpha,k/\sqrt{n}}. \tag{7.101}$$

Applying Corollary 7.22 to the collection $(S_{\alpha,k})_{\alpha \geq 1, k \geq 1}$ with

$$\Delta_{\alpha,k} = C_2(\alpha) k^{2/(2\alpha+1)} \text{ and } x_{\alpha,k} = 4\alpha k^{2/(2\alpha+1)}$$

leads to the penalty

$$\text{pen}(\alpha, k) = \frac{K(\alpha) k^{2/(2\alpha+1)}}{n}$$

where $K(\alpha) = \kappa''(C_2(\alpha) + 4\alpha)$. Noticing that $x_{\alpha,k} \geq \alpha + 2\ln(k)$ one has

$$\Sigma = \sum_{\alpha,k} e^{-x_{\alpha,k}} \leq \left(\sum_{\alpha \geq 1} e^{-\alpha}\right)\left(\sum_{k \geq 1} k^{-2}\right) < 1$$

and it follows from (7.100) that if $\tilde{s}$ denotes the penalized MLE one has whatever the density s

$$\mathbb{E}_s\left[\mathbf{h}^2(s, \tilde{s})\right] \leq C(\alpha) \inf_{\alpha,k}\left(\mathbf{K}(s, S_{\alpha,k}) + \frac{k^{2/(2\alpha+1)}}{n}\right).$$

In particular if $s \in S_{\alpha,R}$ for some integer α and some real number $R \geq 1/\sqrt{n}$, setting $k = \lceil R\sqrt{n}\rceil$ we have $s \in S_{\alpha,k/\sqrt{n}}$ and since $S_{\alpha,k}$ is a $k^{1/(2\alpha+1)}n^{-1/2}$-net of $S_{\alpha,k/\sqrt{n}}$, the previous inequality implies via (7.101) that

$$\sup_{s \in S_{\alpha,R}} \mathbb{E}_s\left[\mathbf{h}^2(s, \tilde{s})\right] \leq 4C(\alpha)\left(\frac{k^{2/(2\alpha+1)}}{n}\right)$$

$$\leq 4C(\alpha) 2^{2/(2\alpha+1)} \frac{(R\sqrt{n})^{2/(2\alpha+1)}}{n}.$$

Since Hellinger distance is bounded by 1, we finally derive that for some constant $C'(\alpha) \geq 1$

$$\sup_{s \in \mathcal{S}_{\alpha,R}} \mathbb{E}_s \left[\mathbf{h}^2 \left(s, \tilde{s} \right) \right] \leq C' \left(\alpha \right) \left(\left(R^{2/(2\alpha+1)} n^{-2\alpha/(2\alpha+1)} \right) \wedge 1 \right) \qquad (7.102)$$

If $\mathcal{H}_{\alpha,R}$ is the class $\mathcal{H}(\alpha, R)$ of Hölder smooth functions defined in Section 7.5.1, it is not difficult to see (using approximations by piecewise polynomials on dyadic regular partitions as in [22] for instance) that the $\mathbb{L}_\infty$-metric entropy of $\sqrt{\mathcal{S}_{\alpha,R}}$, $H_\infty \left(., \sqrt{\mathcal{S}_{\alpha,R}} \right)$ satisfies to

$$H_\infty \left(\delta/2, \sqrt{\mathcal{S}_{\alpha,R}} \right) \leq C_2 \left(\alpha \right) \left(\frac{R}{\delta} \right)^{1/\alpha}$$

which obviously implies that the above entropy with bracketing assumption is satisfied and therefore the previous approach applies to this case. Combining (7.102) with Proposition 7.16 shows that the discretized penalized MLE $\tilde{s}$ is indeed minimax (up to constants) on each compact set $\mathcal{S}_{\alpha,R}$ with $\alpha \in \mathbb{N}^*$ and $R \geq 1/\sqrt{n}$. More generally if $\mathcal{H}_{\alpha,R}$ is a closed ball with radius R centered at 0 with respect to some seminorm $\mathbf{n}_\alpha$ for which $\mathbf{n}_\alpha (1) = 0$ and if we moreover assume that for every $\delta \leq R \wedge 1/2$

$$H \left(\delta, \mathcal{S}_{\alpha,R} \right) \geq C_1 \left(\alpha \right) \left(\frac{\delta}{R} \right)^{-1/\alpha}$$

then, it comes from Proposition 7.19 that for some positive constant $\kappa \left(\alpha \right)$ depending only on α

$$\sup_{s \in S} \mathbb{E}_s \left[\mathbf{h}^2 \left(s, \tilde{s} \right) \right] \geq \kappa \left(\alpha \right) \left(\left(R^{2/(2\alpha+1)} n^{-2\alpha/(2\alpha+1)} \right) \wedge 1 \right),$$

provided that $R \geq 1/\sqrt{n}$ and again the discretized penalized MLE $\tilde{s}$ is approximately minimax on each set $\mathcal{S}_{\alpha,R}$ with $\alpha \in \mathbb{N}^*$ and $R \geq 1/\sqrt{n}$.

7.6 Appendix

The following inequalities are more or less classical and well known. We present some (short) proofs for the sake of completeness.

7.6.1 Kullback–Leibler Information and Hellinger Distance

We begin with the connections between Kullback–Leibler information and Hellinger distance.

Lemma 7.23 *Let P and Q be some probability measures. Then*

$$\mathbf{K} \left(P, \frac{P+Q}{2} \right) \geq \left(2 \ln \left(2 \right) - 1 \right) \mathbf{h}^2 \left(P, Q \right). \qquad (7.103)$$

Moreover whenever $P \ll Q$,

$$2\mathbf{h}^2 \left(P, Q \right) \leq \mathbf{K} \left(P, Q \right) \leq 2 \left(2 + \ln \left(\left\| \frac{dP}{dQ} \right\|_\infty \right) \right) \mathbf{h}^2 \left(P, Q \right) \qquad (7.104)$$

Proof. Let $\mu = (P + Q)/2$, then

$$\frac{dP}{d\mu} = 1 + u \text{ and } \frac{dQ}{d\mu} = 1 - u,$$

where u takes its values in $[-1, +1]$. Hence, setting for every $x \in [-1, +1]$, $\Phi(x) = (1 + x)\ln(1 + x) - x$ and for every $x \in [0, 1]$, $\Psi(x) = 1 - \sqrt{1 - x}$, one has

$$\mathbf{K}(P, \mu) = E_\mu[\Phi(u)] \text{ and } \mathbf{h}^2(P, Q) = E_\mu\left[\Psi(u^2)\right]. \tag{7.105}$$

Now $x \to \Phi(x)/x^2$ is nonincreasing on $[-1, +1]$ and $x \to \Psi(x)/x$ is nondecreasing on $[0, 1]$ so

$$\Phi(x) \geq \Phi(1)x^2 \geq \Phi(1)\Psi(x^2)$$

which leads to (7.103) via (7.105). If $P \ll Q$, introducing for every positive x, $f(x) = x\ln x - x + 1$ and $g(x) = (\sqrt{x} - 1)^2$ we can write

$$\mathbf{K}(P, Q) = E_Q\left[f\left(\frac{dP}{dQ}\right)\right] \text{ and } \mathbf{h}^2(P, Q) = \frac{1}{2}E_Q\left[g\left(\frac{dP}{dQ}\right)\right].$$

On the one hand the identity

$$f(x) - g(x) = 2\sqrt{x}f(\sqrt{x})$$

implies that

$$f(x) \geq g(x) \text{ for all nonnegative } x. \tag{7.106}$$

On the other hand, using elementary calculus, one can check that the function

$$x \to \left(2 + (\ln x)_+\right)g(x) - f(x)$$

achieves a unique minimum at point $x = 1$ and is therefore nonnegative. Since $x \to 2 + (\ln x)_+$ is increasing and $\|dP/dQ\|_\infty \geq 1$, this implies that

$$f(x) \leq \left(2 + \ln\left(\left\|\frac{dP}{dQ}\right\|_\infty\right)\right)g(x) \text{ for all } x \in [0, \|dP/dQ\|_\infty] \tag{7.107}$$

and the proof of (7.104) can be completed by taking expectation (under Q) on both sides of (7.106) and (7.107). ∎

A connection between Kullback–Leibler information and the variance of log-likelihood ratios is also useful. The following Lemma is adapted from Lemma 1 of Barron and Sheu [11] and borrowed to [34].

Lemma 7.24 *For all probability measures P and Q with $P \ll Q$*

$$\frac{1}{2}\int(dP \wedge dQ)\left(\ln\frac{dP}{dQ}\right)^2 \leq \mathbf{K}(P, Q) \leq \frac{1}{2}\int(dP \vee dQ)\left(\ln\frac{dP}{dQ}\right)^2.$$

Proof. Setting for every real number x

$$f(x) = e^{-x} + x - 1,$$

the Kullback–Leibler information can be written as

$$\mathbf{K}(P,Q) = \int \frac{dP}{dQ} f\left(\ln\left(\frac{dP}{dQ}\right)\right) dQ.$$

Since one can check by elementary that

$$\frac{1}{2}x^2\left(1 \wedge e^{-x}\right) \le f(x) \le \frac{1}{2}x^2\left(1 \vee e^{-x}\right),$$

the result immediately follows. ∎

We are also recording below some convexity properties of Hellinger distance.

Lemma 7.25 *Let P, Q and R be probability measures. For all $\theta \in [0,1]$ the following inequalities hold*

$$\mathbf{h}^2\left(P, \theta Q + (1 - \theta) R\right) \le \theta \mathbf{h}^2\left(P, Q\right) + (1 - \theta) \mathbf{h}^2\left(P, R\right) \tag{7.108}$$

and

$$\mathbf{h}^2\left(\theta P + (1 - \theta) R, \theta Q + (1 - \theta) R\right) \le \theta \mathbf{h}^2\left(P, Q\right). \tag{7.109}$$

Proof. Recalling that Hellinger affinity between two probability measures P and Q is defined by

$$\rho(P,Q) = \int \sqrt{dP\,dQ},$$

one has

$$\mathbf{h}^2(P,Q) = 1 - \rho(P,Q).$$

We note that by concavity of the square root

$$\rho(P, \theta Q + (1 - \theta) R) \ge \theta \rho(P,Q) + (1 - \theta) \rho(P,R),$$

which leads to (7.108). To prove (7.109), setting

$$\rho_\theta = \rho(\theta P + (1 - \theta) R, \theta Q + (1 - \theta) R)$$

we use the elementary inequality $2\sqrt{ab} \le a + b$ to derive that

$$\rho_\theta = \int \left(\theta^2 dP\,dQ + (1 - \theta)^2 dR^2 + \theta(1 - \theta)(dP + dQ)\,dR\right)^{1/2}$$

$$\ge \int \left(\theta^2 dP\,dQ + (1 - \theta)^2 dR^2 + 2\theta(1 - \theta)\sqrt{dP\,dQ}\,dR\right)^{1/2}$$

$$\ge \int \left(\theta\sqrt{dP\,dQ} + (1 - \theta)\,dR\right)$$

$$\ge \theta \rho(P,Q) + 1 - \theta,$$

which easily leads to (7.109), achieving the proof of Lemma 7.25. ∎

7.6.2 Moments of Log-Likelihood Ratios

In order to use the maximal inequalities of the previous chapter, it is essential to get good upper bounds for the moments of the log-likelihood ratios. The following Lemma is an improved version (in terms of the absolute constants involved) of a result due to van de Geer (see [118]). The proof is borrowed to [21].

Lemma 7.26 *Let P be some probability measure with density s with respect to some measure μ and t, u be some nonnegative and μ-integrable functions, then, denoting by $\|.\|_2$ the $\mathbb{L}_2(\mu)$-norm, one has for every integer $k \geq 2$*

$$P\left(\left|\ln\left(\sqrt{\frac{s+t}{s+u}}\right)\right|^k\right) \leq \frac{9}{64}k!\left\|\sqrt{t}-\sqrt{u}\right\|^2.$$

Proof. We first establish the elementary inequality

$$x - 1 - \ln(x) \leq \frac{9}{64}\left(x - \frac{1}{x}\right)^2 \tag{7.110}$$

which is valid for every positive x. Indeed setting

$$f(x) = \frac{9}{64}\left(x - \frac{1}{x}\right)^2 - x - 1 - \ln(x)$$

one can check that $x^3 f'(x)$ is nondecreasing on $(0, +\infty)$ and is equal to 0 at point $x = 1$. This implies that f achieves a minimum at point 1 and since $f(1) = 0$, f is nonnegative which means that (7.110) holds. Hence, setting $y = (1/x) \vee x$, we derive from (7.110) that

$$\frac{|\ln(x)|^k}{k!} = \frac{(\ln(y))^k}{k!} \leq y - 1 - \ln(y) \leq \frac{9}{64}\left(x - \frac{1}{x}\right)^2.$$

Plugging $x = \sqrt{(s+t)/s+u}$ in this inequality, we get

$$P\left(\left|\ln\left(\sqrt{\frac{s+t}{s+u}}\right)\right|^k\right) \leq \frac{9}{64}k!P\left(\frac{(t-u)^2}{(s+u)(s+t)}\right).$$

Now

$$\left(\sqrt{t}+\sqrt{u}\right)^2 s \leq (s+u)(s+t),$$

so

$$P\left(\frac{(t-u)^2}{(s+u)(s+t)}\right) = \int \frac{s(t-u)^2}{(s+u)(s+t)}d\mu$$

$$\leq \left\|\sqrt{t}-\sqrt{u}\right\|^2$$

and the result follows. ∎

7.6.3 An Exponential Bound for Log-Likelihood Ratios

Let $\mathbf{h}$ denote Hellinger distance.

Proposition 7.27 *Let* $X_1, \ldots, X_n$ *be independent random variables with common distribution* $P = s\mu$. *Then, for every positive density* f *and any positive number* x

$$\mathbb{P}\left[P_n \left[\ln \left(\frac{f}{s} \right) \right] \geq -2\mathbf{h}^2 \left(s, f \right) + 2\frac{x}{n} \right] \leq e^{-x}.$$

Proof. By Markov's inequality, for every $y \in \mathbb{R}$

$$\mathbb{P}\left[P_n \left[\ln \left(\frac{f}{s} \right) \right] \geq y \right] \leq \left(P \left[\exp \left(\frac{1}{2} \ln \left(\frac{f}{s} \right) \right) \right] \right)^n e^{-ny/2}$$

$$\leq \left(1 - \mathbf{h}^2 \left(s, f \right) \right)^n e^{-ny/2} \leq e^{-n\mathbf{h}^2(s,f)} e^{-ny/2}$$

which implies the result if one takes $y = -2\mathbf{h}^2 \left(s, f \right) + 2x/n$. ∎

8

Statistical Learning

8.1 Introduction

Suppose that one observes independent variables $\xi_1, \ldots, \xi_n$ taking their values in some measurable space Ξ. Let us furthermore assume, for the sake of simplicity, that these variables are identically distributed with common distribution P. The two main frameworks that we have in mind are respectively the classification and the bounded regression frameworks. In those cases, for every i the variable $\xi_i = (X_i, Y_i)$ is a copy of a pair of random variables (X, Y), where X takes its values in some measurable space $\mathcal{X}$ and Y is assumed to take its values in $[0, 1]$. In the classification case, the response variable Y is assumed to belong to $\{0, 1\}$. One defines the regression function η as

$$\eta(x) = \mathbb{E}[Y \mid X = x] \tag{8.1}$$

for every $x \in \mathcal{X}$. In the regression case, one is interested in the estimation of η while in the classification case, one wants to estimate the Bayes classifier s^*, defined for every $x \in \mathcal{X}$ by

$$s^*(x) = 1\!\!1_{\eta(x) \geq 1/2}. \tag{8.2}$$

One of the most commonly used method to estimate η or s^* or more generally to estimate a quantity of interest s depending on the unknown distribution P is the so called empirical risk minimization by Vapnik which is a special instance of minimum contrast estimation.

Empirical Risk Minimization

Basically one considers some set $\mathcal{S}$ which is known to contain s, think of $\mathcal{S}$ as being the set of all measurable functions from $\mathcal{X}$ to $[0, 1]$ in the regression case or to $\{0, 1\}$ in the classification case. Then we consider some *loss (or contrast)* function

$$\gamma \text{ from } \mathcal{S} \times \Xi \text{ to } [0, 1]$$

which is well adapted to our estimation problem of s in the sense that the *expected loss* $\mathbb{E}\left[\gamma\left(t,\xi_1\right)\right]$ achieves a minimum at point s when t varies in $\mathcal{S}$. In other words the *relative expected loss* ℓ defined by

$$\ell\left(s,t\right)=\mathbb{E}\left[\gamma\left(t,\xi_1\right)-\gamma\left(s,\xi_1\right)\right],\quad\text{for all }\ t\in\mathcal{S}\tag{8.3}$$

is nonnegative. In the regression or the classification cases, one can take

$$\gamma\left(t,\left(x,y\right)\right)=\left(y-t\left(x\right)\right)^2$$

since η (resp. s^*) is indeed the minimizer of $\mathbb{E}\left[\left(Y-t\left(X\right)\right)^2\right]$ over the set of measurable functions t taking their values in $[0,1]$ (resp. $\{0,1\}$). The heuristics of empirical risk minimization (or minimum contrast estimation) can be described as follows. If one substitutes the empirical risk

$$\gamma_n\left(t\right)=P_n\left[\gamma\left(t,.\right)\right]=\frac{1}{n}\sum_{i=1}^{n}\gamma\left(t,\xi_i\right),$$

to its expectation $P\left[\gamma\left(t,.\right)\right]=\mathbb{E}\left[\gamma\left(t,\xi_1\right)\right]$ and minimizes γ_n on some subset S of $\mathcal{S}$ (that we call a *model*), there is some hope to get a sensible estimator $\widehat{s}$ of s, at least if s belongs (or is close enough) to the model S.

8.2 Model Selection in Statistical Learning

The purpose of this section is to provide an other look at the celebrated Vapnik's method of *structural risk minimization* (initiated in [121]) based on concentration inequalities. In the next section, we shall present an alternative analysis which can lead to improvements of Vapnik's method for the classification problem. Let us consider some countable or finite (but possibly depending on n) collection of models $\{S_m\}_{m\in\mathcal{M}}$ and the corresponding collection of empirical risk minimizers $\{\widehat{s}_m\}_{m\in\mathcal{M}}$. For every $m\in\mathcal{M}$ an empirical risk minimizer within model S_m is defined by

$$\widehat{s}_m=\operatorname{argmin}_{t\in S_m}\gamma_n\left(t\right).$$

Given some penalty function pen: $\mathcal{M}\to\mathbb{R}_+$ and let us define $\widehat{m}$ as a minimizer of

$$\gamma_n\left(\widehat{s}_m\right)+\operatorname{pen}\left(m\right)\tag{8.4}$$

over $\mathcal{M}$ and finally estimate s by the *penalized estimator*

$$\widetilde{s}=\widehat{s}_{\widehat{m}}.$$

Since some problems can occur with the existence of a solution to the previous minimization problems, it is useful to consider approximate solutions (note that even if $\widehat{s}_m$ does exist, it is relevant from a practical point of view to consider approximate solutions since $\widehat{s}_m$ will typically be approximated by some numerical algorithm). Therefore, given $\rho \geq 0$ (in practice, taking $\rho = n^{-2}$ makes the introduction of an approximate solution painless), we shall consider for every $m \in \mathcal{M}$ some approximate empirical risk minimizer $\widehat{s}_m$ satisfying

$$\gamma_n \left(\widehat{s}_m\right) \leq \gamma_n \left(t\right) + \rho$$

and say that $\widetilde{s}$ is a ρ-penalized estimator of s if

$$\gamma_n \left(\widetilde{s}\right) + \operatorname{pen}\left(\widehat{m}\right) \leq \gamma_n \left(t\right) + \operatorname{pen}\left(m\right) + \rho, \ \forall m \in \mathcal{M} \text{ and } \forall t \in S_m. \quad (8.5)$$

To analyze the statistical performance of this procedure, the key is to take $\ell \left(s, t\right)$ as a loss function and notice that the definition of the penalized procedure leads to a very simple but fundamental control for $\ell \left(s, \widetilde{s}\right)$. Indeed, by the definition of $\widetilde{s}$ we have, whatever $m \in \mathcal{M}$ and $s_m \in S_m$,

$$\gamma_n \left(\widetilde{s}\right) + \operatorname{pen}\left(\widehat{m}\right) \leq \gamma_n \left(s_m\right) + \operatorname{pen}\left(m\right) + \rho,$$

and therefore

$$\gamma_n \left(\widetilde{s}\right) \leq \gamma_n \left(s_m\right) + \operatorname{pen}\left(m\right) - \operatorname{pen}\left(\widehat{m}\right) + \rho. \quad (8.6)$$

If we introduce the centered empirical process

$$\overline{\gamma}_n \left(t\right) = \gamma_n \left(t\right) - \mathbb{E}\left[\gamma \left(t, \xi_1\right)\right], \ t \in \mathcal{S}$$

and notice that $\mathbb{E}\left[\gamma \left(t, \xi_1\right)\right] - \mathbb{E}\left[\gamma \left(u, \xi_1\right)\right] = \ell \left(s, t\right) - \ell \left(s, u\right)$ for all $t, u \in \mathcal{S}$, we readily get from (8.6)

$$\ell \left(s, \widetilde{s}\right) \leq \ell \left(s, s_m\right) + \overline{\gamma}_n \left(s_m\right) - \overline{\gamma}_n \left(\widetilde{s}\right) - \operatorname{pen}\left(\widehat{m}\right) + \operatorname{pen}\left(m\right) + \rho. \quad (8.7)$$

8.2.1 A Model Selection Theorem

Let us first see what can be derived from (8.7) by using only the following boundedness assumption on the contrast function γ

A1 For every t belonging to some set $\mathcal{S}$, one has $0 \leq \gamma \left(t, .\right) \leq 1$.

In order to avoid any measurability problem, let us first assume that each of the models S_m is countable. Given some constant Σ, let us consider some preliminary collection of nonnegative weights $\{x_m\}_{m \in \mathcal{M}}$ such that

$$\sum_{m \in \mathcal{M}} e^{-x_m} \leq \Sigma$$

and let $z > 0$ be given. It follows from (5.7) (which was proved in Chapter 5 to be a consequence of Mc Diarmid's Inequality) that for every $m' \in \mathcal{M}$,

$$\mathbb{P}\left[\sup_{t \in S_{m'}} \left(-\overline{\gamma}_n \left(t\right)\right) \geq \mathbb{E}\left[\sup_{t \in S_{m'}} \left(-\overline{\gamma}_n \left(t\right)\right)\right] + \sqrt{\frac{x_{m'} + z}{2n}}\right] \leq e^{-x_{m'} - z},$$

and therefore, setting $\mathbb{E}\left[\sup_{t \in S_{m'}} (-\overline{\gamma}_n(t))\right] = E_{m'}$, except on a set of probability not larger than Σe^{-z}, one has for every $m' \in \mathcal{M}$,

$$\sup_{t \in S_{m'}} (-\overline{\gamma}_n(t)) \le E_{m'} + \sqrt{\frac{x_{m'} + z}{2n}}.$$

Hence, (8.7) implies that the following inequality holds, except on a set of probability not larger than Σe^{-z}:

$$\ell(s, \widetilde{s}) \le \ell(s, s_m) + \overline{\gamma}_n(s_m) + E_{\widehat{m}} + \sqrt{\frac{x_{\widehat{m}}}{2n}} - \text{pen}(\widehat{m}) + \text{pen}(m) + \sqrt{\frac{z}{2n}} + \rho. \quad (8.8)$$

It is tempting to choose $\text{pen}(m') = E_{m'} + \sqrt{x_{m'}/2n}$ for every $m' \in \mathcal{M}$ but we should not forget that $E_{m'}$ typically depends on the unknown s. Thus, we are forced to consider some upper bound $\widetilde{E}_{m'}$ of $E_{m'}$ which does not depend on s. This upper bound can be either deterministic (we shall discuss below the drawbacks of this strategy) or random and in such a case we shall take benefit of the fact that it is enough to assume that $\widetilde{E}_{m'} \ge E_{m'}$ holds on a set with sufficiently high probability. More precisely, assuming that for some constant K and for every $m' \in \mathcal{M}$

$$\text{pen}(m') \ge E_{m'} + \sqrt{\frac{x_{m'}}{2n}} - K\sqrt{\frac{z}{2n}} \quad (8.9)$$

holds, except on set of probability not larger than $\exp(-x_{m'} - z)$, we derive from (8.8) and (8.9) that

$$\ell(s, \widetilde{s}) \le \ell(s, s_m) + \overline{\gamma}_n(s_m) + \text{pen}(m) + (1 + K)\sqrt{\frac{z}{2n}} + \rho$$

holds except on a set of probability not larger than $2\Sigma e^{-z}$. Thus, integrating with respect to z leads to

$$\mathbb{E}\left[(\ell(s, \widetilde{s}) - \ell(s, s_m) - \overline{\gamma}_n(s_m) - \text{pen}(m) - \rho)^+\right] \le \Sigma(1 + K)\sqrt{\frac{\pi}{2n}}$$

and therefore, since $\overline{\gamma}_n(s_m)$ is centered at expectation

$$\mathbb{E}\left[\ell(s, \widetilde{s})\right] \le \ell(s, s_m) + \mathbb{E}\left[\text{pen}(m)\right] + \Sigma(1 + K)\sqrt{\frac{\pi}{2n}} + \rho.$$

Hence, we have proven the following result.

Theorem 8.1 *Let* $\xi_1, \ldots, \xi_n$ *be independent observations taking their values in some measurable space* Ξ *and with common distribution* P *depending on some unknown parameter* $s \in \mathcal{S}$. *Let* $\gamma : \mathcal{S} \times \Xi \to \mathbb{R}$ *be some contrast function* *satisfying assumption **A1**. Let $\{S_m\}_{m\in\mathcal{M}}$ be some at most countable collection of countable subsets of $\mathcal{S}$ and $\rho \geq 0$ be given. Consider some absolute constant Σ, some family of nonnegative weights $\{x_m\}_{m\in\mathcal{M}}$ such that*

$$\sum_{m\in\mathcal{M}} e^{-x_m} = \Sigma < \infty$$

and some (possibly data-dependent) penalty function pen : $\mathcal{M} \to \mathbb{R}_+$. Let $\widetilde{s}$ be a ρ-penalized estimator of s as defined by (8.5). Then, if for some nonnegative constant K, for every $m \in \mathcal{M}$ and every positive z

$$\text{pen}(m) \geq \mathbb{E}\left[\sup_{t\in S_m}\left(-\overline{\gamma}_n(t)\right)\right] + \sqrt{\frac{x_m}{2n}} - K\sqrt{\frac{z}{2n}}$$

holds with probability larger than $1 - \exp(-x_m - z)$, the following risk bound holds for all $s \in \mathcal{S}$

$$\mathbb{E}\left[\ell(s,\widetilde{s})\right] \leq \inf_{m\in\mathcal{M}}\left(\ell(s,S_m) + \mathbb{E}\left[\text{pen}(m)\right]\right) + \Sigma(1+K)\sqrt{\frac{\pi}{2n}} + \rho, \quad (8.10)$$

where ℓ is defined by (8.3) and $\ell(s,S_m) = \inf_{t\in S_m}\ell(s,t)$.

It is not that easy to discuss whether this result is sharp or not in the generality where it is stated here. Nevertheless we shall see that, at the price of making an extra assumption on the contrast function γ, it is possible to improve on (8.10) by weakening the restriction on the penalty function. This will be the purpose of our next section.

Vapnik's Learning Theory Revisited

We would like here to explain how Vapnik's *structural minimization of the risk method* (as described in [121] and further developed in [122]) fits in the above framework of model selection via penalization. More precisely, we shall consider the *classification* problem and show how to recover (or refine in the spirit of [31]) some of Vapnik's results from Theorem 8.1. The data $\xi_1 = (X_1, Y_1), \ldots, \xi_n = (X_n, Y_n)$ consist of independent, identically distributed copies of the random variable pair (X, Y) taking values in $\mathcal{X} \times \{0,1\}$. Let the models $\{S_m\}_{m\in\mathcal{M}}$ being defined for every $m \in \mathcal{M}$ as

$$S_m = \{\mathbb{1}_C : C \in \mathcal{A}_m\},$$

where $\mathcal{A}_m$ is some countable class of subsets of $\mathcal{X}$. Let $\mathcal{S}$ be the set of measurable functions taking their values in $[0,1]$. In this case, the least squares contrast function fulfills condition **A1**. Indeed, since $\gamma(t,(x,y)) = (y - t(x))^2$,

A1 is fulfilled whenever $t \in \mathcal{S}$ and $y \in [0,1]$. For a function t taking only the two values 0 and 1, the least squares criterion also writes

$$\frac{1}{n} \sum_{i=1}^{n} (Y_i - t(X_i))^2 = \frac{1}{n} \sum_{i=1}^{n} \mathbb{1}_{Y_i \neq t(X_i)}$$

so that minimizing the least squares criterion means minimizing the number of misclassifications on the training sample $\xi_1 = (X_1, Y_1), \ldots, \xi_n = (X_n, Y_n)$. Each estimator $\widehat{s}_m$ represents some possible classification rule and the purpose of model selection is here to select what classification rule is the best according to some risk minimization criterion. At this stage it should be noticed that we have the choice here between two different definitions of the statistical object of interest s. Indeed, we can take s to be the minimizer of $t \to \mathbb{E}[Y - t(X)]^2$ subject or not to the restriction that t takes its values in $\{0,1\}$. On the one hand the Bayes classifier s^* as defined by (8.2) is a minimizer of $\mathbb{E}[Y - t(X)]^2$ under the restriction that t takes its values in $\{0,1\}$ and the corresponding loss function can be written as

$$\ell(s^*, t) = \mathbb{E}[s^*(X) - t(X)]^2 = \mathbb{P}[Y \neq t(X)] - \mathbb{P}[Y \neq s^*(X)].$$

On the other hand, the regression function η as defined by (8.1) minimizes $\mathbb{E}[Y - t(X)]^2$ without the restriction that t takes its values in $\{0,1\}$, and the corresponding loss function is simply $\ell(\eta, t) = \mathbb{E}[\eta(X) - t(X)]^2$. It turns out that the results presented below are valid for both definitions of s simultaneously. In order to apply Theorem 8.1, it remains to upper bound $\mathbb{E}\left[\sup_{t \in S_m}(-\overline{\gamma}_n(t))\right]$. Let us introduce the (random) combinatorial entropy of $\mathcal{A}_m$

$$H_m = \ln |\{A \cap \{X_1, \ldots, X_n\}, A \in \mathcal{A}_m\}|.$$

If we take some independent copy $(\xi_1', \ldots, \xi_n')$ of $(\xi_1, \ldots, \xi_n)$ and consider the corresponding copy γ_n' of γ_n, we can use the following standard symmetrization argument. By Jensen's inequality

$$\mathbb{E}\left[\sup_{t \in S_m}(-\overline{\gamma}_n(t))\right] \leq \mathbb{E}\left[\sup_{t \in S_m}(\gamma_n'(t) - \gamma_n(t))\right],$$

so that, given independent random signs $(\varepsilon_1, \ldots, \varepsilon_n)$, independent of $(\xi_1, \ldots, \xi_n)$, one has,

$$\mathbb{E}\left[\sup_{t \in S_m}(-\overline{\gamma}_n(t))\right] \leq \frac{1}{n}\mathbb{E}\left[\sup_{t \in S_m}\left(\sum_{i=1}^{n}\varepsilon_i\left(\mathbb{1}_{Y_i' \neq t(X_i')} - \mathbb{1}_{Y_i \neq t(X_i)}\right)\right)\right]$$

$$\leq \frac{2}{n}\mathbb{E}\left[\sup_{t \in S_m}\left(\sum_{i=1}^{n}\varepsilon_i\mathbb{1}_{Y_i \neq t(X_i)}\right)\right].$$

Hence, by (6.3) we get

$$\mathbb{E}\left[\sup_{t \in S_m} \left(-\overline{\gamma}_n(t)\right)\right] \leq \frac{2\sqrt{2}}{n} \mathbb{E}\left[\left(H_m \sup_{t \in S_m} \left(\sum_{i=1}^{n} \mathbb{1}_{Y_i \neq t(X_i)}\right)\right)^{1/2}\right],$$

and by Jensen's inequality

$$\mathbb{E}\left[\sup_{t \in S_m} \left(-\overline{\gamma}_n(t)\right)\right] \leq 2\sqrt{\frac{2\mathbb{E}\left[H_m\right]}{n}}. \tag{8.11}$$

The trouble now is that $\mathbb{E}\left[H_m\right]$ is unknown. Two different strategies can be followed to overcome this difficulty. First, one can use the VC-dimension to upper bound $\mathbb{E}\left[H_m\right]$. Assume each $\mathcal{A}_m$ to be a VC-class with VC-dimension V_m (see Definition 6.2), one derives from (6.9) that

$$\mathbb{E}\left[H_m\right] \leq V_m \left(1 + \ln\left(\frac{n}{V_m}\right)\right). \tag{8.12}$$

If $\mathcal{M}$ has cardinality not larger than n, one can take $x_m = \ln(n)$ for each $m \in \mathcal{M}$ which leads to a penalty function of the form

$$\text{pen}(m) = 2\sqrt{\frac{2V_m\left(1 + \ln\left(n/V_m\right)\right)}{n}} + \sqrt{\frac{\ln(n)}{2n}},$$

and to the following risk bound for the corresponding penalized estimator $\widetilde{s}$, since then $\Sigma \leq 1$:

$$\mathbb{E}\left[\ell\left(s, \widetilde{s}\right)\right] \leq \inf_{m \in \mathcal{M}} \left(\ell\left(s, S_m\right) + \text{pen}(m)\right) + \sqrt{\frac{\pi}{2n}} + \rho. \tag{8.13}$$

This approach has two main drawbacks:

- the VC-dimension of a given collection of sets is generally very difficult to compute or even to evaluate (see [6] and [69] for instance);
- even if the VC-dimension is computable (in the case of affine half spaces of $\mathbb{R}^d$ for instance), inequality (8.12) is too pessimistic and it would be desirable to define a penalty function from a quantity which is much closer to $\mathbb{E}\left[H_m\right]$ than the right-hand side of (8.12).

Following [31], the second strategy consists of substituting H_m to $\mathbb{E}\left[H_m\right]$ by using again a concentration argument. Indeed, by (5.22), for any positive z, one has $H_m \geq \mathbb{E}\left[H_m\right] - \sqrt{2\ln(2)\mathbb{E}\left[H_m\right](x_m + z)}$, on a set of probability not less than $1 - \exp(-x_m - z)$. Hence, since

$$\sqrt{2\ln(2)\mathbb{E}\left[H_m\right](x_m + z)} \leq \frac{\mathbb{E}\left[H_m\right]}{2} + \ln(2)(x_m + z),$$

we have on the same set,

$$\mathbb{E}\left[H_m\right] \leq 2H_m + 2\ln(2)(x_m + z),$$

which, by (8.11), yields

$$\mathbb{E}\left[\sup_{t \in S_m} (-\overline{\gamma}_n (t))\right] \leq 4 \left(\sqrt{\frac{H_m}{n}} + \sqrt{\frac{\ln (2) \, x_m}{n}} + \sqrt{\frac{\ln (2) \, z}{n}} \right).$$

Taking $x_m = \ln (n)$ as before leads to the following choice for the penalty function

$$\text{pen} (m) = 4\sqrt{\frac{H_m}{n}} + 4.1\sqrt{\frac{\ln (n)}{n}},$$

which satisfies

$$\text{pen} (m) \geq \mathbb{E}\left[\sup_{t \in S_m} (-\overline{\gamma}_n (t))\right] + \sqrt{\frac{\ln (n)}{2n}} - 4\sqrt{\frac{\ln (2) \, z}{n}}.$$

The corresponding risk bound can be written as

$$\mathbb{E}\left[\ell (s, \widetilde{s})\right] \leq \left[\inf_{m \in \mathcal{M}} (\ell (s, S_m) + \mathbb{E}\left[\text{pen} (m)\right]) + 4\sqrt{\frac{\pi \ln (2)}{n}} + \rho \right],$$

and therefore, by Jensen's inequality

$$\mathbb{E}\left[\ell (s, \widetilde{s})\right] \leq \left[\inf_{m \in \mathcal{M}} \left(\ell (s, S_m) + 4\sqrt{\frac{\mathbb{E}\left[H_m\right]}{n}} \right) + 4.1\sqrt{\frac{\ln (n)}{n}} + \frac{6}{\sqrt{n}} + \rho \right].$$
$$(8.14)$$

Note that if we take $s = s^*$, denoting by L_t the probability of misclassification of the rule t, i.e., $L_t = \mathbb{P}\left[Y \neq t (X)\right]$, the risk bound (8.14) can also be written as

$$\mathbb{E}\left[L_{\widetilde{s}}\right] \leq \inf_{m \in \mathcal{M}} \left(\inf_{t \in S_m} L_t + 4\sqrt{\frac{\mathbb{E}\left[H_m\right]}{n}} \right) + 4.1\sqrt{\frac{\ln (n)}{n}} + \frac{6}{\sqrt{n}} + \rho,$$

which is maybe a more standard way of expressing the performance of a classifier in the statistical learning literature. Of course, if we follow the first strategy of penalization a similar bound can be derived from (8.13), namely

$$\mathbb{E}\left[L_{\widetilde{s}}\right] \leq \inf_{m \in \mathcal{M}} \left(\inf_{t \in S_m} L_t + 2\sqrt{\frac{2V_m (1 + \ln (n/V_m))}{n}} \right)$$
$$+ \sqrt{\frac{\ln (n)}{2n}} + \sqrt{\frac{\pi}{2n}} + \rho.$$

Note that the same conclusions would hold true (up to straightforward modifications of the absolute constants) if instead of the combinatorial entropy H_m, one would take as a random measure of complexity for the class S_m the Rademacher conditional mean

$$\frac{1}{\sqrt{n}} \mathbb{E} \left[\sup_{t \in S_m} \sum_{i=1}^{n} \varepsilon_i \mathbb{1}_{Y_i \neq t(X_i)} \mid (X_i, Y_i)_{1 \leq i \leq n} \right]$$

since we have indeed seen in Chapter 4 that this quantity obeys exactly to the same concentration inequality as H_m. This leads to risk bounds for Rademacher penalties of the same nature as those obtained by Bartlett, Boucheron and Lugosi (see [13]) or Koltchinskii (see [70]).

8.3 A Refined Analysis for the Risk of an Empirical Risk Minimizer

The purpose of this section is to provide a general upper bound for the relative expected loss between $\widehat{s}$ and s, where $\widehat{s}$ denotes the empirical risk minimizer over a given model S.

We introduce the centered empirical process $\overline{\gamma}_n$. In addition to the relative expected loss function ℓ we shall need another way of measuring the closeness between the elements of S which is directly connected to the variance of the increments of $\overline{\gamma}_n$ and therefore will play an important role in the analysis of the fluctuations of $\overline{\gamma}_n$. Let d be some pseudodistance on $S \times S$ (which may perfectly depend on the unknown distribution P) such that

$$P \left((\gamma(t, .) - \gamma(s, .))^2 \right) \leq d^2(s, t), \text{ for every } t \in S.$$

Of course, we can take d as the pseudodistance associated to the variance of γ itself, but it will more convenient in the applications to take d as a more intrinsic distance. For instance, in the regression or the classification setting it is easy to see that d can be chosen (up to constant) as the $\mathbb{L}_2(\mu)$ distance, where μ denotes the distribution of X. Indeed, for classification

$$|\gamma(t, (x, y)) - \gamma(s^*, (x, y))| = \left| \mathbf{I}_{y \neq t(x)} - \mathbf{I}_{y \neq s^*(x)} \right| = |t(x) - s^*(x)|$$

and therefore setting $d^2(s^*, t) = \mathbb{E} \left[(t(X) - s^*(X))^2 \right]$ leads to

$$P \left((\gamma(t, .) - \gamma(s^*, .))^2 \right) \leq d^2(s^*, t).$$

For regression, we write

$$[\gamma(t, (x, y)) - \gamma(\eta, (x, y))]^2 = [t(x) - \eta(x)]^2 [2(y - \eta(x)) - t(x) + \eta(x)]^2.$$

Now

$$\mathbb{E}[Y - \eta(X) \mid X] = 0 \text{ and } \mathbb{E} \left[(Y - \eta(X))^2 \mid X \right] \leq \frac{1}{4},$$

imply that

$$\mathbb{E}\left[\left[2\left(y-\eta\left(x\right)\right)-t\left(x\right)+\eta\left(x\right)\right]^{2}\mid X\right]=4\mathbb{E}\left[\left(Y-\eta\left(X\right)\right)^{2}\mid X\right]$$
$$+\left(-t\left(X\right)+\eta\left(X\right)\right)^{2}\leq2,$$

and therefore

$$P\left(\left(\gamma\left(t,.\right)-\gamma\left(\eta,.\right)\right)^{2}\right)\leq2\mathbb{E}\left(t\left(X\right)-\eta\left(X\right)\right)^{2}. \tag{8.15}$$

Our main result below will crucially depend on two different moduli of uniform continuity:

- the stochastic modulus of uniform continuity of $\overline{\gamma}_n$ over S with respect to d,
- the modulus of uniform continuity of d with respect to ℓ.

The main tool that we shall use is Bousquet's version of Talagrand's inequality for empirical processes (see Chapter 4) which will allow us to control the oscillations of the empirical process $\overline{\gamma}_n$ by the modulus of uniform continuity of $\overline{\gamma}_n$ in expectation. Bousquet's version has the advantage of providing explicit constants and of dealing with one-sided suprema (in the spirit of [91], we could also work with absolute suprema but it is easier and somehow more natural to work with one-sided suprema).

8.3.1 The Main Theorem

We need to specify some mild regularity conditions that we shall assume to be verified by the moduli of continuity involved in our result.

Definition 8.2 *We denote by $\mathcal{C}_1$ the class of nondecreasing and continuous functions ψ from $\mathbb{R}_+$ to $\mathbb{R}_+$ such that $x \to \psi(x)/x$ is nonincreasing on $(0,+\infty)$ and $\psi(1) \geq 1$.*

Note that if ψ is a nondecreasing continuous and concave function on $\mathbb{R}_+$ with $\psi(0) = 0$ and $\psi(1) \geq 1$, then ψ belongs to $\mathcal{C}_1$. In particular, for the applications that we shall study below an example of special interest is $\psi(x) = Ax^{\alpha}$, where $\alpha \in [0,1]$ and $A \geq 1$.

In order to avoid measurability problems and to use the concentration tools, we need to consider some separability condition on S. The following one will be convenient

(M) There exists some countable subset S' of S such that for every $t \in S$, there exists some sequence $(t_k)_{k\geq1}$ of elements of S' such that for every $\xi \in \Xi$, $\gamma(t_k,\xi)$ tends to $\gamma(t,\xi)$ as k tends to infinity.

We are now in a position to state our upper bound for the relative expected loss of any empirical risk minimizer on some given model S.

Theorem 8.3 *Let $\xi_1, \ldots, \xi_n$ be independent observations taking their values in some measurable space Ξ and with common distribution P. Let S be some set, $\gamma : S \times \Xi \to [0, 1]$ be a measurable function such that for every $t \in S$, $x \to \gamma(t, x)$ is measurable. Assume that there exists some minimizer s of $P(\gamma(t, .))$ over S and denote by $\ell(s, t)$ the nonnegative quantity $P(\gamma(t, .)) - P(\gamma(s, .))$ for every $t \in S$. Let γ_n be the empirical risk*

$$\gamma_n(t) = P_n(\gamma(t, .)) = \frac{1}{n} \sum_{i=1}^n \gamma(t, \xi_i), \text{ for every } t \in S$$

and $\overline{\gamma}_n$ be the centered empirical process defined by

$$\overline{\gamma}_n(t) = P_n(\gamma(t, .)) - P(\gamma(t, .)), \text{ for every } t \in S .$$

Let d be some pseudodistance on $S \times S$ such that

$$P\left(\left(\gamma(t, .) - \gamma(s, .)\right)^2\right) \le d^2(s, t), \text{ for every } t \in S. \tag{8.16}$$

Let ϕ and w belong to the class of functions $\mathcal{C}_1$ defined above and let S be a subset of $\mathcal{S}$ satisfying separability condition ($\boldsymbol{M}$). Assume that on the one hand, for every $t \in S$

$$d(s, t) \le w\left(\sqrt{\ell(s, t)}\right) \tag{8.17}$$

and that on the other hand one has for every $u \in S'$

$$\sqrt{n}\mathbb{E}\left[\sup_{t \in S', d(u, t) \le \sigma} [\overline{\gamma}_n(u) - \overline{\gamma}_n(t)]\right] \le \phi(\sigma) \tag{8.18}$$

for every positive σ such that $\phi(\sigma) \le \sqrt{n}\sigma^2$, where S' is given by assumption ($\boldsymbol{M}$). Let ε_* be the unique solution of the equation

$$\sqrt{n}\varepsilon_*^2 = \phi(w(\varepsilon_*)). \tag{8.19}$$

Let ρ be some given nonnegative real number and consider any ρ-empirical risk minimizer, i.e., any estimator $\widehat{s}$ taking its values in S and such that

$$\gamma_n(\widehat{s}) \le \rho + \inf_{t \in S} \gamma_n(t) .$$

Then, setting

$$\ell(s, S) = \inf_{t \in S} \ell(s, t),$$

there exists some absolute constant κ such that for every $y \ge 0$, the following inequality holds

$$\mathbb{P}\left[\ell(s, \widehat{s}) > 2\rho + 2\ell(s, S) + \kappa\left(\varepsilon_*^2 + \frac{(1 \wedge w^2(\varepsilon_*))}{n\varepsilon_*^2} y\right)\right] \le e^{-y}. \tag{8.20}$$

In particular, the following risk bound is available

$$\mathbb{E}[\ell(s, \widehat{s})] \le 2\left(\rho + \ell(s, S) + \kappa\varepsilon_*^2\right) .$$

Comments.

Let us give some first comments about Theorem 8.3.

- The absolute constant 2 appearing in (8.20) has no magic meaning here, it could be replaced by any $C > 1$ at the price of making the constant κ depend on C.
- One can wonder whether an empirical risk minimizer over S does exist or not. Note that condition (**M**) implies that for every positive ρ, there exists some measurable choice of a ρ-empirical risk minimizer since then $\inf_{t \in S'} \gamma_n(t) = \inf_{t \in S} \gamma_n(t)$. If $\rho = 1/n$ for instance, it is clear that, according to (8.20), such an estimator performs as well as a strict empirical risk minimizer.
- For the computation of ϕ satisfying (8.18), since the supremum appearing in the left-hand side of (8.18) is extended to the countable set S' and not S itself, it will allow us to restrict ourself to the case where S is countable.
- It is worth mentioning that, assuming for simplicity that $s \in S$, (8.20) still holds if we consider the relative empirical risk $\gamma_n(s) - \gamma_n(\widehat{s})$ instead of the expected loss $\ell(s, \widehat{s})$. This is indeed a by-product of the proof of Theorem 8.3 below.

Proof. According to measurability condition (**M**), we may without loss of generality assume S to be countable. Suppose, first, for the sake of simplicity that there exists some point $\pi(s)$ belonging to S such that

$$\ell(s, \pi(s)) = \ell(s, S). \tag{8.21}$$

We start from the identity

$$\ell(s, \widehat{s}) = \ell(s, \pi(s)) + \gamma_n(\widehat{s}) - \gamma_n(\pi(s)) + \overline{\gamma}_n(\pi(s)) - \overline{\gamma}_n(\widehat{s}),$$

which, by definition of $\widehat{s}$ implies that

$$\ell(s, \widehat{s}) \leq \rho + \ell(s, \pi(s)) + \overline{\gamma}_n(\pi(s)) - \overline{\gamma}_n(\widehat{s}).$$

Let $x > 0$ with

$$x^2 = \kappa \left(\varepsilon_*^2 + \frac{\left(1 \wedge w^2(\varepsilon_*) \right) y}{n \varepsilon_*^2} \right),$$

where κ is a constant to be chosen later such that $\kappa \geq 1$, and

$$V_x = \sup_{t \in S} \frac{\overline{\gamma}_n(\pi(s)) - \overline{\gamma}_n(t)}{\ell(s, t) + x^2}.$$

Then,

$$\ell(s, \widehat{s}) \leq \rho + \ell(s, \pi(s)) + V_x \left(\ell(s, \widehat{s}) + x^2 \right)$$

and therefore, on the event $V_x < 1/2$, one has

$$\ell(s, \widehat{s}) < 2 \left(\rho + \ell(s, \pi(s)) \right) + x^2$$

yielding

$$\mathbb{P}\left[\ell\left(s,\widehat{s}\right) \geq 2\left(\rho + \ell\left(s, \pi\left(s\right)\right)\right) + x^2\right] \leq \mathbb{P}\left[V_x \geq \frac{1}{2}\right]. \qquad (8.22)$$

Since ℓ is bounded by 1, we may always assume x (and thus ε_*) to be not larger than 1. Assuming that $x \leq 1$, it remains to control the variable V_x via Bousquet's inequality. By (8.16), (8.17), the definition of $\pi\left(s\right)$ and the monotonicity of w, we derive that for every $t \in S$

$$\operatorname{Var}_P\left(-\gamma\left(t,.\right) + \gamma\left(\pi\left(s\right),.\right)\right) \leq \left(d\left(s,t\right) + d\left(s, \pi\left(s\right)\right)\right)^2 \leq 4w^2\left(\sqrt{\ell\left(s,t\right)}\right).$$

Hence, since γ takes its values in $[0,1]$, introducing the function $w_1 = 1 \wedge 2w$, we derive that

$$\sup_{t \in S} \operatorname{Var}_P\left[\frac{\gamma\left(t,.\right) - \gamma\left(\pi\left(s\right),.\right)}{\ell\left(s,t\right) + x^2}\right] \leq \sup_{\varepsilon \geq 0}\frac{w_1^2\left(\varepsilon\right)}{\left(\varepsilon^2 + x^2\right)^2} \leq \frac{1}{x^2}\sup_{\varepsilon \geq 0}\left(\frac{w_1\left(\varepsilon\right)}{\varepsilon \vee x}\right)^2.$$

Now the monotonicity assumptions on w imply that either $w\left(\varepsilon\right) \leq w\left(x\right)$ if $x \geq \varepsilon$ or $w\left(\varepsilon\right)/\varepsilon \leq w\left(x\right)/x$ if $x \leq \varepsilon$, hence one has in any case $w\left(\varepsilon\right)/\left(\varepsilon \vee x\right) \leq w\left(x\right)/x$ which finally yields

$$\sup_{t \in S} \operatorname{Var}_P\left[\frac{\gamma\left(t,.\right) - \gamma\left(\pi\left(s\right),.\right)}{\ell\left(s,t\right) + x^2}\right] \leq \frac{w_1^2\left(x\right)}{x^4}.$$

On the other hand since γ takes its values in $[0,1]$, we have

$$\sup_{t \in S}\left\|\frac{\gamma\left(t,.\right) - \gamma\left(\pi\left(s\right),.\right)}{\ell\left(s,t\right) + x^2}\right\|_\infty \leq \frac{1}{x^2}.$$

We can therefore apply (5.49) with $v = w_1^2\left(x\right)x^{-4}$ and $b = 2x^{-2}$, which gives that, on a set Ω_y with probability larger than $1 - \exp\left(-y\right)$, the following inequality holds

$$V_x < \mathbb{E}\left[V_x\right] + \sqrt{\frac{2\left(w_1^2\left(x\right)x^{-2} + 4\mathbb{E}\left[V_x\right]\right)y}{nx^2}} + \frac{y}{nx^2}$$

$$< 3\mathbb{E}\left[V_x\right] + \sqrt{\frac{2w_1^2\left(x\right)x^{-2}y}{nx^2}} + \frac{2y}{nx^2}. \qquad (8.23)$$

Now since ε_* is assumed to be not larger than 1, one has $w\left(\varepsilon_*\right) \geq \varepsilon_*$ and therefore for every $\sigma \geq w\left(\varepsilon_*\right)$, the following inequality derives from the definition of ε_* by monotonicity

$$\frac{\phi\left(\sigma\right)}{\sigma^2} \leq \frac{\phi\left(w\left(\varepsilon_*\right)\right)}{w^2\left(\varepsilon_*\right)} \leq \frac{\phi\left(w\left(\varepsilon_*\right)\right)}{\varepsilon_*^2} = \sqrt{n}.$$

Hence (8.18) holds for every $\sigma \geq w\left(\varepsilon_*\right)$ and since $u \to \phi\left(2w\left(u\right)\right)/u$ is non-increasing, by assumption (8.17) and (8.21) we can use Lemma 4.23 (and the triangle inequality for d) to get

$$\mathbb{E}\left[V_x\right] \leq \frac{4\phi\left(2w\left(x\right)\right)}{\sqrt{n}x^2}.$$

Hence, by monotonicity of $u \to \phi(u)/u$

$$\mathbb{E}[V_x] \le \frac{8\phi(w(x))}{\sqrt{n}x^2}.$$

Since $\kappa \ge 1$ we note that $x \ge \sqrt{\kappa}\varepsilon_* \ge \varepsilon_*$. Thus, using the monotonicity of $u \to \phi(w(u))/u$, and the definition of ε_*, we derive that

$$\mathbb{E}[V_x] \le \frac{8\phi(w(\varepsilon_*))}{\sqrt{n}x\varepsilon_*} = \frac{8\varepsilon_*}{x} \le \frac{8}{\sqrt{\kappa}}. \tag{8.24}$$

Now, the monotonicity of $u \to w_1(u)/u$ implies that

$$\frac{w_1^2(x)}{x^2} \le \frac{w_1^2(\varepsilon_*)}{\varepsilon_*^2}. \tag{8.25}$$

Plugging (8.24) and (8.25) into (8.23) implies that, on the set Ω_y,

$$V_x < \frac{24}{\sqrt{\kappa}} + \sqrt{\frac{2w_1^2(\varepsilon_*)\varepsilon_*^{-2}y}{nx^2}} + \frac{2y}{nx^2}.$$

Recalling that $\varepsilon_* \le 1$, it remains to use the lower bound $4nx^2 \ge \kappa w_1^2(\varepsilon_*)\varepsilon_*^{-2}y$, noticing that $w_1^2(\varepsilon_*)\varepsilon_*^{-2} \ge 1$ to derive that, on the set Ω_y, the following inequality holds

$$V_x < \frac{24}{\sqrt{\kappa}} + \sqrt{\frac{8}{\kappa}} + \frac{8}{\kappa}.$$

Hence, choosing κ as a large enough numerical constant warrants that $V_x < 1/2$ on Ω_y. Thus

$$\mathbb{P}\left[V_x \ge \frac{1}{2}\right] \le \mathbb{P}[\Omega_y^c] \le e^{-y}, \tag{8.26}$$

and therefore (8.22) leads to

$$\mathbb{P}\left[\ell(s,\widehat{s}) \ge 2(\rho + \ell(s, \pi(s))) + x^2\right] \le e^{-y}.$$

If a point $\pi(s)$ satisfying (8.21) does not exist we can use as well some point $\pi(s)$ satisfying $\ell(s, \pi(s)) \le \ell(s, S) + \delta$ and get the required probability bound (8.20) by letting δ tend to zero. But since $\phi(u)/u \ge \phi(1) \ge 1$ for every $u \in [0, 1]$, we derive from (8.19) and the monotonicity of ϕ and $u \to \phi(u)/u$ that

$$\frac{1 \wedge w^2(\varepsilon_*)}{\varepsilon_*^2} \le \frac{\phi^2(1 \wedge w(\varepsilon_*))}{\varepsilon_*^2} \le \frac{\phi^2(w(\varepsilon_*))}{\varepsilon_*^2}$$

and therefore

$$\frac{1 \wedge w^2(\varepsilon_*)}{\varepsilon_*^2} \le n\varepsilon_*^2. \tag{8.27}$$

The proof can then be easily completed by integrating the tail bound (8.20) to get

$$\mathbb{E}\left[\ell\left(s,\widehat{s}\right)\right] \leq 2\left(\rho + \ell\left(s,S\right)\right) + \kappa\varepsilon_*^2 + \kappa\frac{1 \wedge w^2\left(\varepsilon_*\right)}{n\varepsilon_*^2}.$$

yielding the required upper bound on the expected risk via (8.27). ∎

Even though the main motivation for Theorem 8.3 is the study of classification, it can also be easily applied to bounded regression. We begin the illustration of Theorem 8.3 with this framework which is more elementary than classification since in this case there is a clear connection between the expected loss and the variance of the increments.

8.3.2 Application to Bounded Regression

In this setting, the regression function $\eta : x \to \mathbb{E}\left[Y \mid X = x\right]$ is the target to be estimated, so that here $s = \eta$. We recall that for this framework we can take d to be the $\mathbb{L}_2\left(\mu\right)$ distance times $\sqrt{2}$. The connection between the loss function ℓ and d is especially simple in this case since

$$\left[\gamma\left(t,\left(x,y\right)\right) - \gamma\left(s,\left(x,y\right)\right)\right] = \left[-t\left(x\right) + s\left(x\right)\right]\left[2\left(y - s\left(x\right)\right) - t\left(x\right) + s\left(x\right)\right]$$

which implies since $\mathbb{E}\left[Y - s\left(X\right) \mid X\right] = 0$ that

$$\ell\left(s,t\right) = \mathbb{E}\left[\gamma\left(t,\left(X,Y\right)\right) - \gamma\left(s,\left(X,Y\right)\right)\right] = \mathbb{E}\left(t\left(X\right) - s\left(X\right)\right)^2.$$

Hence $2\ell\left(s,t\right) = d^2\left(s,t\right)$ and in this case the modulus of continuity w can simply be taken as $w\left(\varepsilon\right) = \sqrt{2}\varepsilon$. Note also that in this case, an empirical risk minimizer $\widehat{s}$ over some model S is a LSE. The quadratic risk of $\widehat{s}$ depends only on the modulus of continuity ϕ satisfying (8.18) and one derives from Theorem 8.3 that, for some absolute constant κ',

$$\mathbb{E}\left[d^2\left(s,\widehat{s}\right)\right] \leq 2d^2\left(s,S\right) + \kappa'\varepsilon_*^2$$

where ε_* is the solution of

$$\sqrt{n}\varepsilon_*^2 = \phi\left(\varepsilon_*\right).$$

To be more concrete, let us give an example where this modulus ϕ and the bias term $d^2\left(s,S\right)$ can be evaluated, leading to an upper bound for the minimax risk over some classes of regression functions.

Binary Images

Following [72], our purpose is to study the particular regression framework for which the variables X_i are uniformly distributed on $\left[0,1\right]^2$ and $s\left(x\right) = \mathbb{E}\left[Y \mid X = x\right]$ is of the form

$$s\left(x_1, x_2\right) = b \text{ if } x_2 \leq \partial s\left(x_1\right) \text{ and } a \text{ otherwise},$$

where ∂s is some measurable map from $\left[0,1\right]$ to $\left[0,1\right]$ and $0 < a < b < 1$. The function ∂s should be understood as the parametrization of a boundary fragment corresponding to some portion s of a binary image in the plane (a and b, representing the two level of colors which are taken by the image) and restoring this portion of the image from the noisy data $(X_1, Y_1), \ldots, (X_n, Y_n)$ means estimating s or equivalently ∂s. Let $\mathcal{G}$ be the set of measurable maps from $[0,1]$ to $[0,1]$. For any $f \in \mathcal{G}$, let us denote by χ_f the function defined on $[0,1]^2$ by

$$\chi_f(x_1, x_2) = b \text{ if } x_2 \leq f(x_1) \text{ and } a \text{ otherwise.}$$

From this definition we see that $\chi_{\partial s} = s$ and more generally if we define $\mathcal{S} = \{\chi_f : f \in \mathcal{G}\}$, for every $t \in \mathcal{S}$, we denote by ∂t the element of $\mathcal{G}$ such that $\chi_{\partial t} = t$. It is natural to consider here as an approximate model for s a model S of the form $S = \{\chi_f : f \in \partial S\}$, where ∂S denotes some subset of $\mathcal{G}$. In what follows, we shall assume condition (**M**) to be fulfilled which allows us to make as if S was countable. Denoting by $\|.\|_1$ (resp. $\|.\|_2$) the Lebesgue $\mathbb{L}_1$-norm (resp. $\mathbb{L}_2$-norm), one has for every $f, g \in \mathcal{G}$

$$\|\chi_f - \chi_g\|_1 = (b-a)\|f - g\|_1 \text{ and } \|\chi_f - \chi_g\|_2^2 = (b-a)^2 \|f - g\|_1$$

or equivalently for every $s, t \in \mathcal{S}$,

$$\|s - t\|_1 = (b-a)\|\partial s - \partial t\|_1 \text{ and } \|s - t\|_2^2 = (b-a)^2 \|\partial s - \partial t\|_1.$$

Given $u \in S$ and setting $S_\sigma = \{t \in S, d(t, u) \leq \sigma\}$, we have to compute some function ϕ fulfilling (8.18) and therefore to upper bound $\mathbb{E}[W(\sigma)]$, where

$$W(\sigma) = \sup_{t \in S_\sigma} \overline{\gamma}_n(u) - \overline{\gamma}_n(t).$$

This can be done using entropy with bracketing arguments. Indeed, let us notice that if g belongs to some ball with radius δ in $\mathbb{L}_\infty[0,1]$, then for some function $f \in \mathbb{L}_\infty[0,1]$, one has $f - \delta \leq g \leq f + \delta$ and therefore, defining $f_L = \sup(f - \delta, 0)$ and $f_U = \inf(f + \delta, 1)$

$$\chi_{f_L} \leq \chi_g \leq \chi_{f_U}$$

with $\|\chi_{f_L} - \chi_{f_U}\|_1 \leq 2(b-a)\delta$. This means that, defining $H_\infty(\delta, \partial S, \rho)$ as the supremum over $g \in \partial S$ of the $\mathbb{L}_\infty$-metric entropy for radius δ of the $\mathbb{L}_1$ ball centered at g with radius ρ in ∂S, one has for every positive ε

$$H_{[.]}(\varepsilon, S_\sigma, \mu) \leq H_\infty\left(\frac{\varepsilon}{2(b-a)}, \partial S, \frac{\sigma^2}{2(b-a)^2}\right).$$

Moreover if $[t_L, t_U]$ is a bracket with extremities in $\mathcal{S}$ and $\mathbb{L}_1(\mu)$ diameter not larger than δ and if $t \in [t_L, t_U]$, then

$$y^2 - 2t_U(x)y + t_L^2(x) \leq (y - t(x))^2 \leq y^2 - 2t_L(x)y + t_U^2(x),$$

which implies that $\gamma(t,.)$ belongs to a bracket with $\mathbb{L}_1(P)$-diameter not larger than

$$2\mathbb{E}\left[(t_U(X) - t_L(X))\left(Y + \frac{t_U(X) + t_L(X)}{2}\right)\right] \le 4\delta.$$

Hence, if $\mathcal{F} = \{\gamma(t,.), t \in S \text{ and } d(t,u) \le \sigma\}$, then

$$H_{[.]}(x, \mathcal{F}, P) \le H_\infty\left(\frac{x}{8(b-a)}, \partial S, \frac{\sigma^2}{2(b-a)^2}\right)$$

and furthermore, if $d(t,u) \le \sigma$

$$\mathbb{E}\left[\left|(Y - t(X))^2 - (Y - u(X))^2\right|\right] \le 2\|u - t\|_1 = \frac{2\|u - t\|_2^2}{(b-a)} \le \frac{\sigma^2}{(b-a)}.$$

We can therefore apply Lemma 6.5 to the class $\mathcal{F}$ and derive that, setting

$$\varphi(\sigma) = \int_0^{\sigma/\sqrt{b-a}}\left(H_\infty\left(\frac{x^2}{8(b-a)}, \partial S, \frac{\sigma^2}{2(b-a)^2}\right)\right)^{1/2} dx,$$

one has

$$\sqrt{n}\mathbb{E}[W(\sigma)] \le 12\varphi(\sigma),$$

provided that

$$4\varphi(\sigma) \le \sqrt{n}\frac{\sigma^2}{(b-a)}. \tag{8.28}$$

The point now is that, whenever ∂S is part of a linear finite dimensional subspace of $\mathbb{L}_\infty[0,1]$, $H_\infty(\delta, \partial S, \rho)$ is typically bounded by $D[B + \ln(\rho/\delta)]$ for some appropriate constants D and B. If it is so then

$$\varphi(\sigma) \le \sqrt{D}\int_0^{\sigma/\sqrt{b-a}}\left(B + \ln\left(\frac{4\sigma^2}{x^2(b-a)}\right)\right)^{1/2} dx$$

$$\le \frac{\sqrt{D}\sigma}{\sqrt{b-a}}\int_0^1\sqrt{B + 2|\ln(2\delta)|}d\delta,$$

which implies that for some absolute constant κ

$$\varphi(\sigma) \le \kappa\sigma\sqrt{\frac{(1+B)D}{(b-a)}}.$$

The restriction (8.28) is a fortiori satisfied if $\sigma\sqrt{b-a} \ge 4\kappa\sqrt{(1+B)D/n}$. Hence if we take

$$\phi(\sigma) = 12\kappa\sigma\sqrt{\frac{(1+B)D}{(b-a)}},$$

assumption (8.18) is satisfied. To be more concrete let us consider the example where ∂S is taken to be the set of piecewise constant functions on a regular partition with D pieces on $[0, 1]$ with values in $[0, 1]$. Then, it is shown in [12] that

$$H_\infty\left(\delta, \partial S, \rho\right) \leq D\left[\ln\left(\rho/\delta\right)\right]$$

and therefore the previous analysis can be used with $B = 0$. As a matter of fact this extends to piecewise polynomials with degree not larger than r via some adequate choice of B as a function of r but we just consider the histogram case here to be simple. As a conclusion, Theorem 8.3 yields in this case for the LSE $\widehat{s}$ over S

$$\mathbb{E}\left[\|\partial s - \partial \widehat{s}\|_1\right] \leq 2 \inf_{t \in S}\|\partial s - \partial t\|_1 + C\frac{D}{(b-a)^3 n}$$

for some absolute constant C. In particular, if ∂s is Hölder smooth,

$$|\partial s\left(x\right) - \partial s\left(x'\right)| \leq R\left|x - x'\right|^\alpha \tag{8.29}$$

with $R > 0$ and $\alpha \in (0, 1]$, then

$$\inf_{t \in S}\|\partial s - \partial t\|_1 \leq RD^{-\alpha}$$

leading to

$$\mathbb{E}\left[\|\partial s - \partial \widehat{s}\|_1\right] \leq 2RD^{-\alpha} + C\frac{D}{(b-a)^3 n}.$$

Hence, if $\mathcal{H}\left(R, \alpha\right)$ denotes the set of functions from $[0, 1]$ to $[0, 1]$ satisfying (8.29), an adequate choice of D yields for some constant C' depending only on a and b

$$\sup_{\partial s \in \mathcal{H}(R,\alpha)} \mathbb{E}\left[\|\partial s - \partial \widehat{s}\|_1\right] \leq C'\left(R \vee \frac{1}{n}\right)^{\frac{1}{\alpha+1}} n^{-\frac{\alpha}{1+\alpha}}.$$

As a matter of fact, this upper bound is unimprovable (up to constants) from a minimax point of view (see [72] for the corresponding minimax lower bound).

8.3.3 Application to Classification

Our purpose is to apply Theorem 8.3 to the classification setting, assuming that the Bayes classifier is the target to be estimated, so that here $s = s^*$. We recall that for this framework we can take d to be the $\mathbb{L}_2\left(\mu\right)$ distance and $S = \{\mathbb{1}_A, A \in \mathcal{A}\}$, where $\mathcal{A}$ is some class of measurable sets. Our main task is to compute the moduli of continuity ϕ and w. In order to evaluate w, we need some margin type condition. For instance we can use Tsybakov's margin condition

$$\ell\left(s, t\right) \geq h^\theta d^{2\theta}\left(s, t\right), \text{ for every } t \in S, \tag{8.30}$$

where h is some positive constant (that we can assume to be smaller than 1 since we can always change h into $h \wedge 1$ without violating (8.30))and $\theta \geq 1$. As explained by Tsybakov in [115], this condition is fulfilled if the distribution of $\eta(X)$ is well behaved around $1/2$. A simple situation is the following. Assume that, for some positive number h, one has for every $x \in \mathcal{X}$

$$|2\eta(x) - 1| \geq h. \tag{8.31}$$

Then

$$\ell(s,t) = \mathbb{E}\left[|2\eta(X) - 1| \, |s(X) - t(X)|\right] \geq hd^2(s,t)$$

which means that Tsybakov's condition is satisfied with $\theta = 1$. Of course, Tsybakov's condition implies that the modulus of continuity w can be taken as

$$w(\varepsilon) = h^{-1/2}\varepsilon^{1/\theta}. \tag{8.32}$$

In order to evaluate ϕ, we shall consider two different kinds of assumptions on S which are well known to imply the Donsker property for the class of functions $\{\gamma(t,.), t \in S\}$ and therefore the existence of a modulus ϕ which tends to 0 at 0, namely a VC-condition or an entropy with bracketing assumption. Given $u \in S$, in order to bound the expectation of

$$W(\sigma) = \sup_{t \in S; d(u,t) \leq \sigma} \left(-\overline{\gamma}_n(t) + \overline{\gamma}_n(u)\right),$$

we shall use the maximal inequalities for empirical processes which are established in Chapter 6 via slightly different techniques according to the way the "size" of the class $\mathcal{A}$ is measured.

The VC-Case

Let us first assume for the sake of simplicity that $\mathcal{A}$ is countable. We use the definitions, notations and results of Section 6.1.2, to express ϕ in terms of the random combinatorial entropy or the VC-dimension of $\mathcal{A}$. Indeed, we introduce the classes of sets

$$\mathcal{A}_+ = \left\{\left\{(x,y): \mathbb{1}_{y \neq t(x)} \leq \mathbb{1}_{y \neq u(x)}\right\}, t \in S\right\}$$

and

$$\mathcal{A}_- = \left\{\left\{(x,y): \mathbb{1}_{y \neq t(x)} \geq \mathbb{1}_{y \neq u(x)}\right\}, t \in S\right\}$$

and define for every class of sets $\mathcal{B}$ of $\mathcal{X} \times \{0,1\}$

$$W_{\mathcal{B}}^+(\sigma) = \sup_{B \in \mathcal{B}, P(B) \leq \sigma^2} (P_n - P)(B), \quad W_{\mathcal{B}}^-(\sigma) = \sup_{B \in \mathcal{B}, P(B) \leq \sigma^2} (P - P_n)(B).$$

Then,

$$\mathbb{E}[W(\sigma)] \leq \mathbb{E}\left[W_{\mathcal{A}_+}^+(\sigma)\right] + \mathbb{E}\left[W_{\mathcal{A}_-}^-(\sigma)\right] \tag{8.33}$$

and it remains to control $\mathbb{E}\left[W_{\mathcal{A}_+}^+(\sigma)\right]$ and $\mathbb{E}\left[W_{\mathcal{A}_-}^-(\sigma)\right]$ via Lemma 6.4.

Since the VC-dimension of $\mathcal{A}_+$ and $\mathcal{A}_-$ are not larger than that of $\mathcal{A}$, and that similarly, the combinatorial entropies of $\mathcal{A}_+$ and $\mathcal{A}_-$ are not larger than the combinatorial entropy of $\mathcal{A}$, denoting by $V_{\mathcal{A}}$ the VC-dimension of $\mathcal{A}$ (assuming that $V_{\mathcal{A}} \geq 1$), we derive from (8.33) and Lemma 6.4 that

$$\sqrt{n}\mathbb{E}\left[W\left(\sigma\right)\right] \leq \phi\left(\sigma\right)$$

provided that $\phi\left(\sigma\right) \leq \sqrt{n}\sigma^2$, where ϕ can be taken either as

$$\phi\left(\sigma\right) = K\sigma\sqrt{\left(1 \vee \mathbb{E}\left[H_{\mathcal{A}}\right]\right)} \tag{8.34}$$

or as

$$\phi\left(\sigma\right) = K\sigma\sqrt{V\left(1 + \ln\left(\sigma^{-1} \vee 1\right)\right)}. \tag{8.35}$$

In both cases, assumption (8.18) is satisfied and we can apply Theorem 8.3 with $w \equiv 1$ or w defined by (8.32). When ϕ is given by (8.34) the solution ε_* of equation (8.19) can be explicitly computed when w is given by (8.32) or $w \equiv 1$. Hence the conclusion of Theorem 8.3 holds with

$$\varepsilon_*^2 = \left(\frac{K^2\left(1 \vee \mathbb{E}\left[H_{\mathcal{A}}\right]\right)}{nh}\right)^{\theta/(2\theta-1)} \wedge \sqrt{\frac{K^2\left(1 \vee \mathbb{E}\left[H_{\mathcal{A}}\right]\right)}{n}}.$$

In the second case i.e., when ϕ is given by (8.35), $w \equiv 1$ implies by (8.19) that

$$\varepsilon_*^2 = K\sqrt{\frac{V}{n}}$$

while if $w\left(\varepsilon_*\right) = h^{-1/2}\varepsilon_*^{1/\theta}$ then

$$\varepsilon_*^2 = K\varepsilon_*^{1/\theta}\sqrt{\frac{V}{nh}}\sqrt{1 + \ln\left(\left(\sqrt{h}\varepsilon_*^{-1/\theta}\right) \vee 1\right)}. \tag{8.36}$$

Since $1 + \ln\left(\left(\sqrt{h}\varepsilon_*^{-1/\theta}\right) \vee 1\right) \geq 1$ and $K \geq 1$, we derive from (8.36) that

$$\varepsilon_*^2 \geq \left(\frac{V}{nh}\right)^{\theta/(2\theta-1)}. \tag{8.37}$$

Plugging this inequality in the logarithmic factor of (8.36) yields

$$\varepsilon_*^2 \leq K\varepsilon_*^{1/\theta}\sqrt{\frac{V}{nh}}\sqrt{1 + \frac{1}{2\left(2\theta-1\right)}\ln\left(\left(\frac{nh^{2\theta}}{V}\right) \vee 1\right)}$$

and therefore, since $\theta \geq 1$

$$\varepsilon_*^2 \leq K\varepsilon_*^{1/\theta}\sqrt{\frac{V}{nh}}\sqrt{1 + \ln\left(\left(\frac{nh^{2\theta}}{V}\right) \vee 1\right)}.$$

Hence

$$\varepsilon_*^2 \leq \left(\frac{K^2 V \left(1 + \ln \left(\left(n h^{2\theta}/V\right) \vee 1\right)\right)}{nh} \right)^{\theta/(2\theta-1)}$$

$$\leq K^2 \left(\frac{V \left(1 + \ln \left(\left(n h^{2\theta}/V\right) \vee 1\right)\right)}{nh} \right)^{\theta/(2\theta-1)}$$

and therefore the conclusion of Theorem 8.3 holds with

$$\varepsilon_*^2 = K^2 \left[\left(\frac{V \left(1 + \ln \left(\left(n h^{2\theta}/V\right) \vee 1\right)\right)}{nh} \right)^{\theta/(2\theta-1)} \wedge \sqrt{\frac{V}{n}} \right].$$

Of course, if S (and therefore $\mathcal{A}$) is not countable but fulfills condition (**M**), the previous arguments still apply for a conveniently countable subclass of $\mathcal{A}$ so that we have a fortiori obtained the following result.

Corollary 8.4 Let $\mathcal{A}$ be a VC-class with dimension $V \geq 1$ and assume that s^* belongs to $S = \{\mathbb{1}_A, A \in \mathcal{A}\}$. Assuming that S satisfies to (**M**), there exists an absolute constant C such that if $\widehat{s}$ denotes an empirical risk minimizer over S, the following inequality holds

$$\mathbb{E}\left[\ell\left(s^*, \widehat{s}\right)\right] \leq C \sqrt{\frac{V \wedge \left(1 \vee \mathbb{E}\left[H_\mathcal{A}\right]\right)}{n}}. \tag{8.38}$$

Moreover if $\theta \geq 1$ is given and one assumes that the margin condition (8.30) holds, then the following inequalities are also available

$$\mathbb{E}\left[\ell\left(s^*, \widehat{s}\right)\right] \leq C \left(\frac{\left(1 \vee \mathbb{E}\left[H_\mathcal{A}\right]\right)}{nh} \right)^{\theta/(2\theta-1)} \tag{8.39}$$

and

$$\mathbb{E}\left[\ell\left(s^*, \widehat{s}\right)\right] \leq C \left(\frac{V \left(1 + \ln \left(n h^{2\theta}/V\right)\right)}{nh} \right)^{\theta/(2\theta-1)}, \tag{8.40}$$

provided that $h \geq (V/n)^{1/2\theta}$.

Comments.

- The risk bound (8.38) is well known. Our purpose was just here to show how it can be derived from our approach.
- The risk bounds (8.39) and (8.40) are new and they perfectly fit with (8.38) when one considers the borderline case $h = (V/n)^{1/2\theta}$. They look very similar but are not strictly comparable since roughly speaking they differ from a logarithmic factor. Indeed it may happen that $\mathbb{E}\left[H_\mathcal{A}\right]$ turns out to be of the order of V (without any extra log-factor). This the case

when $\mathcal{A}$ is the family of all subsets of a given finite set with cardinality V. In such a case, $\mathbb{E}[H_{\mathcal{A}}] \leq V$ and (8.39) is sharper than (8.40). On the contrary, for some arbitrary VC-class, let us remember that the consequence (6.9) of Sauer's lemma tells us that $H_{\mathcal{A}} \leq V(1 + \ln(n/V))$. The logarithmic factor $1 + \ln(n/V)$ is larger than $1 + \ln(nh^{2\theta}/V)$ and turns out to be especially over pessimistic when h is close to the borderline value $(V/n)^{1/2\theta}$.

- For the sake of simplicity we have assumed s^* to belong to S in the above statement. Of course this assumption is not necessary (since our main Theorem does not require it). The price to pay if s^* does not belong to S is simply to add $2\ell(s^*, S)$ to the right hand side of the risk bounds above.

In [92] the optimality of (8.40) from a minimax point of view is discussed in the case where $\theta = 1$, showing that it is essentially unimprovable in that sense.

Bracketing Conditions

For the same reasons as in the previous section, let us make the preliminary assumption that S is countable (the final result will easily extend to the case where S satisfies (**M**) anyway). If t_1 and t_2 are measurable functions such that $t_1 \leq t_2$, the collection of measurable functions t such that $t_1 \leq t \leq t_2$ is denoted by $[t_1, t_2]$ and called bracket with lower extremity t_1 and upper extremity t_2. Recalling that μ denotes the probability distribution of the explanatory variable X, the $\mathbb{L}_1(\mu)$-diameter of a bracket $[t_1, t_2]$ is given by $\mu(t_2) - \mu(t_1)$. Recall that the $\mathbb{L}_1(\mu)$-entropy with bracketing of S is defined for every positive δ, as the logarithm of the minimal number of brackets with $\mathbb{L}_1(\mu)$-diameter not larger than δ which are needed to cover S and is denoted by $H_{[.]}(\delta, S, \mu)$. The point is that if $\mathcal{F}$ denotes the class of functions $\mathcal{F} = \{\gamma(t, .), t \in S$ with $d(u, t) \leq \sigma\}$, one has

$$H_{[.]}(\delta, \mathcal{F}, P) \leq H_{[.]}(\delta, S, \mu)$$

hence, we derive from (8.33) and Lemma 6.5 that, setting

$$\varphi(\sigma) = \int_0^\sigma H_{[.]}^{1/2}(x^2, S, \mu)\, dx,$$

the following inequality is available

$$\sqrt{n}\mathbb{E}[W(\sigma)] \leq 12\varphi(\sigma)$$

provided that $4\varphi(\sigma) \leq \sigma^2\sqrt{n}$. Hence, we can apply Theorem 8.3 with $\phi = 12\varphi$ and if we assume Tsybakov's margin condition (8.30) to be satisfied, then we can also take w according to (8.32) and derive from that the conclusions of Theorem 8.3 hold with ε_* solution of the equation

$$\sqrt{n}\varepsilon_*^2 = \phi\left(h^{-1/2}\varepsilon_*^{1/\theta}\right).$$

In particular, if

$$H_{[.]}(x, S, \mu) \leq C x^{-r} \text{ with } 0 < r < 1, \tag{8.41}$$

then for some constant C' depending only on C, one has

$$\varepsilon_*^2 \leq C' \left[(1-r)^2 \, nh^{1-r} \right]^{-\frac{\theta}{2\theta - 1 + r}}. \tag{8.42}$$

If S' is taken as a δ_n-net (with respect to the $\mathbb{L}_2(\mu)$-distance d) of a bigger class S to which the target s^* is assumed to belong, then we can also apply Theorem 8.3 to the empirical risk minimizer over S' and since $H_{[.]}(x, S', \mu) \leq H_{[.]}(x, S, \mu)$, we still get the conclusions of Theorem 8.3 with ε_* satisfying (8.42) and $\ell(s^*, S') \leq \delta_n^2$. This means that if δ_n is conveniently chosen (in a way that δ_n is of lower order as compared to ε_*), for instance $\delta_n^2 = n^{-1/(1+r)}$, then, for some constant C'' depending only on C, one has

$$\mathbb{E}[\ell(s^*, \widehat{s})] \leq C'' \left[(1-r)^2 \, nh^{1-r} \right]^{-\frac{\theta}{2\theta - 1 + r}}.$$

This means that we have recovered Tsybakov's Theorem 1 in [115] (as a matter of fact our result is slightly more precise since it also provides the dependency of the risk bound with respect to the margin parameter h and not only on θ as in Tsybakov's Theorem). We refer to [85] for concrete examples of classes of sets with smooth boundaries satisfying (8.41) when μ is equivalent to the Lebesgue measure on some compact set of $\mathbb{R}^d$.

8.4 A Refined Model Selection Theorem

It is indeed quite easy to formally derive from (8.20) the following model selection version of Theorem 8.3.

Theorem 8.5 *Let $\xi_1, \ldots, \xi_n$ be independent observations taking their values in some measurable space Ξ and with common distribution P. Let S be some set, $\gamma : S \times \Xi \rightarrow [0, 1]$ be a measurable function such that for every $t \in S$, $x \rightarrow \gamma(t, x)$ is measurable. Assume that there exists some minimizer s of $P(\gamma(t, .))$ over S and denote by $\ell(s, t)$ the nonnegative quantity $P(\gamma(t, .)) - P(\gamma(s, .))$ for every $t \in S$. Let γ_n be the empirical risk*

$$\gamma_n(t) = P_n(\gamma(t, .)) = \frac{1}{n} \sum_{i=1}^{n} \gamma(t, \xi_i), \text{ for every } t \in S$$

and $\overline{\gamma}_n$ be the centered empirical process defined by

$$\overline{\gamma}_n(t) = P_n(\gamma(t, .)) - P(\gamma(t, .)), \text{ for every } t \in S .$$

Let d be some pseudodistance on $S \times S$ such that (8.16) holds. Let $\{S_m\}_{m \in \mathcal{M}}$ be some at most countable collection of subsets of S, each model S_m admitting *some countable subset S'_m such that S_m satisfies to separability condition (**M**). Let w and ϕ_m belong to the class of functions $\mathcal{C}_1$ defined above for every $m \in \mathcal{M}$. Assume that on the one hand assumption (8.17) holds and that on the other hand one has for every $m \in \mathcal{M}$ and $u \in S'_m$*

$$\sqrt{n}\mathbb{E}\left[\sup_{t \in S'_m, d(u,t) \leq \sigma} [\overline{\gamma}_n(u) - \overline{\gamma}_n(t)]\right] \leq \phi_m(\sigma) \tag{8.43}$$

for every positive σ such that $\phi_m(\sigma) \leq \sqrt{n}\sigma^2$. Let ε_m be the unique solution of the equation

$$\sqrt{n}\varepsilon_m^2 = \phi_m(w(\varepsilon_m)). \tag{8.44}$$

Let ρ be some given nonnegative real number and consider $\widehat{s}_m$ taking its values in S_m and such that

$$\gamma_n(\widehat{s}_m) \leq \inf_{t \in S_m} \gamma_n(t) + \rho.$$

Let $\{x_m\}_{m \in \mathcal{M}}$ be some family of nonnegative weights such that

$$\sum_{m \in \mathcal{M}} e^{-x_m} = \Sigma < \infty,$$

pen $: \mathcal{M} \to \mathbb{R}_+$ *such that for every $m \in \mathcal{M}$*

$$\mathrm{pen}(m) \geq K\left(\varepsilon_m^2 + \frac{w^2(\varepsilon_m)x_m}{n\varepsilon_m^2}\right).$$

Then, if K is large enough, there almost surely exists some minimizer $\widehat{m}$ of

$$\gamma_n(\widehat{s}_m) + \mathrm{pen}(m). \tag{8.45}$$

and some constant $C(K)$ such that the penalized estimator $\widetilde{s} = \widehat{s}_{\widehat{m}}$ satisfies the following inequality

$$\mathbb{E}\left[\ell(s,\widetilde{s})\right] \leq C(K)\left[\inf_{m \in \mathcal{M}}(\ell(s, S_m) + \mathrm{pen}(m)) + \frac{(\Sigma + 1)}{n} + \rho\right].$$

Concerning the proof of Theorem 8.5, the hard work has been already done to derive (8.20). The proof of Theorem 8.5 can indeed be sketched as follows: start from exponential bound (8.26) (which as a matter of fact readily implies (8.20)) and use a union bound argument. The calculations are quite similar to those of the proof of Theorem 4.18. At this stage they can be considered as routine and we shall therefore skip them. From the point of view of model selection for classification, Theorem 8.5 is definitely disappointing and far from producing the result we could expect anyway. In this classification context, it should be considered as a formal exercise. Indeed, the classification framework was the main motivation for introducing the "margin" function w in Theorem 8.3. The major drawback of Theorem 8.5 is that the penalization procedure involved *does require the knowledge* of w. Hence, apart from the situation where w can "legally" be assumed to be known (like for bounded regression where one can take $w(\varepsilon) = \sqrt{2}\varepsilon$), we cannot freely use Theorem 8.5 to build adaptive estimators as we did with the related model selection theorems in the other functional estimation frameworks that we have studied in the previous chapters (Gaussian white noise or density estimation). We shall come back to the classification framework in Section 8.5 below to design "margin adaptive" model selection strategies. For the moment we may at least use Theorem 8.5 in the bounded regression framework (note that more generally, when $w(\varepsilon) = C\varepsilon$ for a known absolute constant C, Theorem 8.5 is nothing more than Theorem 4.2. in [91]).

8.4.1 Application to Bounded Regression

As mentioned above, bounded regression is a typical framework for which the previous model selection theorem (Theorem 8.5) is relevant. Indeed, let us recall that in this setting, the regression function $\eta : x \to \mathbb{E}[Y \mid X = x]$ is the target s to be estimated and d may be taken as the $\mathbb{L}_2(\mu)$ distance times $\sqrt{2}$. The connection between the loss function ℓ and d is trivial since

$$\ell(s, t) = \mathbb{E}(t(X) - s(X))^2 = d^2(s, t)/2$$

and therefore w can simply be taken as $w(\varepsilon) = \sqrt{2}\varepsilon$. The penalized criterion given by (8.45) is a *penalized least squares criterion* and the corresponding penalized estimator $\tilde{s}$ is merely a *penalized LSE* . It is not very difficult to study again the example of boundary images, showing this time that some adequate choice of the collection of models leads to adaptive properties for the penalized LSE on classes of binary images with smooth boundaries.

Binary Images

We consider the same framework as in Section 8.3.2, i.e., the variables X_i are uniformly distributed on $[0, 1]^2$ and the regression function s is of the form

$$s(x_1, x_2) = b \text{ if } x_2 \le \partial s(x_1) \text{ and } a \text{ otherwise,}$$

where ∂s is some measurable map from $[0, 1]$ to $[0, 1]$ and $0 < a < b < 1$. Let $\mathcal{G}$ be the set of measurable maps from $[0, 1]$ to $[0, 1]$ and, for any $f \in \mathcal{G}$, χ_f denotes the function defined on $[0, 1]^2$ by

$$\chi_f(x_1, x_2) = b \text{ if } x_2 \le f(x_1) \text{ and } a \text{ otherwise.}$$

Setting $\mathcal{S} = \{\chi_f : f \in \mathcal{G}\}$, for every $t \in \mathcal{S}$, ∂t denotes the element of $\mathcal{G}$ such that $\chi_{\partial t} = t$. Consider for every positive integer m, ∂S_m to be the set of piecewise constant functions on a regular partition of $[0, 1]$ by m intervals and define $S_m = \{\chi_f : f \in \partial S_m\}$. We take $\{S_m\}_{m \in \mathbb{N}^*}$ as a collection of models. In order to apply Theorem 8.5, given $u \in S_m$, we need to upper bound $\mathbb{E}\left[W_m\left(\sigma\right)\right]$ where

$$W_m\left(\sigma\right) = \sup_{t \in S_m; d(u,t) \leq \sigma} \overline{\gamma}_n\left(u\right) - \overline{\gamma}_n\left(t\right).$$

We derive from the calculations of Section 8.3.2 that for some absolute numerical constant κ'

$$\sqrt{n}\mathbb{E}\left[W_m\left(\sigma\right)\right] \leq \kappa'\sigma\sqrt{\frac{m}{(b-a)}}$$

so that we can take

$$\phi_m\left(\sigma\right) = \kappa'\sigma\sqrt{\frac{m}{(b-a)}}.$$

Hence the solution ε_m of (8.44) is given by

$$\varepsilon_m^2 = \frac{2m\kappa'^2}{n\left(b-a\right)}.$$

Choosing $x_m = m$, leads to $\Sigma < 1$ and therefore, applying Theorem 8.5, we know that for some adequate numerical constants K' and C', one can take

$$\text{pen}\left(m\right) = K'\frac{m}{n\left(b-a\right)}$$

and the resulting penalized LSE $\widetilde{s}$ satisfies to

$$\mathbb{E}\left[\|\partial s - \partial\widetilde{s}\|_1\right] \leq C' \inf_{m \geq 1}\left\{ \inf_{t \in S_m}\|\partial s - \partial t\|_1 + \frac{m}{(b-a)^3 n} \right\}.$$

Assuming now that ∂s is Hölder smooth

$$|\partial s\left(x\right) - \partial s\left(x'\right)| \leq R\left|x - x'\right|^\alpha$$

with $R > 0$ and $\alpha \in (0,1]$, then

$$\inf_{t \in S_m}\|\partial s - \partial t\|_1 \leq Rm^{-\alpha},$$

leading to

$$\mathbb{E}\left[\|\partial s - \partial\widetilde{s}\|_1\right] \leq C' \inf_{m \geq 1}\left\{ Rm^{-\alpha} + \frac{m}{(b-a)^3 n} \right\}.$$

Hence, provided that $R \geq 1/n$, optimizing this bound with respect to m implies that

$$\sup_{\partial s \in \mathcal{H}(R,\alpha)} \mathbb{E}\left[\|\partial s - \partial\widetilde{s}\|_1\right] \leq C' R^{\frac{1}{\alpha+1}} n^{-\frac{\alpha}{1+\alpha}}.$$

Taking into account that the minimax risk is indeed of order $R^{\frac{1}{\alpha+1}} n^{-\frac{\alpha}{1+\alpha}}$ according to [72], this proves that the penalized LSE $\tilde{s}$ is adaptive on each of the Hölder classes $\mathcal{H}(R, \alpha)$ such that $R \geq 1/n$ and $\alpha \in (0, 1]$. Of course, with a little more efforts, the same kind of results could be obtained with collections of piecewise polynomials with variable degree, leading to adaptive estimators on Hölder classes $\mathcal{H}(R, \alpha)$ such that $R \geq 1/n$, for any positive value of α.

Selecting Nets

We can try to mimic the discretization strategies that we have developed in Chapter 4 and Chapter 7. As compared to the density estimation problem for instance, there is at least one noticeable difference. Indeed for density estimation, the dominating probability measure μ is assumed to be known. Here the role of this dominating measure is played by the distribution of the explanatory variables X_is. For some specific problems it makes sense to assume that μ is known (as we did in the previous boundary images estimation problem above), but most of the time one cannot make such an assumption. In such a situation there are at least two possibilities to overcome this difficulty: use $\mathbb{L}_\infty$ nets or empirical nets based on the empirical distribution of the variables X_is. Even if the second approach is more general than the first one, it would lead us to use extra technicalities that we prefer to avoid here. Constructing $\mathbb{L}_\infty$ nets concretely means that if the variables X_is take their values in $\mathbb{R}^d$ for instance, one has to assume that they are compactly supported and that we know their support. Moreover it also means that we have in view to estimate a rather smooth regression function s. Let us first state a straightforward consequence of Theorem 8.5 when applied to the selection of finite models problem in the regression framework.

Corollary 8.6 *Let $\{S_m\}_{m \in \mathcal{M}}$ be some at most countable collection of models, where for each $m \in \mathcal{M}$, S_m is assumed to be a finite set of functions taking their values in $[0, 1]$. We consider a corresponding collection $(\widehat{s}_m)_{m \in \mathcal{M}}$ of LSE, which means that for every $m \in \mathcal{M}$*

$$\sum_{i=1}^{n} (Y_i - \widehat{s}_m(X_i))^2 = \inf_{t \in S_m} \sum_{i=1}^{n} (Y_i - t(X_i))^2.$$

Let $\{\Delta_m\}_{m \in \mathcal{M}}, \{x_m\}_{m \in \mathcal{M}}$ be some families of nonnegative numbers such that $\Delta_m \geq \ln(|S_m|)$ for every $m \in \mathcal{M}$ and

$$\sum_{m \in \mathcal{M}} e^{-x_m} = \Sigma < \infty.$$

Define

$$\operatorname{pen}(m) = \frac{\kappa''(\Delta_m + x_m)}{n} \quad \text{for every } m \in \mathcal{M} \tag{8.46}$$

for some suitable numerical constant κ''. Then, if κ'' is large enough, there almost surely exists some minimizer $\widehat{m}$ of

$$\sum_{i=1}^{n} (Y_i - \widehat{s}_m (X_i))^2 + \mathrm{pen}\,(m)$$

over $\mathcal{M}$. Moreover, for such a minimizer, the following inequality is valid whatever the regression function s

$$\mathbb{E}_s \left[d^2 \left(s, \widehat{s}_{\widehat{m}} \right) \right] \leq C'' \left(\inf_{m \in \mathcal{M}} \left(d^2 \left(s, S_m \right) + \frac{\kappa'' \left(\Delta_m + x_m \right)}{n} \right) + \frac{(1 + \Sigma)}{n} \right).$$

$$(8.47)$$

Proof. It suffices to apply Theorem 8.5 with $w\,(\varepsilon) = \sqrt{2}\varepsilon$ and for each model $m \in \mathcal{M}$, check that by (6.4) the function ϕ_m defined by

$$\phi_m\,(\sigma) = 2\sigma\sqrt{\Delta_m}$$

does satisfy to (8.43). The result easily follows. ∎

Let us see what kind of result is achievable when working with nets by considering the same type of example as in the Gaussian or the density estimation frameworks. Let us consider some collection of compact subsets $\{\mathcal{S}_{\alpha,R}, \alpha \in \mathbb{N}^*, R > 0\}$ of $\mathcal{S}$ with the following structure:

$$\mathcal{S}_{\alpha,R} = \mathcal{S} \cap R\mathcal{H}_{\alpha,1},$$

where $\mathcal{H}_{\alpha,1}$ is star-shaped at 0 and satisfies for some positive constant $C_2\,(\alpha)$ to

$$H_\infty\,(\delta, \mathcal{H}_{\alpha,1}) \leq C_2\,(\alpha)\,\delta^{-1/\alpha}$$

for every $\delta \leq 1$. We consider for every positive integers α and k some $\mathbb{L}_\infty$-net $S_{\alpha,k}$ of $\mathcal{S}_{\alpha,k/\sqrt{n}}$ with radius $\delta_{\alpha,k} = k^{1/(2\alpha+1)}/\sqrt{n}$, so that

$$\ln |S_{\alpha,k}| \leq C_2\,(\alpha) \left(\frac{k}{\sqrt{n}\delta_{\alpha,k}} \right)^{1/\alpha} \leq C_2\,(\alpha)\,k^{2/(2\alpha+1)}$$

and

$$d^2\,(s, S_{\alpha,k}) \leq 2\delta_{\alpha,k}^2, \text{ for all } s \in \mathcal{S}_{\alpha,k/\sqrt{n}}. \qquad (8.48)$$

Applying Corollary 8.6 to the collection $(S_{\alpha,k})_{\alpha \geq 1, k \geq 1}$ with

$$\Delta_{\alpha,k} = C_2\,(\alpha)\,k^{2/(2\alpha+1)} \text{ and } x_{\alpha,k} = 4\alpha k^{2/(2\alpha+1)}$$

leads to the penalty

$$\mathrm{pen}\,(\alpha, k) = K\,(\alpha)\,k^{2/(2\alpha+1)}\varepsilon^2$$

where $K\,(\alpha) = \kappa''\,(C_2\,(\alpha) + 4\alpha)$. Noticing that $x_{\alpha,k} \geq \alpha + 2\ln\,(k)$ one has

$$\Sigma = \sum_{\alpha,k} e^{-x_{\alpha,k}} \le \left(\sum_{\alpha \ge 1} e^{-\alpha}\right)\left(\sum_{k \ge 1} k^{-2}\right) < 1$$

and it follows from (8.47) that if $\tilde{s}$ denotes the penalized LSE one has whatever the regression function s

$$\mathbb{E}_s\left[d^2\left(s,\tilde{s}\right)\right] \le C\left(\alpha\right) \inf_{\alpha,k}\left(d^2\left(s, S_{\alpha,k}\right) + \frac{k^{2/(2\alpha+1)}}{n}\right).$$

In particular if $s \in S_{\alpha,R}$ for some integer α and some real number $R \ge 1/\sqrt{n}$, setting $k = \lceil R\sqrt{n}\rceil$ we have $s \in S_{\alpha,k/\sqrt{n}}$ and since $S_{\alpha,k}$ is a $k^{1/(2\alpha+1)}\varepsilon$-net of $S_{\alpha,k\varepsilon}$, the previous inequality implies via (8.48) that

$$\sup_{s \in S_{\alpha,R}} \mathbb{E}_s\left[d^2\left(s,\tilde{s}\right)\right] \le 3C\left(\alpha\right)\left(\frac{k^{2/(2\alpha+1)}}{n}\right)$$

$$\le 3C\left(\alpha\right) 2^{2/(2\alpha+1)} \frac{\left(R\sqrt{n}\right)^{2/(2\alpha+1)}}{n}.$$

Since d^2 is upper bounded by 2, we finally derive that for some constant $C'\left(\alpha\right) \ge 1$

$$\sup_{s \in S_{\alpha,R}} \mathbb{E}_s\left[d^2\left(s,\tilde{s}\right)\right] \le C'\left(\alpha\right)\left(\left(R^{2/(2\alpha+1)}n^{-2\alpha/(2\alpha+1)}\right) \wedge 1\right). \qquad (8.49)$$

If $\mathcal{H}_{\alpha,R}$ is the Hölder class $\mathcal{H}\left(\alpha, R\right)$ defined in Section 7.5.1, we have already used the following property

$$H_\infty\left(\delta, \mathcal{H}\left(\alpha,R\right)\right) \le C_2\left(\alpha\right)\left(\frac{R}{\delta}\right)^{1/\alpha}.$$

Hence the previous approach applies to this case. Of course, nothing warrants that the above upper bound for the risk is minimax for arbitrary probability measures μ. For Hölder classes , it would not be difficult to show that this is indeed the case provided that one restricts to probability measures μ which are absolutely continuous with respect to Lebesgue measure with density f satisfying $0 < a \le f \le b < \infty$, for given positive constants a and b.

8.5 Advanced Model Selection Problems

All along the preceding Chapters, we have focused on model selection via penalization. It is worth noticing however, that some much simpler procedure can be used if one is ready to split the data into two parts, using the first half of the original simple to build the collection of estimators on each model and the second half to select among the family. This is the so-called *hold-out*. It should be seen as some primitive version of the V-fold cross-validation method which is commonly used in practice when one deals with i.i.d. data as it is the case in this Section. The advantage of hold-out is that it is very easy to study from a mathematical point of view. Of course it would be very interesting to derive similar results for V-fold cross-validation but we do not see how to do it for the moment.

8.5.1 Hold-Out as a Margin Adaptive Selection Procedure

Our purpose is here to show that the hold-out is a naturally margin adaptive selection procedure for classification. More generally, for i.i.d. data we wish to understand what is the performance of the hold-out as a model selection procedure. Our analysis will be based on the following abstract selection theorem among some family of functions $\{f_m, m \in \mathcal{M}\}$. The reason for introducing an auxiliary family of functions $\{g_m, m \in \mathcal{M}\}$ in the statement of Theorem 8.7 below will become clear in the section devoted to the study of MLEs. At first reading it is better to consider the simplest case where $g_m = f_m$ for every $m \in \mathcal{M}$, which is indeed enough to deal with the applications to bounded regression or classification that we have in view.

Theorem 8.7 *Let* $\{f_m, m \in \mathcal{M}\}$ *be some at most countable collection of real-valued measurable functions defined on some measurable space* $\mathcal{X}$. *Let* $\xi_1, \ldots, \xi_n$ *be some i.i.d. random variables with common distribution* P *and denote by* P_n *the empirical probability measure based on* $\xi_1, \ldots, \xi_n$. *Assume that for some family of positive numbers* $\{\sigma_m, m \in \mathcal{M}\}$ *and some positive constant* c, *one has for every integer* $k \geq 2$

$$P\left(|f_m - f_{m'}|^k\right) \leq \frac{k!}{2} c^{k-2} \left(\sigma_m + \sigma_{m'}\right)^2 \text{ for every } m \in \mathcal{M}, m' \in \mathcal{M}. \quad (8.50)$$

Assume furthermore that $P(f_m) \geq 0$ *for every* $m \in \mathcal{M}$ *and let* w *be some nonnegative and nondecreasing continuous function on* $\mathbb{R}_+$ *such that* $w(x)/x$ *is nonincreasing on* $(0, +\infty)$ *and*

$$\sigma_m \leq w\left(\sqrt{P(f_m)}\right) \text{ for every } m \in \mathcal{M}. \quad (8.51)$$

Let $\{g_m, m \in \mathcal{M}\}$ *be a family of functions such that* $f_m \leq g_m$ *and* $\{x_m\}_{m \in \mathcal{M}}$ *some family of nonnegative numbers such that*

$$\sum_{m \in \mathcal{M}} e^{-x_m} = \Sigma < \infty.$$

Let pen $: \mathcal{M} \to \mathbb{R}_+$ *and consider some random variable* $\widehat{m}$ *such that*

$$P_n\left(g_{\widehat{m}}\right) + \text{pen}\left(\widehat{m}\right) = \inf_{m \in \mathcal{M}} \left(P_n\left(g_m\right) + \text{pen}\left(m\right)\right).$$

Define δ_* *as the unique positive solution of the equation*

$$w\left(\delta\right) = \sqrt{n}\delta^2$$

and suppose that for some constant $\theta \in (0,1)$

$$\text{pen}\left(m\right) \geq x_m \left(\frac{\delta_*^2}{\theta} + \frac{c}{n}\right) \quad \text{for every } m \in \mathcal{M}. \tag{8.52}$$

Then, setting

$$\mathcal{R}_{\min} = \inf_{m \in \mathcal{M}} \left(P\left(g_m\right) + \text{pen}\left(m\right)\right),$$

one has

$$(1-\theta)\,\mathbb{E}\left[P\left(f_{\widehat{m}}\right)\right] \leq (1+\theta)\,\mathcal{R}_{\min} + \delta_*^2 \left(2\theta + \Sigma\theta^{-1}\right) + \frac{c\Sigma}{n}. \tag{8.53}$$

Moreover, if $f_m = g_m$ *for every* $m \in \mathcal{M}$, *the following exponential bound holds for every positive real number* x

$$\mathbb{P}\left[(1-\theta)\,P\left(f_{\widehat{m}}\right) > (1+\theta)\,\mathcal{R}_{\min} + \delta_*^2 \left(2\theta + x\theta^{-1}\right) + \frac{cx}{n}\right] \leq \Sigma e^{-x}. \tag{8.54}$$

Proof. We may always assume that the infimum of $P\left(g_m\right) + \text{pen}\left(m\right)$ is achieved on $\mathcal{M}$ (otherwise we can always take m_ε such that $P\left(g_{m_\varepsilon}\right) + \text{pen}\left(m_\varepsilon\right) \leq \inf_{m \in \mathcal{M}} \left(P\left(g_m\right) + \text{pen}\left(m\right)\right) + \varepsilon$ and make ε tend to 0 in the resulting tail bound). So let m such that

$$P\left(g_m\right) + \text{pen}\left(m\right) = \inf_{m' \in \mathcal{M}} \left(P\left(g_{m'}\right) + \text{pen}\left(m'\right)\right).$$

Our aim is to prove that, except on a set of probability less than Σe^{-x}, one has

$$(1-\theta)\,P\left(f_{\widehat{m}}\right) + U_m \leq (1+\theta)\left(P\left(g_m\right) + \text{pen}\left(m\right)\right) + \delta_*^2 \left(2\theta + x\theta^{-1}\right) + \frac{cx}{n}, \tag{8.55}$$

where $U_m = (P - P_n)\left(g_m - f_m\right)$. Noticing that U_m is centered at expectation and is equal to 0 whenever $f_m = g_m$, this will achieve the proof of Theorem 8.7. Indeed (8.55) leads to (8.53) by integrating with respect to x and (8.55) means exactly (8.54) whenever $f_m = g_m$. To prove (8.55), let us notice that by definition of $\widehat{m}$

$$P_n\left(g_{\widehat{m}}\right) + \text{pen}\left(\widehat{m}\right) \leq P_n\left(g_m\right) + \text{pen}\left(m\right),$$

hence, since $f_{\widehat{m}} \leq g_{\widehat{m}}$

$$P\left(f_{\widehat{m}}\right) = (P - P_n)\left(f_{\widehat{m}}\right) + P_n\left(f_{\widehat{m}}\right)$$
$$\leq P_n\left(g_m\right) + \text{pen}\left(m\right) + (P - P_n)\left(f_{\widehat{m}}\right) - \text{pen}\left(\widehat{m}\right)$$

and therefore

$$P\left(f_{\widehat{m}}\right) + U_m \leq P\left(g_m\right) + \text{pen}\left(m\right) + (P - P_n)\left(f_{\widehat{m}} - f_m\right) - \text{pen}\left(\widehat{m}\right). \tag{8.56}$$

It comes from Bernstein's inequality that for every $m' \in \mathcal{M}$ and every positive number $y_{m'}$, the following inequality holds, except on a set with probability less than $e^{-y_{m'}}$

$$(P - P_n)(f_{m'} - f_m) \leq \sqrt{\frac{2y_{m'}}{n}}(\sigma_m + \sigma_{m'}) + \frac{cy_{m'}}{n}.$$

Choosing $y_{m'} = x_{m'} + x$ for every $m' \in \mathcal{M}$, this implies that, except on some set Ω_x with probability less than Σe^{-x},

$$(P - P_n)(f_{\widehat{m}} - f_m) \leq \sqrt{\frac{2y_{\widehat{m}}}{n}}(\sigma_m + \sigma_{\widehat{m}}) + \frac{cy_{\widehat{m}}}{n}. \tag{8.57}$$

If u is some nonnegative real number, we derive from the monotonicity assumptions on w that

$$w(\sqrt{u}) \leq w\left(\sqrt{u + \delta_*^2}\right) \leq \sqrt{u + \delta_*^2}\frac{w(\delta_*)}{\delta_*}.$$

Hence, for every positive number y, we get by definition of δ_*

$$\sqrt{\frac{2y}{n}}w(\sqrt{u}) \leq \theta(u + \delta_*^2) + \theta^{-1}\frac{yw^2(\delta_*)}{2n\delta_*^2} \leq \theta(u + \delta_*^2) + \frac{y\delta_*^2\theta^{-1}}{2}.$$

Using this inequality with $y = y_{\widehat{m}}$ and successively $u = P(f_m)$ and $u = P(f_{\widehat{m}})$, it comes from (8.51) that

$$\sqrt{\frac{2y_{\widehat{m}}}{n}}(\sigma_m + \sigma_{\widehat{m}}) \leq \delta_*^2(2\theta + y_{\widehat{m}}\theta^{-1}) + \theta P(f_m) + \theta P(f_{\widehat{m}}).$$

Combining this inequality with (8.57) and (8.52) implies that, except on Ω_x

$$(P - P_n)(f_{\widehat{m}} - f_m) \leq \mathrm{pen}(\widehat{m}) + \delta_*^2(2\theta + x\theta^{-1}) + \frac{cx}{n} + \theta P(f_m) + \theta P(f_{\widehat{m}}).$$

Plugging this inequality in (8.56) yields since $f_m \leq g_m$

$$(1 - \theta)P(f_{\widehat{m}}) + U_m \leq (1 + \theta)P(g_m) + \mathrm{pen}(m) + \delta_*^2(2\theta + x\theta^{-1}) + \frac{cx}{n}$$

which a fortiori implies that (8.55) holds. ∎

Theorem 8.7 has a maybe more easily understandable corollary directly orientated towards the hold-out procedure *without* penalization in statistical learning.

Corollary 8.8 *Let $\{f_m, m \in \mathcal{M}\}$ be some finite collection of real-valued measurable functions defined on some measurable space $\mathcal{X}$. Let $\xi_1, \ldots, \xi_n$ be some i.i.d. random variables with common distribution P and denote by P_n the empirical probability measure based on $\xi_1, \ldots, \xi_n$. Assume that $f_m - f_{m'} \leq 1$*

for every $m, m' \in \mathcal{M}$. Assume furthermore that $P(f_m) \geq 0$ for every $m \in \mathcal{M}$ and let w be some nonnegative and nondecreasing continuous function on $\mathbb{R}_+$ such that $w(x)/x$ is nonincreasing on $(0, +\infty)$, $w(1) \geq 1$ and

$$P(f_m^2) \leq w^2\left(\sqrt{P(f_m)}\right) \quad \text{for every } m \in \mathcal{M}. \tag{8.58}$$

Consider some random variable $\widehat{m}$ such that

$$P_n(f_{\widehat{m}}) = \inf_{m \in \mathcal{M}} P_n(f_m).$$

Define δ_* as the unique positive solution of the equation

$$w(\delta) = \sqrt{n}\delta^2.$$

Then, for every $\theta \in (0,1)$

$$(1-\theta)\mathbb{E}\left[P\left(f_{\widehat{m}}\right)\right] \leq (1+\theta)\inf_{m \in \mathcal{M}} P(f_m) + \delta_*^2\left(2\theta + \ln(e|\mathcal{M}|)\left(\frac{1}{3} + \theta^{-1}\right)\right). \tag{8.59}$$

Proof. Noticing that since $w(1) \geq 1$, one has $\delta_*^2 \geq 1/n$, we simply apply Theorem 8.7 with $c = 1/3$, $x_m = \ln(|\mathcal{M}|)$ and

$$\text{pen}(m) = \delta_*^2 \ln(|\mathcal{M}|)\left(\theta^{-1} + (1/3)\right).$$

Since $\Sigma = 1$, (8.53) leads to (8.59). ∎

Hold-Out for Bounded Contrasts

Let us consider again the empirical risk minimization procedure. Assume that we observe $N + n$ random variables with common distribution P depending on some parameter s to be estimated. The first N observations $\xi'_1, \ldots, \xi'_N$ are used to build some preliminary collection of estimators $\{\widehat{s}_m\}_{m \in \mathcal{M}}$ and we use the remaining observations $\xi_1, \ldots, \xi_n$ to select some estimator $\widehat{s}_m$ among the collection $\{\widehat{s}_m\}_{m \in \mathcal{M}}$. We more precisely consider here the situation where there exists some (bounded) loss or contrast

$$\gamma \text{ from } \mathcal{S} \times \Xi \text{ to } [0,1]$$

which is well adapted to our estimation problem of s in the sense that the expected loss $P[\gamma(t, .)]$ achieves a minimum at point s when t varies in $\mathcal{S}$. We recall that the relative expected loss is defined by

$$\ell(s,t) = P[\gamma(t,.) - \gamma(s,.)], \quad \text{for all } t \in \mathcal{S}.$$

In the bounded regression or the classification cases, we have already seen that one can take

$$\gamma\left(t,\left(x,y\right)\right) = \left(y - t\left(x\right)\right)^{2}$$

since η (resp. s^*) is indeed the minimizer of $\mathbb{E}\left[\left(Y - t\left(X\right)\right)^{2}\right]$ over the set of measurable functions t taking their values in $[0,1]$ (resp.$\{0,1\}$). The idea is now to apply the results of the previous section conditionally on the training sample $\xi'_1, \ldots, \xi'_N$. For instance, we can apply Corollary 8.8 to the collection of functions $\{f_m = \gamma\left(\widehat{s}_m, .\right) - \gamma\left(s, .\right), m \in \mathcal{M}\}$. Let us consider the case where $\mathcal{M}$ is finite and define $\widehat{m}$ as a minimizer of the empirical risk $P_n\left(\gamma\left(\widehat{s}_m, .\right)\right)$ over $\mathcal{M}$. If $w \in \mathcal{C}_1$ is such that for all $t \in \mathcal{S}$

$$P\left(\left(\gamma\left(t, .\right) - \gamma\left(s, .\right)\right)^{2}\right) \le w^{2}\left(\sqrt{\ell\left(s,t\right)}\right),$$

we derive from (8.59) that conditionally on $\xi'_1, \ldots, \xi'_N$, one has for every $\theta \in (0,1)$

$$\left(1 - \theta\right)\mathbb{E}\left[\ell\left(s, \widehat{s}_{\widehat{m}}\right) \mid \xi'\right] \le \left(1 + \theta\right)\inf_{m \in \mathcal{M}}\ell\left(s, \widehat{s}_m\right)$$
$$+ \delta_*^2\left(2\theta + \ln\left(e\left|\mathcal{M}\right|\right)\left(\frac{1}{3} + \theta^{-1}\right)\right), \qquad (8.60)$$

where δ_* satisfies to $\sqrt{n}\delta_*^2 = w\left(\delta_*\right)$. The striking feature of this result is that the hold-out selection procedure provides an oracle type inequality involving the modulus of continuity w which is not known in advance. This is especially interesting in the classification framework for which w can be of very different natures according to the difficulty of the classification problem. The main issue is therefore to understand whether the term $\delta_*^2\left(1 + \ln\left(\left|\mathcal{M}\right|\right)\right)$ appearing in (8.60) is indeed a remainder term or not. We cannot exactly answer to this question because it is hard to compare δ_*^2 with $\inf_{m \in \mathcal{M}}\ell\left(s, \widehat{s}_m\right)$. However, if $\widehat{s}_m$ is itself an empirical risk minimizer over some model S_m, we can compare δ_*^2 with $\inf_{m \in \mathcal{M}}\varepsilon_m^2$, where ε_m^2 is (up to constant) the upper bound for the expected risk $\mathbb{E}\left[\ell\left(s, \widehat{s}_m\right)\right]$ provided by Theorem 8.3. More precisely, taking for instance $\theta = 1/2$, we derive from (8.60) that

$$\mathbb{E}\left[\ell\left(s, \widehat{s}_{\widehat{m}}\right)\right] \le 3\inf_{m \in \mathcal{M}}\mathbb{E}\left[\ell\left(s, \widehat{s}_m\right)\right] + \delta_*^2\left(3 + 2\ln\left(\left|\mathcal{M}\right|\right)\right).$$

By Theorem 8.3, setting $\ell\left(s, S_m\right) = \inf_{t \in S_m}\ell\left(s, t\right)$, one has for some absolute constant κ

$$\mathbb{E}\left[\ell\left(s, \widehat{s}_{\widehat{m}}\right)\right] \le 6\inf_{m \in \mathcal{M}}\left(\ell\left(s, S_m\right) + \kappa\varepsilon_m^2\right) + \delta_*^2\left(3 + 2\ln\left(\left|\mathcal{M}\right|\right)\right), \qquad (8.61)$$

where ε_m is defined by the equation

$$\sqrt{N}\varepsilon_m^2 = \phi_m\left(w\left(\varepsilon_m\right)\right).$$

Let ϕ_m belong to $\mathcal{C}_1$ controlling the modulus of continuity of the empirical process $\left(P'_N - P\right)\left(\gamma\left(t, .\right)\right)$ over model S_m with respect to some pseudodistance d satisfying to (8.16) and let w satisfy to (8.17). If N and n are of the same order of magnitude, $N = n$ say to be as simple as possible, then, since one can always assume that $w \leq 1$ (otherwise one can change w into $1 \wedge w$) one has $\phi_m(w(\varepsilon_m)) \geq w(\varepsilon_m)$ and therefore $\delta_* \leq \varepsilon_m$. This shows that in full generality, the risk of $\widehat{s}_{\widehat{m}}$ is at most of order

$$\ln(e|\mathcal{M}|) \inf_{m \in \mathcal{M}} \left(\ell(s, S_m) + \kappa \varepsilon_m^2\right). \tag{8.62}$$

Up to the unpleasant logarithmic factor $\ln(e|\mathcal{M}|)$, this is exactly what one could expect of a clever model selection procedure, i.e., it performs as well as if the margin function w was known. This is of course especially interesting in the classification setting. We were in fact over pessimistic when deriving (8.62) from (8.61). To see this, let us consider the classification framework and consider the VC case with margin function $w(\varepsilon) = 1 \wedge h^{-1/2}\varepsilon$, assuming that $|\mathcal{M}| \leq n$. Then, if V_m denotes the VC-dimension of S_m, combining (8.61) with Theorem 8.3 (in the spirit of Corollary 8.4) yields

$$\mathbb{E}\left[\ell(s, \widehat{s}_{\widehat{m}})\right] \leq 6 \inf_{m \in \mathcal{M}} \left(\ell(s, S_m) + C\ln(n)\left(\sqrt{\frac{V_m}{n}}\right) \wedge \left(\frac{V_m}{nh}\right)\right).$$

Hold-Out for the Maximum Likelihood Criterion

We consider here the maximum likelihood criterion. We can derive from Theorem 8.7 the following general result for penalized log-likelihood hold-out procedures. We recall that $\mathbf{K}$ (resp. $\mathbf{h}$) denote the Kullback–Leibler information number (resp. the Hellinger distance) as defined at the beginning of Chapter 7.

Theorem 8.9 *Assume that we observe $N+n$ random variables with common distribution P with density s with respect to some given positive σ-finite measure μ. The first N observations $\xi'_1, \ldots, \xi'_N$ are used to build some preliminary collection of estimators $\{\widehat{s}_m\}_{m \in \mathcal{M}}$ and we use the remaining observations $\xi_1, \ldots, \xi_n$ to select some estimator $\widehat{s}_m$ among the collection $\{\widehat{s}_m\}_{m \in \mathcal{M}}$. Let* pen $: \mathcal{M} \to \mathbb{R}_+$ *and denoting by P_n the empirical probability measure based on $\xi_1, \ldots, \xi_n$ consider some random variable $\widehat{m}$ such that*

$$P_n\left(-\ln(\widehat{s}_{\widehat{m}})\right) + \mathrm{pen}(\widehat{m}) = \inf_{m \in \mathcal{M}}\left(P_n(-\ln(\widehat{s}_m)) + \mathrm{pen}(m)\right).$$

Let $\{x_m\}_{m \in \mathcal{M}}$ be some family of nonnegative numbers such that

$$\sum_{m \in \mathcal{M}} e^{-x_m} = \Sigma < \infty,$$

and suppose that for some constant $\theta \in (0,1)$

$$\mathrm{pen}(m) \geq \frac{x_m}{n}\left(\frac{3}{\theta} + 2\right) \quad \text{for every } m \in \mathcal{M}. \tag{8.63}$$

Then,

$$(1-\theta)\,\mathbb{E}\left[\mathbf{K}\left(s,\frac{s+\widehat{s_m}}{2}\right)\right] \leq (1+\theta)\inf_{m\in\mathcal{M}}\left(\mathbb{E}\left[\mathbf{K}\left(s,\widehat{s}_m\right)\right]+\text{pen}\left(m\right)\right)$$
$$+\frac{3\left(2\theta+\Sigma\theta^{-1}\right)+2\Sigma}{n}. \tag{8.64}$$

Proof. We work conditionally to $\xi'_1,\ldots,\xi'_N$ and apply Theorem 8.7 to the family of functions

$$g_m = -\frac{1}{2}\ln\left(\frac{\widehat{s}_m}{s}\right) \text{ and } f_m = -\ln\left(\frac{s+\widehat{s}_m}{2s}\right),\; m\in\mathcal{M}.$$

By concavity of the logarithm, we indeed have $f_m \leq g_m$ for every $m \in \mathcal{M}$. Now we must check the moment condition (8.50). It comes from Lemma 7.26 that given two probability densities u and t, by the triangle inequality, the following moment inequality is available for every integer $k \geq 2$

$$P\left(\left|\ln\left(\frac{s+u}{s+t}\right)\right|^k\right) \leq 2^{k-2}k! \times \frac{9}{8}\left(\mathbf{h}\left(s,u\right)+\mathbf{h}\left(s,t\right)\right)^2.$$

Since $9/\left(8\left(2\ln\left(2\right)-1\right)\right) \leq 3$, combining this inequality with (7.103) leads to

$$P\left(\left|\ln\left(\frac{s+u}{s+t}\right)\right|^k\right) \leq 2^{k-2}k! \times 3\left(\sqrt{\mathbf{K}\left(s,\frac{s+u}{2}\right)}+\sqrt{\mathbf{K}\left(s,\frac{s+t}{2}\right)}\right)^2,$$

which means that (8.50) holds with $c = 2$ and $\sigma_m^2 = 3\mathbf{K}\left(s,\left(s+\widehat{s}_m\right)/2\right)$. Hence, since

$$P\left(f_m\right) = \mathbf{K}\left(s,\frac{s+\widehat{s}_m}{2}\right),$$

we derive from the definition of σ_m^2 that assumption (8.51) holds true with $w\left(x\right) = \sqrt{3}x$. Hence, setting

$$\delta_*^2 = \frac{3}{n}$$

(8.53) is valid (conditionally to $\xi'_1,\ldots,\xi'_N$), provided that condition (8.52) is satisfied. This clearly yields (8.64). ∎

The oracle inequality above is expressed in terms of the unusual loss function $\mathbf{K}\left(s,\left(s+t\right)/2\right)$. It comes from Lemma 7.23 that this loss function is also linked to the square Hellinger distance, so that, up to some absolute constant (8.64) remains true for the square Hellinger loss $\mathbf{h}^2\left(s,t\right)$.

8.5.2 Data-Driven Penalties

It could seem a bit disappointing to discover that a very crude method like hold-out is working so well. This is especially true in the classification framework. It is indeed a really hard work in this context to design margin adaptive penalties. Of course recent works on the topic (see [71] for a review), involving local Rademacher penalties for instance, provide at least some theoretical solution to the problem but still if one carefully looks at the penalties which are proposed in these works, they systematically involve constants which are typically unknown. In some cases, these constants are absolute constants which should nevertheless considered as unknown just because the numerical values coming from the theory are obviously over pessimistic. In some other cases, it is even worse since they also depend on nuisance parameters related to the unknown distribution (like for instance the infimum of the density of the explanatory variables). In any case these penalization methods are not ready to be implemented and remain far from being competitive with simple methods like hold out (or more generally with cross-validation methods).

Hence, two natural and connected questions emerge:

- Is there some room left for penalization methods?
- How to calibrate penalties to design efficient penalization criteria?

There are at least two reasons for which despite of the arguments against penalization that we have developed at the beginning of this Section, one should however keep interest for penalization methods. The first one is that for independent but not identically distributed observations (we typically think of Gaussian regression on a fixed design), hold out or more generally cross-validation may become irrelevant. The second one is that, talking about hold-out, since one uses part of the original data as testing data, one looses a bit of efficiency. A close inspection of the oracle inequalities presented in the preceding section shows that in the situation of half-sampling for instance one typically looses some factor 2 in the oracle inequality. Moreover hold-out is also known to be quite unstable and this is the reason why V-fold cross-validation is preferred to hold-out and widely used in practice. But now, concerning V-fold cross-validation, the question becomes how to choose V and what is the influence of this choice on the statistical performance of the method. This means that on the one hand, it remains to better understand cross-validation from a theoretical point of view and on the other hand that there is some room left for improvements. One can indeed hope to do better when using some direct method like penalization. But of course, since the opponent is strong, beating it requires to calibrate penalties sharply. This leads us to the second question raised above. We would like to conclude this Chapter by providing some possible answers to this last question, partly based on theoretical results which are already available and partly based on heuristics and thoughts which lead to some empirical rules and new theoretical problems.

A Practical Rule for Calibrating Penalties from the Data

To explain our idea which consists in guessing what is the right penalty to be used from the data themselves, let us come back to Gaussian model selection.

If we consider again the Gaussian model selection theorem for linear models, the following points can be made

- Mallows' C_p can underpenalize.
- Condition $K > 1$ in the statement of Theorem 4.2 is sharp.
- What penalty should be recommended? One can try to optimize the oracle inequality. The result is that roughly speaking, $K = 2$ is a good choice (see [24]).

In practice, the level of noise is unknown, but one can retain from the Gaussian theory the rule of thumb:

$$\text{``optimal'' penalty } = 2 \times \text{``minimal'' penalty.}$$

Interestingly the minimal penalty can be evaluated from the data because when the penalty is not heavy enough one systematically chooses models with large dimension. It remains to multiply by 2 to produce the desired (nearly) optimal penalty. This is a strategy for designing a data-driven penalty without knowing in advance the level of noise.

Practical implementation of penalization methods involves the extension to non Gaussian frameworks of the data-driven penalty choice strategy suggested above in the Gaussian case. It can roughly be described as follows

- Compute the minimum contrast estimator $\widehat{s}_D$ on the union of models defined by the same number D of parameters.
- Use the theory to guess the shape of the penalty $\text{pen}(D)$, typically $\text{pen}(D) = \alpha D$ (but other forms are also possible, like $\text{pen}(D) = \alpha D (1 + \ln(n/D))$).
- Estimate α from the data by multiplying by 2 the smallest value for which the corresponding penalized criterion does not explode.

In the context of change points detection, this data-driven calibration method for the penalty has been successfully implemented and tested by Lebarbier (see [74]). In the non-Gaussian case, we believe that this procedure remains valid but theoretical justification is far from being trivial and remains open. As already mentioned at the beginning of this Section, this problem is especially challenging in the classification context since it is connected to the question of defining optimal classifiers *without* knowing in advance the margin condition on the underlying distribution, which is a topic attracting much attention in the statistical learning community at this moment (see [115], [116], [14] for instance and [71] for a review).

Some Heuristics

More generally, defining proper data-driven strategies for choosing a penalty offers a new field of mathematical investigation since future progress on the topic requires to understand in depth the behavior of $\gamma_n(\widehat{s}_D)$. Recent advances involve new concentration inequalities. A first step in this direction is made in [32] and a joint work in progress with S. Boucheron is building upon the new moment inequalities proved in [30]. If one wants to better understand how to penalize optimally and the role that concentration inequalities could play in this matter, one has to come back to the root of the topic of model selection via penalization i.e., to Mallows' and Akaike's heuristics which are both based on the idea of estimating the risk in an unbiased way (at least asymptotically as far as Akaike's heuristics is concerned). The idea is the following.

Let us consider, in each model S_m some minimizer s_m of $t \to \mathbb{E}[\gamma_n(t)]$ over S_m (assuming that such a point does exist). Defining for every $m \in \mathcal{M}$,

$$\widehat{b}_m = \gamma_n(s_m) - \gamma_n(s) \text{ and } \widehat{v}_m = \gamma_n(s_m) - \gamma_n(\widehat{s}_m),$$

minimizing some penalized criterion

$$\gamma_n(\widehat{s}_m) + \text{pen}(m)$$

over $\mathcal{M}$ amounts to minimize

$$\widehat{b}_m - \widehat{v}_m + \text{pen}(m).$$

The point is that since $\widehat{b}_m$ is an unbiased estimator of the bias term $\ell(s, s_m)$. If we have in mind to use concentration arguments, one can hope that minimizing the quantity above will be approximately equivalent to minimize

$$\ell(s, s_m) - \mathbb{E}[\widehat{v}_m] + \text{pen}(m).$$

Since the purpose of the game is to minimize the risk $\mathbb{E}[\ell(s, \widehat{s}_m)]$, an ideal penalty would therefore be

$$\text{pen}(m) = \mathbb{E}[\widehat{v}_m] + \mathbb{E}[\ell(s_m, \widehat{s}_m)].$$

In the Mallows' C_p case, the models S_m are linear and $\mathbb{E}[\widehat{v}_m] = \mathbb{E}[\ell(s_m, \widehat{s}_m)]$ are explicitly computable (at least if the level of noise is assumed to be known). For Akaike's penalized log-likelihood criterion, this is similar, at least asymptotically. More precisely, one uses the fact that

$$\mathbb{E}[\widehat{v}_m] \approx \mathbb{E}[\ell(s_m, \widehat{s}_m)] \approx D_m/(2n),$$

where D_m stands for the number of parameters defining model S_m. The conclusion of these considerations is that Mallows' C_p as well as Akaike's criterion are indeed both based on the unbiased risk estimation principle.

Our guess is that we can go further in this direction and that the approximation $\mathbb{E}[\widehat{v}_m] \approx \mathbb{E}[\ell(s_m, \widehat{s}_m)]$ remains generally valid. If we believe in it then a good penalty becomes $2\mathbb{E}[\widehat{v}_m]$ or equivalently (having still in mind concentration arguments) $2\widehat{v}_m$. This in some sense explains the rule of thumb which is given in the preceding Section: the minimal penalty is $\widehat{v}_m$ while the optimal penalty should be $\widehat{v}_m + \mathbb{E}\left[\ell\left(s_m, \widehat{s}_m\right)\right]$ and their ratio is approximately equal to 2.

Of course, concentration arguments will work only if the list of models is not too rich. In practice this means that starting from a given list of models, one has first to decide to penalize in the same way the models which are defined by the same number of parameters. Then one considers a new list of models $(S_D)_{D \geq 1}$, where for each integer D, S_D is the union of those among the initial models which are defined by D parameters and then apply the preceding heuristics to this new list.

References

1. ADLER, R.J. An introduction to continuity, extrema and related topics for general Gaussian processes. *Institute of Mathematical Statistics Lecture Notes-Monograph Series*, **12** (1990).
2. AKAIKE, H. Information theory and an extension of the maximum likelihood principle. In P.N. Petrov and F. Csaki, editors, *Proceedings 2nd International Symposium on Information Theory*, pages 267–281. Akademia Kiado, Budapest, 1973.
3. ALDOUS, D.J. Exchangeability and related topics. In *Ecole d'Eté de Probabilités de St-Flour 1983*, 1–198 Springer Verlag Lecture Notes in Mathematics **1117** (1985).
4. ALEXANDER, K.S. Rates of growth and sample moduli for weighted empirical processes indexed by sets. *Probab. Theory Rel. Fields* **75** n°3, 379–423 (1987).
5. ANÉ, C. et al. Sur les inégalités de Sobolev logarithmiques. *Panoramas et Synthèses*, vol. **10**. Soc. Math. de France (2000).
6. ASSOUAD, P. Densité et dimension. *Ann. Inst. Fourier* **33**, n°3, 233–282 (1983).
7. BAHADUR, R.R. Examples of inconsistency of maximum likelihood estimates. *Sankhya Ser.A* **20**, 207–210 (1958).
8. BARAUD, Y. Model selection for regression on a fixed design. *Probability Theory and Related Fields* **117**, n°4 467–493 (2000).
9. BARAUD, Y., COMTE, F. and VIENNET, G. Model selection for (auto-)regression with dependent data. *ESAIM: Probability and Statistics* **5**, 33–49 (2001) http://www.emath.fr/ps/.
10. BARRON, A.R. and COVER, T.M. Minimum complexity density estimation. *IEEE Transactions on Information Theory* **37**, 1034–1054 (1991).
11. BARRON, A.R. and SHEU, C.H. Approximation of density functions by sequences of exponential families. *Ann. Statist.* **19**, 1054–1347 (1991).
12. BARRON, A.R., BIRGÉ, L. and MASSART, P. Risk bounds for model selection via penalization. *Probab. Th. Rel. Fields.* **113**, 301–415 (1999).
13. BARTLETT, P., BOUCHERON, S. and LUGOSI, G. Model selection and error estimation. *Machine Learning* **48**, 85–113 (2001).
14. BARTLETT, P., BOUSQUET, O. and MENDELSON, S. Local Rademacher Complexities (submitted) (2005).
15. BENNETT, G. Probability inequalities for the sum of independent random variables. *Journal of the American Statistical Association* **57**, 33–45 (1962).

16. BIRGÉ, L. Approximation dans les espaces métriques et théorie de l'estimation. *Z. Wahrscheinlichkeitstheorie Verw. Geb.* **65**, 181–237 (1983).

17. BIRGÉ, L. A new lower bound for multiple hypothesis testing. *IEEE Trans. Inform. Theory.* **51**, 1611–1615 (2005).

18. BIRGÉ, L. Model selection via testing: an alternative to (penalized) maximum likelihood estimators. *Ann. Inst. Henri Poincaré* **42**, 273–325 (2006).

19. BIRGÉ, L. and MASSART, P. Rates of convergence for minimum contrast estimators. *Probab. Th. Relat. Fields* **97**, 113–150 (1993).

20. BIRGÉ, L. and MASSART, P. From model selection to adaptive estimation. In *Festschrift for Lucien Lecam: Research Papers in Probability and Statistics* (D. Pollard, E. Torgersen and G. Yang, eds.), 55–87 (1997) Springer-Verlag, NewYork.

21. BIRGÉ, L. and MASSART, P. Minimum contrast estimators on sieves: exponential bounds and rates of convergence. *Bernoulli*, **4** (3), 329–375 (1998).

22. BIRGÉ, L. and MASSART, P. An adaptive compression algorithm in Besov spaces. *Constructive Approximation* **16**, 1–36 (2000).

23. BIRGÉ, L. and MASSART, P. Gaussian model selection. *Journal of the European Mathematical Society*, n°3, 203–268 (2001).

24. BIRGÉ, L. and MASSART, P. A generalized C_p criterion for Gaussian model selection. *Prépublication*, n°647, Universités de Paris 6 & Paris 7 (2001).

25. BIRGÉ, L. and MASSART, P. Minimal penalties for Gaussian model selection. *Probab. Th. Relat. Fields* (*to appear*).

26. BIRGÉ, L. and ROZENHOLC, Y. How many bins should be put in a regular histogram. *ESAIM: Probability and Statistics* **10**, 24–45 (2006).

27. BOBKOV, S. On Gross' and Talagrand's inequalities on the discrete cube. *Vestnik of Syktyvkar University* **Ser. 1,1** 12–19 (1995) (in Russian).

28. BOBKOV, S. and GOETZE, F. Exponential integrability and transportation cost under logarithmic Sobolev inequalities. *J. Funct. Anal.* **163**, n°1 1–28 (1999).

29. BORELL, C. The Brunn-Minkowski inequality in Gauss space. Invent. Math. **30**, 207–216 (1975).

30. BOUCHERON, S., BOUSQUET, O., LUGOSI, G. and MASSART, P. Moment inequalities for functions of independent random variables. *Ann. of Probability* **33**, n°2, 514–560 (2005).

31. BOUCHERON, S., LUGOSI, G. and MASSART, P. A sharp concentration inequality with applications. *Random Structures and Algorithms* **16**, n°3, 277–292 (2000).

32. BOUCHERON, S., LUGOSI, G. and MASSART, P. Concentration inequalities using the entropy method. *Ann. Probab.* **31**, n°3, 1583–1614 (2003).

33. BOUSQUET, O. A Bennett concentration inequality and its application to suprema of empirical processes. *C.R. Math. Acad. Sci. Paris* **334** n°6, 495–500 (2002).

34. CASTELLAN, G. Modified Akaike's criterion for histogram density estimation. Technical report #99.61, (1999) Université de Paris-Sud.

35. CASTELLAN, G. Density estimation via exponential model selection. *IEEE Trans. Inform. Theory* **49** n°8, 2052–2060 (2003).

36. CIREL'SON, B.S., IBRAGIMOV, I.A. and SUDAKOV, V.N. Norm of Gaussian sample function. In *Proceedings of the 3rd Japan-U.S.S.R. Symposium on Probability Theory*, Lecture Notes in Mathematics **550**, 20–41 (1976) Springer-Verlag, Berlin.

37. CIREL'SON, B.S. and SUDAKOV, V.N. Extremal properties of half spaces for spherically invariant measures. *J. Soviet. Math.* **9**, 9–18 (1978); translated from *Zap. Nauch. Sem. L.O.M.I.* **41**, 14–24 (1974).

38. COHEN, A., DAUBECHIES, I. and VIAL, P. Wavelets and fast wavelet transform on an interval. *Appl. Comput. Harmon. Anal.* **1**, 54–81 (1993).

39. COVER, T.M. and THOMAS, J.A. *Elements of Information Theory.* Wiley series in telecommunications. Wiley (1991).

40. CSISZAR, I. and KÖRNER, J. *Information Theory: Coding Theorems for discrete Memory-less Systems.* Academic Press, New York (1981).

41. DANIEL, C. and WOOD, F.S. *Fitting Equations to Data.* Wiley, New York (1971).

42. DEMBO, A. Information inequalities and concentration of measure. *Ann. Probab.* **25**, 927–939 (1997).

43. DEVORE, R.A., KYRIZIAKIS, G., LEVIATAN, D. and TIKHOMIROV, V.M. Wavelet compression and nonlinear n-widths. *Adv. Computational Math.* **1**, 197–214 (1993).

44. DEVORE, R.A. and LORENTZ, G.G. *Constructive Approximation.* Springer-Verlag, Berlin (1993).

45. DEVROYE, L. and LUGOSI, G. Lower bounds in pattern recognition. *Pattern recognition* **28**, 1011–1018 (1995).

46. DOBRUSHIN, R.L. Prescribing a system of random variables by conditional distributions. *Theor. Prob. Appl.* **15**, 458–486 (1970).

47. DONOHO, D.L. and JOHNSTONE, I.M. Ideal spatial adaptation by wavelet shrinkage. *Biometrika* **81**, 425–455 (1994).

48. DONOHO, D.L. and JOHNSTONE, I.M. Minimax risk over ℓ_p-balls for ℓ_q-error. *Probab. Theory Relat. Fields* **99**, 277–303 (1994).

49. DONOHO, D.L. and JOHNSTONE, I.M. Ideal denoising in an orthonormal basis chosen from a library of bases. *C. R. Acad. Sc. Paris Sér. I Math.* **319**, 1317–1322 (1994).

50. DONOHO, D.L. and JOHNSTONE, I.M. Adapting to unknown smoothness via wavelet shrinkage. *JASA.* **90**, 1200–1224 (1995).

51. DONOHO, D.L. and JOHNSTONE, I.M. Neo-classical minimax problems, thresholding and adaptive function estimation. *Bernoulli* **2**, 39–62 (1996).

52. DONOHO, D.L. and JOHNSTONE, I.M. Minimax estimation via wavelet shrinkage. *Ann. Statist.* **26**, 879–921 (1998).

53. DONOHO, D.L. and JOHNSTONE, I.M., KERKYACHARIAN, G. and PICARD, D. Wavelet shrinkage: Asymptopia? *J.R. Statist. Soc. B* **57**, 301–369 (1995).

54. DONOHO, D.L. and JOHNSTONE, I.M., KERKYACHARIAN, G. and PICARD, D. Density estimation by wavelet thresholding. *Ann. Statist.* **24**, 508–539 (1996).

55. DRAPER, N.R. and SMITH, H. Applied Regression Analysis, second edition. Wiley, New York (1981).

56. DUDLEY, R.M. The sizes of compact subsets of Hilbert space and continuity of Gaussian processes. *J. Funct. Anal.* **1**, 290–330 (1967).

57. DUDLEY, R.M. *Uniform Central Limit Theorems.* Cambridge Studies in advanced mathematics **63**, Cambridge University Press (1999).

58. EFROIMOVITCH, S.Yu. and PINSKER, M.S. Learning algorithm for nonparametric filtering. *Automat. Remote Control* **11**, 1434–1440 (1984), translated from Avtomatika i Telemekhanika **11**, 58–65.

59. EFRON, B. and STEIN, C. The Jacknife estimate of variance. *Ann. Statist.* **9**, 586–596 (1981).

60. FEDERER, H. *Geometric measure theory*. Springer (1969).
61. FERNIQUE, X. Régularité des trajectoires des fonctions aléatoires gaussiennes. Ecole d'Eté de Probabilités de St Flour 1974. *Lecture Notes in Mathematics* (Springer), **480** (1975).
62. FERNIQUE, X. Fonctions aléatoires gaussiennes, vecteurs aléatoires gaussiens. Les publications CRM, Montréal (1997).
63. FIGIEL, T., LINDENSTRAUSS, J., MILMAN, V.D. The dimensions of almost spherical sections of convex bodies. *Acta Math.* **139**, 52–94 (1977).
64. GROMOV, M. Paul Lévy's isoperimetric inequality. Preprint I.H.E.S. (1980).
65. GROSS, L. Logarithmic Sobolev inequalities. *Amer. J. Math.* **97**, 1061–1083 (1975).
66. HAUSSLER, D. Sphere packing numbers for subsets of the boolean n-cube with bounded Vapnik-Chervonenkis dimension. *Journal of Combinatorial Theory Series A*, **69**, 217–232 (1995).
67. HAUSSLER, D., LITTLESTONE, N. and WARMUTH, M. Predicting $\{0,1\}$ − functions on randomly drawn points. *Information and Computation*, **115**, 248–292 (1994).
68. HOEFFDING, W. Probability inequalities for sums of bounded random variables. *Journal of the American Statistical Association* **58**, 13–30 (1963).
69. KARPINSKI, M. and MACINTYRE, A. Polynomial bounds for VC dimension of sigmoidal neural networks. *Proceedings of the 27th annual ACM symposium on the theory of computing (STOC)*. Las Vegas. NV, USA, May 29 – June 1, 1995, NY: ACM, 200–208 (1995).
70. KOLTCHINSKII, V. Rademacher penalties and structural risk minimization. *IEEE Transactions on Information Theory* **47**, 1902–1914 (2001).
71. KOLTCHINSKII, V. 2004 IMS Medallion Lecture: Local Rademacher complexities and oracle inequalities in risk minimization. *Annals of Statistics (to appear)*.
72. KOROSTELEV, A.P. and TSYBAKOV, A.B. Minimax theory of image reconstruction. *Lectures notes in Statistics* **82**, Springer Verlag, New York (1993).
73. LATALA, R. and OLESZKIEWICZ, C. Between Sobolev and Poincaré. In *Geometric aspects of Functional Analysis, Israel Seminar (GAFA), 1996–2000*, pages 147–168 Springer, 2000. Lecture Notes in Mathematics **1745**.
74. LEBARBIER, E. Detecting multiple change points in the mean of Gaussian process by model selection. *Signal Processing* **85**, n°4, 717–736 (2005).
75. LE CAM, L. M. Convergence of estimates under dimensionality restrictions. *Ann. Statist.*, **1**, 38–53 (1973).
76. LEDOUX, M. Isoperimetry and Gaussian Analysis. In *Lectures on Probability Theory and Statistics, Ecole d'Eté de Probabilités de St-Flour XXIV-1994* (P. Bernard, ed.), 165–294 (1996) Springer, Berlin.
77. LEDOUX, M. On Talagrand deviation inequalities for product measures. *ESAIM: Probability and Statistics* **1**, 63–87 (1996) http://www.emath.fr/ps/.
78. LEDOUX, M. *The concentration of measure phenomenon*. Mathematical Surveys and Monographs **89**, American Mathematical Society.
79. LEDOUX, M. and TALAGRAND, M. *Probability in Banach spaces (Isoperimetry and processes)*. Ergebnisse der Mathematik und ihrer Grenzgebiete (1991) Springer-Verlag.
80. LEPSKII, O.V. On a problem of adaptive estimation in Gaussian white noise. *Theory Probab. Appl.* **36**, 454–466 (1990).

81. LEPSKII, O.V. Asymptotically minimax adaptive estimation I: Upper bounds. Optimally adaptive estimates. *Theory Probab. Appl.* **36**, 682–697 (1991).

82. LÉVY, P. *Problèmes concrets d'analyse fonctionnelle.* Gauthier-Villars (1951).

83. LUGOSI, G. Pattern classification and learning theory. *Principles of Nonparametric Learning* (L. Györfi ed.) Springer, Wien, New York, 1–56 (2002).

84. MALLOWS, C.L. Some comments on C_p. *Technometrics* **15**, 661–675 (1973).

85. MAMMEN, E. and TSYBAKOV, A.B. Smooth discrimination analysis. *Ann. Statist.* **27**, N°6, 1808–1829 (1999).

86. MARTON, K. A simple proof of the blowing up lemma. IEEE Trans. Inform. Theory **IT-32**, 445–446 (1986).

87. MARTON, K. Bounding $\bar{d}$-distance by information divergence: a method to prove measure concentration. *Ann. Probab.* **24**, 927–939 (1996).

88. MARTON, K. A measure concentration inequality for contracting Markov chains. *Geom. Funct. Anal.* **6**, n°3, 556–571 (1996).

89. MASON, D.M. and VAN ZWET, W.R. A refinement of the KMT inequality for the uniform empirical process. *Ann. Probab.* **15**, 871–884 (1987).

90. MASSART, P. About the constants in Talagrand's concentration inequalities for empirical processes. *Ann. Probab.* **28**, n°2, 863–884 (2000).

91. MASSART, P. Some applications of concentration inequalities to Statistics. Probability Theory. *Annales de la Faculté des Sciences de Toulouse (6)* **9**, n°2, 245–303 (2000).

92. MASSART, P. and NÉDÉLEC, E. Risk bounds for statistical learning. *Ann. Statist.* **34**, n°5, 2326–2366.

93. MCDIARMID, C. On the method of bounded differences. In *Surveys in Combinatorics 1989*, pages 148–188. Cambridge University Press, Cambridge (1989).

94. MEYER, Y. *Ondelettes et Opérateurs* I. Hermann, Paris (1990).

95. MILMAN, V.D. and SCHECHTMAN, G. Asymptotic theory of finite dimensional normed spaces. *Lecture Notes in Mathematics* (Springer), **1200** (1986).

96. MISITI, M., MISITI, Y., OPPENHEIM, G. and POGGI, J.M. *Matlab Wavelet Toolbox.* The Math Works Inc., Natick (1996).

97. PINELIS, I. Optimum bounds on moments of sums of independent random vectors. *Siberian Advances in Mathematics,* **5**, 141–150 (1995).

98. PINSKER, M.S. *Information and Information Stability of Random Variables and Processes.* Holden-Day, San Francisco (1964).

99. PINSKER, M.S. Optimal filtration of square-integrable signals in Gaussian noise. *Problems of Information Transmission* **16**, 120–133 (1980).

100. PISIER, G. Some applications of the metric entropy condition to harmonic analysis. In *Banach spaces, Harmonic analysis and Probability, Univ. of Connecticut 1980–81.* Springer, Berlin Heidelberg 1983, p.123–159. Lecture Notes in Mathematics, **995**.

101. RACHEV, S.T. *Probability metrics and the stability of stochastic models.* Wiley Series in Probability and Mathematical Statistics, Wiley Sons Ltd XIV 494 p. (1991).

102. REYNAUD-BOURET, P. Adaptive estimation of the intensity of inhomogeneous Poisson processes via concentration inequalities. *Probab. Theory Relat. Fields* **126**, n°1, 103–153 (2003).

103. RIO, E. Une inégalité de Bennett pour les maxima de processus empiriques. *Ann. I. H. Poincaré.* **38**, n°6 1053–1057 (2002).

104. RUDEMO, M. Empirical choice of histograms and kernel density estimators. *Scand. J. Statist.* **9**, 65–78 (1982).

105. SAMSON, P.M. Concentration of measure inequalities for Markov chains and Φ-mixing processes. *Ann. Probab.* **28**, 416–461 (2000).

106. SCHMIDT, E. Die Brunn-Minkowskische Ungleichung und ihr Spiegelbild sowie die isoperimetrische Eigenschaft der Kugel in der euklidischen und nichteuklidischen Geometrie. *Math. Nach.* **1**, 81–15 (1948).

107. SCHUMAKER, L.L. *Spline Functions: Basic Theory.* Wiley, New York (1981).

108. STRASSEN, V. The existence of probability measures with given marginals. *Ann. Math. Statist.* **36**, 423–439 (1965).

109. TALAGRAND, M. Regularity of Gaussian processes. *Acta Math.* **159**, 99–149 (1987).

110. TALAGRAND, M. An isoperimetric theorem on the cube and the Khintchine-Kahane inequalities in product spaces. *Proc. Amer. Math. Soc.* **104**, 905–909 (1988).

111. TALAGRAND, M. Sharper bounds for empirical processes. *Ann. Probab.* **22**, 28–76 (1994).

112. TALAGRAND, M. Concentration of measure and isoperimetric inequalities in product spaces. *Publications Mathématiques de l'I.H.E.S.* **81**, 73–205 (1995).

113. TALAGRAND, M. New concentration inequalities in product spaces. *Invent. Math.* **126**, 505–563 (1996).

114. TALAGRAND, M. Transportation cost for Gaussian and other product measures. *Geometric and Functional Analysis* **6**, 587–600 (1996).

115. TSYBAKOV, A.B. Optimal Aggregation of Classifiers in Statistical Learning. *Ann. Statist.* **32**, n°1 (2004).

116. TSYBAKOV, A.B. and VAN DE GEER, S. Square root penalty: adaptation to the margin in classification and in edge estimation. Prépublication PMA-820, Laboratoire de Probabilités et Modèles Aléatoires, Université Paris VI, (2003).

117. USPENSKY, J.V. *Introduction to Mathematical Probability.* New York: McGraw-Hill (1937).

118. VAN DE GEER, S. The method of sieves and minimum contrast estimators. *Math. Methods Statist.* **4**, 20–38 (1995).

119. VAN DER VAART, A. *Asymptotic Statistics.* Cambridge University Press (1998).

120. VAN DER VAART, A. and WELLNER J. *Weak Convergence and Empirical Processes.* Springer, New York (1996).

121. VAPNIK, V.N. *Estimation of dependencies based on empirical data.* Springer, New York (1982).

122. VAPNIK, V.N. *Statistical learning theory.* J. Wiley, New York (1990).

123. VAPNIK, V.N. and CHERVONENKIS A. Ya. *Theory of pattern recognition.* Nauka, Moscow, 1974. (in Russian); German translation; *Theorie der Zeichenerkennung*, Akademie Verlag, Berlin, 1979.

124. WAHBA, G. *Spline Models for Observational Data.* S.I.A.M., Philadelphia (1990).

125. WHITTAKER, E.T., WATSON, G.N. *A Course of Modern Analysis.* Cambridge University Press, London (1927).

126. YANG, Y. and BARRON, A.R. Information-theoretic determination of minimax rates of convergence. *Ann. Statist.* **27**, n°5, 1564–1599 (1999).

Index

Adaptive
  estimation, 85, 102, 139, 144, 219, 251, 267, 303
  estimator, 98, 102, 142, 203, 204, 270, 271, 303, 305
Akaike, 7, 89, 226, 227, 236
Akaike's criterion, **7**, 225–227, 236
Aldous, 108
Alexander, 139
Approximation theory, 80, 88, 111, 251, 255, 266

Baraud, 181
Barron, 274
Bartlett, 287
Bayes, 2, 106, 279, 284, 296
Besov
  ball, 99, 115, 121, **144**
  body, 107, **121**, 122, 125, 144, 146, 263, 266
  ellipsoid, 111, 115, **132**, 133, 134, 136, 138, 139, 143
  semi-norm, **144**
Bias term, **6**, 89, 91, 92, 102, 116, 120, 124, 134, 244, 245, 266, 267, 269, 270, 293
Birgé, 32, 143, 144
Birgé's lemma, **32**, 34, 102, 109, 253
BirgéE, 31
Borel–Cantelli lemma, 54, 82
Borell's theorem, **60**
Boucheron, 287, 317
Bounded regression, 279, 293, 303, 308, 311

Bousquet, 170, 208, 288
Brownian
  motion, 61, 79, 83
  sheet, 3, 5, 84

Central Limit Theorem, 1, 2, 64
Chaining argument, 17, 72, 183, 186, 189, 190
Change points detection, **8**, 91, 94, 316
Chernoff, 15, 21, 159
Chi-square statistics, 170, 171, **172**, 207–209, 211, 227, 228, 230, 231, 234, 237, 251
Classification, **2**, 5, 279, 280, 283, 284, 287, 293, 296, 302, 303, 308, 311–313, 316
Classifier, 2, **5**, 186, 279, 284, 286, 296, 316
Combinatorial entropy, 157, **160**, 187, 188, 284, 286, 297, 298
Concentration Inequalities, 9
Concentration inequalities, 2, 8, 9, 12, 26, 27, 39, 40, 58, 60, 65, 77, 147, 157, 167, 170, 171, 227, 251, 280, 317
Coupling, 35–37, 39, 51, 168
Cramér, 15, 21, 159
Cramér transform, **16**, 19, 32, 34, 108
Cross-validation, 203, 204, **205**, 215

Data-driven penalty, 157, 208, 215, 314, 316
Density estimation, **2**, 4, 7, 201, 204, 205, 251, 257, 303, 305, 306

Donoho, 7, 85, 89, 92, 99, 102, 205, 267
Donsker, 183, 297
Duality formula
  for entropy, 12, **28**, 29–32, 40, 41, 48, 166
  for ϕ-entropy, 44, **46**, 48, 49
Dudley, 1, 70, 74, 81
Dudley's criterion, 71, **74**, 75, 76, 80

Efron, 50
Ehrhardt, 60
Ellipsoid, **80**, 82, 85, 86, 107, 113, 115–117, 119, 121, 122, 131, 132, 135–137, 219, 254, 255, 257, 263, 264, 266, 267
Empirical coefficients, **93**, 204, 205
Empirical contrast, 3, **3**, 7, 202
Empirical criterion, 3, **3**, 4, 6
Empirical measure, **171**, 183, 203, 208, 209, 305, 308, 310, 313
Empirical process, 1, 9, **9**, 11, 12, 17, 19, 74, 148, 151, 153, 157, 162, 167, 168, 170, 180, 183–186, 192, 206, 227, 281, 287–289, 297, 301, 312
Empirical projection, **88**, 203
Empirical risk, **280**, 289, 290, 301, 312
  minimization, **5**, 279, **280**, 311
  minimizer, 280, **280**, 287–290, 299, 301, 312
Entropy, 12, **27**, 29, 30, 40, 43, 46, 48, 62, 154, 157
Entropy method, 12, 148, 154, 170
Entropy with bracketing, 183, 184, **190**, 238, 239, 245, 247, 251, 257, 271, 273, 294, 297, 300

Fernique, 56, 70, 76

Gaussian process, **53**, 70, 72, 74, 77, 78, 183
Gaussian regression, **83**, 86, 87
Gaussian sequence, **85**, 87, 99, 115, 122, 125, 132, 136
Girsanov's formula, 102
Gross, 62

Hölder class, **252**, 268, 305, 307
Hölder smooth, 79, 80, 273, **296**, 304
Hamming distance, **10**, 11, 105, 107, 147, 151
Haussler's bound, **187**, 189
Herbst argument, **31**, 62
Hilbert space, 78, 79, 83, 87, 115
Hilbert–Schmidt ellipsoid, 76, **80**, 131
Histogram, 171, 172, 201, 204, 225, **225**, 232, 236–238, 250, 296
Hoeffding, 21, 36, 40, 148, 153, 162
Hoeffding's lemma, **21**, 31
Hold-out, 307, 308, **310**, 312, 313
Hypothesis testing, 31

Ideal model, 6, 204
Inequalities
  Bennett, **23**, 26, 30, 34
  Bernstein, 22, **24**, 26, 30, 36, 148, 168, 170, 185, 192, 211, 217, 222, 235, 243, 310
  Bienaymé–Chebycheff, 54
  Borell, **56**
  Bousquet, **170**, 291
  Burkholder, 172
  Cauchy–Schwarz, 37, 40, 47, 65, 81, 130, 149, 171, 188, 196, 207, 209, 210, 213, 224, 228, 230, 235
  Chernoff, **16**, 18–20, 22, 24–26, 62, 105, 108, 149, 159
  Cirelson–Ibragimov–Sudakov, **10**, 56, 61
  Efron–Stein, **12**, 13, 51, **51**, 155, 167, 168, 173, 176
  Gross, **62**
  Hölder, 16, 173, 177
  Han, **42**, 158, 160, 161
  Hoeffding, 21, 22, **22**, 26
  Jensen, 17, 27, 38, 43, 46, 47, 108, 130, 188, 209, 264, 284–286
  Logarithmic Sobolev, 30, 56, 61, 62, 165
  Markov, 15, 18, 75, 277
  Marton, **37**, 148, 149
  maximal, 139, 183–187, 190, 241, 276, 297
  Mc Diarmid, **148**, 281
  moment, 155, 166, 174, 314, 317
  Pinsker, 31, **32**, 35–38

Sobolev, 155, 162
sub-Gaussian, **10**, 11, 188
symmetrization
    for empirical processes, 170, **188**,
        189
    for phi-entropy, **156**, 162
Talagrand, 9, **11**, 13, 169, 170, 180,
    208, 288
tensorization for entropy, 27, **40**, 42,
    51, 62, 63, 157
tensorization for ϕ-entropy, 43, 44,
    49, 154, 155, 162–165
tensorization for the variance, **50**,
    155
Isonormal process, 56, 76, **77**, 79–81,
    83, 84, 103, 115, 125, 131

Johnstone, 7, 85, 89, 92, 99, 102, 205,
    267

Kerkyacharian, 102, 205, 267
Koltchinskii, 287
Kullback–Leibler
    bias, 236, 267
    information, 4, **27**, 102, 104, 158,
        201, 254, 257, 273–275, 313
    loss, 202, 225–227, 231, 236, 244, 245,
        253, 256, 257
    projection, 225
    risk, 225, 232, 233, 236

Lévy, 58, 61
Lévy's isoperimetric theorem, 58
Latala, 44, 46, 154
Latala–Oleszkiewicz condition, 43
Least squares criterion
    in the classification framework, **5**, 284
    in the density framework, **4**, 202, 203
    in the Gaussian framework, 5, **87**
    in the regression framework, **4**, 283
Least squares estimator (LSE)
    in the density framework, **202**, 203,
        204, 212
    in the Gaussian framework, 6, **87**, 88,
        89, 106, 126, 136, 141

in the regression framework, **293**,
    296, 305
Lebarbier, 316
Ledoux, 12, 18, 40, 44, 46, 58, 153, 157,
    168
Lepskii's method, 102
Linear model, 83, 88, 89, 91, 99, 131,
    201, 204, 316
Lipschitz function, 55–62, 147, 150
Loss function, 3–5, 9, 87, 202, 225, 281,
    284, 287, 293, 303, 314
ℓ_p-body, **85**, 86, 105–107, 115, 116, 119,
    121, 144, 146
Lugosi, 287

Mallows, 7, 89
Mallows' C_p, **7**, 89, 97, 98, 316
Margin condition, **296**, 300
Marton, 12, 36, 148
Maximum likelihood
    criterion, **4**, 104, 201, 204, 226, 313
    estimator (MLE), 1, 24, 143, 144,
        225, 240, 257, 267, 270, 308
Metric entropy, 56, 70, **71**, 74–76, 80,
    110, 111, 139, 141, 183, 257, 260,
    273, 294
Minimax
    estimator, 99, **102**, 106, 110, 117,
        119, 122, 125, 134, 138, 142, 143,
        251, 269, 271, 273, 307
    lower bound, 12, 27, 31, 105, 106, 110,
        111, 117, 122, 134, 252, 257, 260,
        263, 269, 296
    point of view, **101**, 296, 300
    risk, 32, **101**, 102, 106, 107, 122, 142,
        144, 237, 263, 265, 267, 271, 272,
        293, 305
Minimum contrast
    estimation, 3, 5, 279, 280
    estimator, **3**, 6, 7, 316
Model selection, 1, 2, 8, 11, 56, 283,
    301, 303, 308, 313
    in the classification framework, 157,
        284, 302, 303
    in the density framework, 201, 202,
        204, 206–208, 211, 212, 238,
        251

in the Gaussian framework, 86, 88, 89, 98, 102, 115, 122, 125, 131, 201, 316
in the regression framework, 181, 303, 305
Moment generating function, **15**, 17, 23, 27, 29, 30, 32, 51, 56, 61, 108, 148, 157, 168, 169

Net, **71**, 76, 110, 112, 114, 142–144, 261, 262, 272, 301, 306, 307

Oleszkiewicz, 44, 46, 154
Oracle, **7**, 9, 89, 91, 92, 225, 226, 236, 237
inequality, 92, 101, 116, 133, 236, 244, 312, 314, 316

Packing number, **71**, 111
Penalized criterion, **7**, 8, 11, 201, 303, 316
Penalized estimator, **8**, 115, 203, 280, 281, 283, 285, 302, 303
Penalized least squares criterion
in the density framework, 201, **203**
in the Gaussian framework, **90**, 94, 100, 125, 136
in the regression framework, 7, **303**
Penalized log-likelihood criterion, 7, 201, **227**, 232, 237, 238, 240, 313
Penalized LSE
in the density framework, **202**, 203–206, 212, 216, 219, 220, 251, 263, 265
in the Gaussian framework, **90**, 92–94, 98, 100–102, 110, 116, 118, 120, 125, 126, 133, 134, 136, 142, 143, 204
in the regression framework, 303, **303**, 304, 305, 307
Penalized MLE, 204, **227**, 230, 232, 235–237, 250, 267, 268, 270–273
Penalty function, 7, **7**, 8, 90, 95, 99, 116, 123, 136, 202–208, 212, 215–223, 225, 230, 232, 235–237, 263, 265, 268, 280, 283, 285, 286
ϕ-entropy, **43**, 46, 48–50, 154, 155, 162
Picard, 102, 205, 267

Piecewise polynomial, 8, 91, 122, 145, 225, 238, 246, 247, 250, 268, 273, 296, 305
Pinelis, 181
Pinsker, 85
Pisier, 17
Product measure, 56, 59, 147
Product space, 35–37, 39
Prohorov distance, 36

Quadratic risk, 6, 88, **88**, 89, 98, 106, 221, 224, 293

Random variables
Bernoulli, 20, 32, 34, 44, 62, 63, 160
binomial, 20, 105, 107, 108, 160
Gaussian, 18, 19, 53, 55, 69
hypergeometric, 107, 108
Poisson, 19, 23, 24, 157, 159, 160
Rademacher, 22, 39, 63, 161, 184, 185
Regression function, 2, **2**, 3, 5, 279, 284, 293, 303, 305–307
Rio, 26, 170
Risk bound, 91, 94, 98, 100, 126, 130, 133, 136, 142, 204, 206, 208, 220, 227, 230, 235–237, 263, 265, 271, 283, 285–287, 289, 299–301

Sauer's lemma, **187**, 300
Schmidt, 58
Self-bounding condition, **157**, 160, 161
Shannon entropy, **41**
Sheu, 274
Slepian's lemma, 55, **66**, 69
Sobolev ball, **85**, 99, 102
Statistical learning, 2, 161, 280, 286, 310, 316
Stein, 50
Sudakov's minoration, 56, **69**, 80

Talagrand, 2, 18, 56, 59, 76, 147, 148, 150, 151, 168
Thresholding estimator, **92**, 98, 99, 205, 206, 265, 267
Total variation distance, **31**, 35
Transportation
cost, 30, 31, 36, **36**, 37, 40, 148–150
method, 36, 39, 149
Tsybakov, 296, 297, 300, 301

Universal entropy, 183, **187**, 188

Van der Vaart, 1
Vapnik, 5, 279, 280, 283
Variable selection, **86**, 87, 91, 92, 94, 97–99, 107, 115, 116
Variance term, **6**, 89
Varshamov–Gilbert's lemma, **105**, 106, 107

VC-class, **186**, 285, 299, 300
VC-dimension, **186**, 285, 297, 298, 313

Wavelet, 8, 86, 91, 99, 102, 122, 144, 145, 205, 211, 247, 255, 265
Wellner, 1
White noise, **3**, 5, 6, 79, 80, 83, 84, 86, 92, 102, 104, 143, 303

List of Participants

Lecturers

DEMBO Amir	Stanford Univ., USA
FUNAKI Tadahisa	Univ. Tokyo, Japan
MASSART Pascal	Univ. Paris-Sud, Orsay, F

Participants

AILLOT Pierre	Univ. Rennes 1, F
ATTOUCH Mohammed Kadi	Univ. Djillali Liabès, Sidi Bel Abbès, Algérie
AUDIBERT Jean-Yves	Univ. Pierre et Marie Curie, Paris, F
BAHADORAN Christophe	Univ. Blaise Pascal, Clermont-Ferrand, F
BEDNORZ Witold	Warsaw Univ., Poland
BELARBI Faiza	Univ. Djillali Liabès, Sidi Bel Abbès, Algérie
BEN AROUS Gérard	Courant Institute, New York, USA
BLACHE Fabrice	Univ. Blaise Pascal, Clermont-Ferrand, F
BLANCHARD Gilles	Univ. Paris-Sud, Orsay, F
BOIVIN Daniel	Univ. Brest, F
CHAFAI Djalil	Ecole Vétérinaire Toulouse, F
CHOUAF Benamar	Univ. Djillali Liabès, Sidi Bel Abbès, Algérie
DACHIAN Serguei	Univ. Blaise Pascal, Clermont-Ferrand, F
DELMOTTE Thierry	Univ. Paul Sabatier, Toulouse, F
DJELLOUT Hacène	Univ. Blaise Pascal, Clermont-Ferrand, F
DUROT Cécile	Univ. Paris-Sud, Orsay, F

FLORESCU Ionut	Purdue Univ., West Lafayette, USA
FONTBONA Joaquin	Univ. Pierre et Marie Curie, Paris, F
FOUGERES Pierre	Imperial College, London, UK
FROMONT Magalie	Univ. Paris-Sud, Orsay, F
GAIFFAS Stéphane	Univ. Pierre et Marie Curie, Paris, F
GUERIBALLAH Abdelkader	Univ. Djillali Liabès, Sidi Bel Abbès, Algérie
GIACOMIN Giambattista	Univ. Paris 7, F
GOLDSCHMIDT Christina	Univ. Cambridge, UK
GUSTO Gaelle	INRA, Jouy en Josas, F
HARIYA Yuu	Kyoto Univ., Japan
JIEN Yu-Juan	Purdue Univ., West Lafayette, USA
JOULIN Aldéric	Univ. La Rochelle, F
KLEIN Thierry	Univ. Versailles, F
KLUTCHNIKOFF Nicolas	Univ. Provence, Marseille, F
LEBARBIER Emilie	INRIA Rhône-Alpes, Saint-Ismier, F
LEVY-LEDUC Céline	Univ. Paris-Sud, Orsay, F
MAIDA Mylène	ENS Lyon, F
MALRIEU Florent	Univ. Rennes 1, F
MARTIN James	CNRS, Univ. Paris 7, F
MARTY Renaud	Univ. Paul Sabatier, Toulouse, F
MEREDITH Mark	Univ. Oxford, UK
MERLE Mathieu	ENS Paris, F
MOCIOALCA Oana	Purdue Univ., West Lafayette, USA
NISHIKAWA Takao	Univ. Tokyo, Japan
OBLOJ Jan	Univ. Pierre et Marie Curie, Paris, F
OSEKOWSKI Adam	Warsaw Univ., Poland
PAROUX Katy	Univ. Besançon, F
PASCU Mihai	Purdue Univ., West Lafayette, USA
PICARD Jean	Univ. Blaise Pascal, Clermont-Ferrand, F
REYNAUD-BOURET Patricia	Georgia Instit. Technology, Atlanta, USA
RIOS Ricardo	Univ. Central, Caracas, Venezuela
ROBERTO Cyril	Univ. Marne-la-Vallée, F
ROITERSHTEIN Alexander	Technion, Haifa, Israel

ROUX Daniel	Univ. Blaise Pascal, Clermont-Ferrand, F
ROZENHOLC Yves	Univ. Paris 7, F
SAINT-LOUBERT BIE Erwan	Univ. Blaise Pascal, Clermont-Ferrand, F
SCHAEFER Christin	Fraunhofer Institut FIRST, Berlin, D
STOLTZ Gilles	Univ. Paris-Sud, Orsay, F
TOUZILLIER Brice	Univ. Pierre et Marie Curie, Paris, F
TURNER Amanda	Univ. Cambridge, UK
VERT Régis	Univ. Paris-Sud, Orsay, F
VIENS Frederi	Purdue Univ., West Lafayette, USA
VIGON Vincent	INSA Rouen, F
YOR Marc	Univ. Pierre et Marie Curie, Paris, F
ZACHARUK Mariusz	Univ. Wrocław, Poland
ZEITOUNI Ofer	Univ. Minnesota, Minneapolis, USA
ZHANG Tao	Purdue Univ., West Lafayette, USA
ZWALD Laurent	Univ. Paris-Sud, Orsay, F

List of Short Lectures

Jean-Yves AUDIBERT	Aggregated estimators and empirical complexity for least squares regression
Christophe BAHADORAN	Convergence and local equilibrium for the one-dimensional asymmetric exclusion process
Fabrice BLACHE	Backward stochastic differential equations on manifolds
Gilles BLANCHARD	Some applications of model selection to statistical learning procedures
Serguei DACHIAN	Description of specifications by means of probability distributions in small volumes under condition of very weak positivity
Thierry DELMOTTE	How to use the stationarity of a reversible random environment to estimate derivatives of the annealed diffusion
Ionut FLORESCU	Pricing the implied volatility surface
Joaquin FONTBONA	Probabilistic interpretation and stochastic particle approximations of the 3-dimensional Navier-Stokes equation
Pierre FOUGERES	Curvature-dimension inequality and projections; some applications to Sobolev inequalities

Magalie FROMONT — Tests adaptatifs d'adéquation dans un modèle de densité

Giambattista GIACOMIN — On random co-polymers and disordered wetting models

Christina GOLDSCHMIDT — Critical random hypergraphs: a stochastic process approach

Yuu HARIYA — Large time limiting laws for Brownian motion perturbed by normalized exponential weights (Part II)

Aldéric JOULIN — Isoperimetric and functional inequalities in discrete settings: application to the geometric distribution

Thierry KLEIN — Processus empirique et concentration autour de la moyenne

Céline LEVY-LEDUC — Estimation de périodes de fonctions périodiques bruitées et de forme inconnue; applications à la vibrométrie laser

James MARTIN — Particle systems, queues, and geodesics in percolation

Renaud MARTY — Théorème limite pour une équation différentielle à coefficient aléatoire à mémoire longue

Oana MOCIOALCA — Additive summable processes and their stochastic integral

Takao NISHIKAWA — Dynamic entropic repulsion for the Ginzburg-Landau $\nabla\phi$ interface model

Jan OBLOJ — The Skorokhod embedding problem for functionals of Brownian excursions

Katy PAROUX — Convergence locale d'un modèle booléen de couronnes

Mihai N. PASCU — Maximum principles for Neumann/mixed Dirichlet-Neumann eigenvalue problem

Patricia REYNAUD-BOURET — Adaptive estimation of the Aalen multiplicative intensity by model selection

Cyril ROBERTO	Sobolev inequalities for probability measures on the real line
Alexander ROITERSHTEIN	Limit theorems for one-dimensional transient random walks in Markov environments
Gilles STOLTZ	Internal regret in on-line prediction of individual sequences and in on-line portfolio selection
Amanda TURNER	Convergence of Markov processes near hyperbolic fixed points
Vincent VIGON	Les abrupts et les érodés
Marc YOR	Large time limiting laws for Brownian motion perturbed by normalized exponential weights (Part I)
Ofer ZEITOUNI	Recursions and tightness

Lecture Notes in Mathematics

For information about earlier volumes
please contact your bookseller or Springer
LNM Online archive: springerlink.com

Vol. 1711: W. Ricker, Operator Algebras Generated by Commuting Projections: A Vector Measure Approach (1999)

Vol. 1712: N. Schwartz, J. J. Madden, Semi-algebraic Function Rings and Reflectors of Partially Ordered Rings (1999)

Vol. 1713: F. Bethuel, G. Huisken, S. Müller, K. Steffen, Calculus of Variations and Geometric Evolution Problems. Cetraro, 1996. Editors: S. Hildebrandt, M. Struwe (1999)

Vol. 1714: O. Diekmann, R. Durrett, K. P. Hadeler, P. K. Maini, H. L. Smith, Mathematics Inspired by Biology. Martina Franca, 1997. Editors: V. Capasso, O. Diekmann (1999)

Vol. 1715: N. V. Krylov, M. Röckner, J. Zabczyk, Stochastic PDE's and Kolmogorov Equations in Infinite Dimensions. Cetraro, 1998. Editor: G. Da Prato (1999)

Vol. 1716: J. Coates, R. Greenberg, K. A. Ribet, K. Rubin, Arithmetic Theory of Elliptic Curves. Cetraro, 1997. Editor: C. Viola (1999)

Vol. 1717: J. Bertoin, F. Martinelli, Y. Peres, Lectures on Probability Theory and Statistics. Saint-Flour, 1997. Editor: P. Bernard (1999)

Vol. 1718: A. Eberle, Uniqueness and Non-Uniqueness of Semigroups Generated by Singular Diffusion Operators (1999)

Vol. 1719: K. R. Meyer, Periodic Solutions of the N-Body Problem (1999)

Vol. 1720: D. Elworthy, Y. Le Jan, X-M. Li, On the Geometry of Diffusion Operators and Stochastic Flows (1999)

Vol. 1721: A. Iarrobino, V. Kanev, Power Sums, Gorenstein Algebras, and Determinantal Loci (1999)

Vol. 1722: R. McCutcheon, Elemental Methods in Ergodic Ramsey Theory (1999)

Vol. 1723: J. P. Croisille, C. Lebeau, Diffraction by an Immersed Elastic Wedge (1999)

Vol. 1724: V. N. Kolokoltsov, Semiclassical Analysis for Diffusions and Stochastic Processes (2000)

Vol. 1725: D. A. Wolf-Gladrow, Lattice-Gas Cellular Automata and Lattice Boltzmann Models (2000)

Vol. 1726: V. Marić, Regular Variation and Differential Equations (2000)

Vol. 1727: P. Kravanja M. Van Barel, Computing the Zeros of Analytic Functions (2000)

Vol. 1728: K. Gatermann Computer Algebra Methods for Equivariant Dynamical Systems (2000)

Vol. 1729: J. Azéma, M. Émery, M. Ledoux, M. Yor (Eds.) Séminaire de Probabilités XXXIV (2000)

Vol. 1730: S. Graf, H. Luschgy, Foundations of Quantization for Probability Distributions (2000)

Vol. 1731: T. Hsu, Quilts: Central Extensions, Braid Actions, and Finite Groups (2000)

Vol. 1732: K. Keller, Invariant Factors, Julia Equivalences and the (Abstract) Mandelbrot Set (2000)

Vol. 1733: K. Ritter, Average-Case Analysis of Numerical Problems (2000)

Vol. 1734: M. Espedal, A. Fasano, A. Mikelić, Filtration in Porous Media and Industrial Applications. Cetraro 1998. Editor: A. Fasano. 2000.

Vol. 1735: D. Yafaev, Scattering Theory: Some Old and New Problems (2000)

Vol. 1736: B. O. Turesson, Nonlinear Potential Theory and Weighted Sobolev Spaces (2000)

Vol. 1737: S. Wakabayashi, Classical Microlocal Analysis in the Space of Hyperfunctions (2000)

Vol. 1738: M. Émery, A. Nemirovski, D. Voiculescu, Lectures on Probability Theory and Statistics (2000)

Vol. 1739: R. Burkard, P. Deuflhard, A. Jameson, J.-L. Lions, G. Strang, Computational Mathematics Driven by Industrial Problems. Martina Franca, 1999. Editors: V. Capasso, H. Engl, J. Periaux (2000)

Vol. 1740: B. Kawohl, O. Pironneau, L. Tartar, J.-P. Zolesio, Optimal Shape Design. Tróia, Portugal 1999. Editors: A. Cellina, A. Ornelas (2000)

Vol. 1741: E. Lombardi, Oscillatory Integrals and Phenomena Beyond all Algebraic Orders (2000)

Vol. 1742: A. Unterberger, Quantization and Non-holomorphic Modular Forms (2000)

Vol. 1743: L. Habermann, Riemannian Metrics of Constant Mass and Moduli Spaces of Conformal Structures (2000)

Vol. 1744: M. Kunze, Non-Smooth Dynamical Systems (2000)

Vol. 1745: V. D. Milman, G. Schechtman (Eds.), Geometric Aspects of Functional Analysis. Israel Seminar 1999-2000 (2000)

Vol. 1746: A. Degtyarev, I. Itenberg, V. Kharlamov, Real Enriques Surfaces (2000)

Vol. 1747: L. W. Christensen, Gorenstein Dimensions (2000)

Vol. 1748: M. Ruzicka, Electrorheological Fluids: Modeling and Mathematical Theory (2001)

Vol. 1749: M. Fuchs, G. Seregin, Variational Methods for Problems from Plasticity Theory and for Generalized Newtonian Fluids (2001)

Vol. 1750: B. Conrad, Grothendieck Duality and Base Change (2001)

Vol. 1751: N. J. Cutland, Loeb Measures in Practice: Recent Advances (2001)

Vol. 1752: Y. V. Nesterenko, P. Philippon, Introduction to Algebraic Independence Theory (2001)

Vol. 1753: A. I. Bobenko, U. Eitner, Painlevé Equations in the Differential Geometry of Surfaces (2001)

Vol. 1754: W. Bertram, The Geometry of Jordan and Lie Structures (2001)

Vol. 1755: J. Azéma, M. Émery, M. Ledoux, M. Yor (Eds.), Séminaire de Probabilités XXXV (2001)

Vol. 1756: P. E. Zhidkov, Korteweg de Vries and Nonlinear Schrödinger Equations: Qualitative Theory (2001)

Vol. 1757: R. R. Phelps, Lectures on Choquet's Theorem (2001)

Vol. 1758: N. Monod, Continuous Bounded Cohomology of Locally Compact Groups (2001)

Vol. 1759: Y. Abe, K. Kopfermann, Toroidal Groups (2001)

Vol. 1760: D. Filipović, Consistency Problems for Heath-Jarrow-Morton Interest Rate Models (2001)

Vol. 1761: C. Adelmann, The Decomposition of Primes in Torsion Point Fields (2001)

Vol. 1762: S. Cerrai, Second Order PDE's in Finite and Infinite Dimension (2001)

Vol. 1763: J.-L. Loday, A. Frabetti, F. Chapoton, F. Goichot, Dialgebras and Related Operads (2001)

Vol. 1764: A. Cannas da Silva, Lectures on Symplectic Geometry (2001)

Vol. 1765: T. Kerler, V. V. Lyubashenko, Non-Semisimple Topological Quantum Field Theories for 3-Manifolds with Corners (2001)

Vol. 1766: H. Hennion, L. Hervé, Limit Theorems for Markov Chains and Stochastic Properties of Dynamical Systems by Quasi-Compactness (2001)

Vol. 1767: J. Xiao, Holomorphic Q Classes (2001)

Vol. 1768: M. J. Pflaum, Analytic and Geometric Study of Stratified Spaces (2001)

Vol. 1769: M. Alberich-Carramiñana, Geometry of the Plane Cremona Maps (2002)

Vol. 1770: H. Gluesing-Luerssen, Linear Delay-Differential Systems with Commensurate Delays: An Algebraic Approach (2002)

Vol. 1771: M. Émery, M. Yor (Eds.), Séminaire de Probabilités 1967-1980. A Selection in Martingale Theory (2002)

Vol. 1772: F. Burstall, D. Ferus, K. Leschke, F. Pedit, U. Pinkall, Conformal Geometry of Surfaces in S^4 (2002)

Vol. 1773: Z. Arad, M. Muzychuk, Standard Integral Table Algebras Generated by a Non-real Element of Small Degree (2002)

Vol. 1774: V. Runde, Lectures on Amenability (2002)

Vol. 1775: W. H. Meeks, A. Ros, H. Rosenberg, The Global Theory of Minimal Surfaces in Flat Spaces. Martina Franca 1999. Editor: G. P. Pirola (2002)

Vol. 1776: K. Behrend, C. Gomez, V. Tarasov, G. Tian, Quantum Comohology. Cetraro 1997. Editors: P. de Bartolomeis, B. Dubrovin, C. Reina (2002)

Vol. 1777: E. García-Río, D. N. Kupeli, R. Vázquez-Lorenzo, Osserman Manifolds in Semi-Riemannian Geometry (2002)

Vol. 1778: H. Kiechle, Theory of K-Loops (2002)

Vol. 1779: I. Chueshov, Monotone Random Systems (2002)

Vol. 1780: J. H. Bruinier, Borcherds Products on O(2,1) and Chern Classes of Heegner Divisors (2002)

Vol. 1781: E. Bolthausen, E. Perkins, A. van der Vaart, Lectures on Probability Theory and Statistics. Ecole d' Eté de Probabilités de Saint-Flour XXIX-1999. Editor: P. Bernard (2002)

Vol. 1782: C.-H. Chu, A. T.-M. Lau, Harmonic Functions on Groups and Fourier Algebras (2002)

Vol. 1783: L. Grüne, Asymptotic Behavior of Dynamical and Control Systems under Perturbation and Discretization (2002)

Vol. 1784: L. H. Eliasson, S. B. Kuksin, S. Marmi, J.-C. Yoccoz, Dynamical Systems and Small Divisors. Cetraro, Italy 1998. Editors: S. Marmi, J.-C. Yoccoz (2002)

Vol. 1785: J. Arias de Reyna, Pointwise Convergence of Fourier Series (2002)

Vol. 1786: S. D. Cutkosky, Monomialization of Morphisms from 3-Folds to Surfaces (2002)

Vol. 1787: S. Caenepeel, G. Militaru, S. Zhu, Frobenius and Separable Functors for Generalized Module Categories and Nonlinear Equations (2002)

Vol. 1788: A. Vasil'ev, Moduli of Families of Curves for Conformal and Quasiconformal Mappings (2002)

Vol. 1789: Y. Sommerhäuser, Yetter-Drinfel'd Hopf algebras over groups of prime order (2002)

Vol. 1790: X. Zhan, Matrix Inequalities (2002)

Vol. 1791: M. Knebusch, D. Zhang, Manis Valuations and Prüfer Extensions I: A new Chapter in Commutative Algebra (2002)

Vol. 1792: D. D. Ang, R. Gorenflo, V. K. Le, D. D. Trong, Moment Theory and Some Inverse Problems in Potential Theory and Heat Conduction (2002)

Vol. 1793: J. Cortés Monforte, Geometric, Control and Numerical Aspects of Nonholonomic Systems (2002)

Vol. 1794: N. Pytheas Fogg, Substitution in Dynamics, Arithmetics and Combinatorics. Editors: V. Berthé, S. Ferenczi, C. Mauduit, A. Siegel (2002)

Vol. 1795: H. Li, Filtered-Graded Transfer in Using Noncommutative Gröbner Bases (2002)

Vol. 1796: J.M. Melenk, hp-Finite Element Methods for Singular Perturbations (2002)

Vol. 1797: B. Schmidt, Characters and Cyclotomic Fields in Finite Geometry (2002)

Vol. 1798: W.M. Oliva, Geometric Mechanics (2002)

Vol. 1799: H. Pajot, Analytic Capacity, Rectifiability, Menger Curvature and the Cauchy Integral (2002)

Vol. 1800: O. Gabber, L. Ramero, Almost Ring Theory (2003)

Vol. 1801: J. Azéma, M. Émery, M. Ledoux, M. Yor (Eds.), Séminaire de Probabilités XXXVI (2003)

Vol. 1802: V. Capasso, E. Merzbach, B. G. Ivanoff, M. Dozzi, R. Dalang, T. Mountford, Topics in Spatial Stochastic Processes. Martina Franca, Italy 2001. Editor: E. Merzbach (2003)

Vol. 1803: G. Dolzmann, Variational Methods for Crystalline Microstructure – Analysis and Computation (2003)

Vol. 1804: I. Cherednik, Ya. Markov, R. Howe, G. Lusztig, Iwahori-Hecke Algebras and their Representation Theory. Martina Franca, Italy 1999. Editors: V. Baldoni, D. Barbasch (2003)

Vol. 1805: F. Cao, Geometric Curve Evolution and Image Processing (2003)

Vol. 1806: H. Broer, I. Hoveijn. G. Lunther, G. Vegter, Bifurcations in Hamiltonian Systems. Computing Singularities by Gröbner Bases (2003)

Vol. 1807: V. D. Milman, G. Schechtman (Eds.), Geometric Aspects of Functional Analysis. Israel Seminar 2000-2002 (2003)

Vol. 1808: W. Schindler, Measures with Symmetry Properties (2003)

Vol. 1809: O. Steinbach, Stability Estimates for Hybrid Coupled Domain Decomposition Methods (2003)

Vol. 1810: J. Wengenroth, Derived Functors in Functional Analysis (2003)

Vol. 1811: J. Stevens, Deformations of Singularities (2003)

Vol. 1812: L. Ambrosio, K. Deckelnick, G. Dziuk, M. Mimura, V. A. Solonnikov, H. M. Soner, Mathematical Aspects of Evolving Interfaces. Madeira, Funchal, Portugal 2000. Editors: P. Colli, J. F. Rodrigues (2003)

Vol. 1813: L. Ambrosio, L. A. Caffarelli, Y. Brenier, G. Buttazzo, C. Villani, Optimal Transportation and its

Applications. Martina Franca, Italy 2001. Editors: L. A. Caffarelli, S. Salsa (2003)

Vol. 1814: P. Bank, F. Baudoin, H. Föllmer, L.C.G. Rogers, M. Soner, N. Touzi, Paris-Princeton Lectures on Mathematical Finance 2002 (2003)

Vol. 1815: A. M. Vershik (Ed.), Asymptotic Combinatorics with Applications to Mathematical Physics. St. Petersburg, Russia 2001 (2003)

Vol. 1816: S. Albeverio, W. Schachermayer, M. Talagrand, Lectures on Probability Theory and Statistics. Ecole d'Eté de Probabilités de Saint-Flour XXX-2000. Editor: P. Bernard (2003)

Vol. 1817: E. Koelink, W. Van Assche (Eds.), Orthogonal Polynomials and Special Functions. Leuven 2002 (2003)

Vol. 1818: M. Bildhauer, Convex Variational Problems with Linear, nearly Linear and/or Anisotropic Growth Conditions (2003)

Vol. 1819: D. Masser, Yu. V. Nesterenko, H. P. Schlickewei, W. M. Schmidt, M. Waldschmidt, Diophantine Approximation. Cetraro, Italy 2000. Editors: F. Amoroso, U. Zannier (2003)

Vol. 1820: F. Hiai, H. Kosaki, Means of Hilbert Space Operators (2003)

Vol. 1821: S. Teufel, Adiabatic Perturbation Theory in Quantum Dynamics (2003)

Vol. 1822: S.-N. Chow, R. Conti, R. Johnson, J. Mallet-Paret, R. Nussbaum, Dynamical Systems. Cetraro, Italy 2000. Editors: J. W. Macki, P. Zecca (2003)

Vol. 1823: A. M. Anile, W. Allegretto, C. Ringhofer, Mathematical Problems in Semiconductor Physics. Cetraro, Italy 1998. Editor: A. M. Anile (2003)

Vol. 1824: J. A. Navarro González, J. B. Sancho de Salas, $\mathscr{C}^\infty$ – Differentiable Spaces (2003)

Vol. 1825: J. H. Bramble, A. Cohen, W. Dahmen, Multiscale Problems and Methods in Numerical Simulations, Martina Franca, Italy 2001. Editor: C. Canuto (2003)

Vol. 1826: K. Dohmen, Improved Bonferroni Inequalities via Abstract Tubes. Inequalities and Identities of Inclusion-Exclusion Type. VIII, 113 p, 2003.

Vol. 1827: K. M. Pilgrim, Combinations of Complex Dynamical Systems. IX, 118 p, 2003.

Vol. 1828: D. J. Green, Gröbner Bases and the Computation of Group Cohomology. XII, 138 p, 2003.

Vol. 1829: E. Altman, B. Gaujal, A. Hordijk, Discrete-Event Control of Stochastic Networks: Multimodularity and Regularity. XIV, 313 p, 2003.

Vol. 1830: M. I. Gil', Operator Functions and Localization of Spectra. XIV, 256 p, 2003.

Vol. 1831: A. Connes, J. Cuntz, E. Guentner, N. Higson, J. E. Kaminker, Noncommutative Geometry, Martina Franca, Italy 2002. Editors: S. Doplicher, L. Longo (2004)

Vol. 1832: J. Azéma, M. Émery, M. Ledoux, M. Yor (Eds.), Séminaire de Probabilités XXXVII (2003)

Vol. 1833: D.-Q. Jiang, M. Qian, M.-P. Qian, Mathematical Theory of Nonequilibrium Steady States. On the Frontier of Probability and Dynamical Systems. IX, 280 p, 2004.

Vol. 1834: Yo. Yomdin, G. Comte, Tame Geometry with Application in Smooth Analysis. VIII, 186 p, 2004.

Vol. 1835: O.T. Izhboldin, B. Kahn, N.A. Karpenko, A. Vishik, Geometric Methods in the Algebraic Theory of Quadratic Forms. Summer School, Lens, 2000. Editor: J.-P. Tignol (2004)

Vol. 1836: C. Năstăsescu, F. Van Oystaeyen, Methods of Graded Rings. XIII, 304 p, 2004.

Vol. 1837: S. Tavaré, O. Zeitouni, Lectures on Probability Theory and Statistics. Ecole d'Eté de Probabilités de Saint-Flour XXXI-2001. Editor: J. Picard (2004)

Vol. 1838: A.J. Ganesh, N.W. O'Connell, D.J. Wischik, Big Queues. XII, 254 p, 2004.

Vol. 1839: R. Gohm, Noncommutative Stationary Processes. VIII, 170 p, 2004.

Vol. 1840: B. Tsirelson, W. Werner, Lectures on Probability Theory and Statistics. Ecole d'Eté de Probabilités de Saint-Flour XXXII-2002. Editor: J. Picard (2004)

Vol. 1841: W. Reichel, Uniqueness Theorems for Variational Problems by the Method of Transformation Groups (2004)

Vol. 1842: T. Johnsen, A. L. Knutsen, K_3 Projective Models in Scrolls (2004)

Vol. 1843: B. Jefferies, Spectral Properties of Noncommuting Operators (2004)

Vol. 1844: K.F. Siburg, The Principle of Least Action in Geometry and Dynamics (2004)

Vol. 1845: Min Ho Lee, Mixed Automorphic Forms, Torus Bundles, and Jacobi Forms (2004)

Vol. 1846: H. Ammari, H. Kang, Reconstruction of Small Inhomogeneities from Boundary Measurements (2004)

Vol. 1847: T.R. Bielecki, T. Björk, M. Jeanblanc, M. Rutkowski, J.A. Scheinkman, W. Xiong, Paris-Princeton Lectures on Mathematical Finance 2003 (2004)

Vol. 1848: M. Abate, J. E. Fornaess, X. Huang, J. P. Rosay, A. Tumanov, Real Methods in Complex and CR Geometry, Martina Franca, Italy 2002. Editors: D. Zaitsev, G. Zampieri (2004)

Vol. 1849: Martin L. Brown, Heegner Modules and Elliptic Curves (2004)

Vol. 1850: V. D. Milman, G. Schechtman (Eds.), Geometric Aspects of Functional Analysis. Israel Seminar 2002-2003 (2004)

Vol. 1851: O. Catoni, Statistical Learning Theory and Stochastic Optimization (2004)

Vol. 1852: A.S. Kechris, B.D. Miller, Topics in Orbit Equivalence (2004)

Vol. 1853: Ch. Favre, M. Jonsson, The Valuative Tree (2004)

Vol. 1854: O. Saeki, Topology of Singular Fibers of Differential Maps (2004)

Vol. 1855: G. Da Prato, P.C. Kunstmann, I. Lasiecka, A. Lunardi, R. Schnaubelt, L. Weis, Functional Analytic Methods for Evolution Equations. Editors: M. Iannelli, R. Nagel, S. Piazzera (2004)

Vol. 1856: K. Back, T.R. Bielecki, C. Hipp, S. Peng, W. Schachermayer, Stochastic Methods in Finance, Bressanone/Brixen, Italy, 2003. Editors: M. Fritelli, W. Runggaldier (2004)

Vol. 1857: M. Émery, M. Ledoux, M. Yor (Eds.), Séminaire de Probabilités XXXVIII (2005)

Vol. 1858: A.S. Cherny, H.-J. Engelbert, Singular Stochastic Differential Equations (2004)

Vol. 1859: E. Letellier, Fourier Transforms of Invariant Functions on Finite Reductive Lie Algebras (2005)

Vol. 1860: A. Borisyuk, G.B. Ermentrout, A. Friedman, D. Terman, Tutorials in Mathematical Biosciences I. Mathematical Neurosciences (2005)

Vol. 1861: G. Benettin, J. Henrard, S. Kuksin, Hamiltonian Dynamics – Theory and Applications, Cetraro, Italy, 1999. Editor: A. Giorgilli (2005)

Vol. 1862: B. Helffer, F. Nier, Hypoelliptic Estimates and Spectral Theory for Fokker-Planck Operators and Witten Laplacians (2005)

Vol. 1863: H. Führ, Abstract Harmonic Analysis of Continuous Wavelet Transforms (2005)

Vol. 1864: K. Efstathiou, Metamorphoses of Hamiltonian Systems with Symmetries (2005)

Vol. 1865: D. Applebaum, B.V. R. Bhat, J. Kustermans, J. M. Lindsay, Quantum Independent Increment Processes I. From Classical Probability to Quantum Stochastic Calculus. Editors: M. Schürmann, U. Franz (2005)

Vol. 1866: O.E. Barndorff-Nielsen, U. Franz, R. Gohm, B. Kümmerer, S. Thorbjønsen, Quantum Independent Increment Processes II. Structure of Quantum Lévy Processes, Classical Probability, and Physics. Editors: M. Schürmann, U. Franz, (2005)

Vol. 1867: J. Sneyd (Ed.), Tutorials in Mathematical Biosciences II. Mathematical Modeling of Calcium Dynamics and Signal Transduction. (2005)

Vol. 1868: J. Jorgenson, S. Lang, $Pos_n(R)$ and Eisenstein Series. (2005)

Vol. 1869: A. Dembo, T. Funaki, Lectures on Probability Theory and Statistics. Ecole d'Eté de Probabilités de Saint-Flour XXXIII-2003. Editor: J. Picard (2005)

Vol. 1870: V.I. Gurariy, W. Lusky, Geometry of Müntz Spaces and Related Questions. (2005)

Vol. 1871: P. Constantin, G. Gallavotti, A.V. Kazhikhov, Y. Meyer, S. Ukai, Mathematical Foundation of Turbulent Viscous Flows, Martina Franca, Italy, 2003. Editors: M. Cannone, T. Miyakawa (2006)

Vol. 1872: A. Friedman (Ed.), Tutorials in Mathematical Biosciences III. Cell Cycle, Proliferation, and Cancer (2006)

Vol. 1873: R. Mansuy, M. Yor, Random Times and Enlargements of Filtrations in a Brownian Setting (2006)

Vol. 1874: M. Yor, M. Émery (Eds.), In Memoriam Paul-André Meyer - Séminaire de probabilités XXXIX (2006)

Vol. 1875: J. Pitman, Combinatorial Stochastic Processes. Ecole d'Eté de Probabilités de Saint-Flour XXXII-2002. Editor: J. Picard (2006)

Vol. 1876: H. Herrlich, Axiom of Choice (2006)

Vol. 1877: J. Steuding, Value Distributions of L-Functions (2007)

Vol. 1878: R. Cerf, The Wulff Crystal in Ising and Percolation Models, Ecole d'Eté de Probabilités de Saint-Flour XXXIV-2004. Editor: Jean Picard (2006)

Vol. 1879: G. Slade, The Lace Expansion and its Applications, Ecole d'Eté de Probabilités de Saint-Flour XXXIV-2004. Editor: Jean Picard (2006)

Vol. 1880: S. Attal, A. Joye, C.-A. Pillet, Open Quantum Systems I, The Hamiltonian Approach (2006)

Vol. 1881: S. Attal, A. Joye, C.-A. Pillet, Open Quantum Systems II, The Markovian Approach (2006)

Vol. 1882: S. Attal, A. Joye, C.-A. Pillet, Open Quantum Systems III, Recent Developments (2006)

Vol. 1883: W. Van Assche, F. Marcellàn (Eds.), Orthogonal Polynomials and Special Functions, Computation and Application (2006)

Vol. 1884: N. Hayashi, E.I. Kaikina, P.I. Naumkin, I.A. Shishmarev, Asymptotics for Dissipative Nonlinear Equations (2006)

Vol. 1885: A. Telcs, The Art of Random Walks (2006)

Vol. 1886: S. Takamura, Splitting Deformations of Degenerations of Complex Curves (2006)

Vol. 1887: K. Habermann, L. Habermann, Introduction to Symplectic Dirac Operators (2006)

Vol. 1888: J. van der Hoeven, Transseries and Real Differential Algebra (2006)

Vol. 1889: G. Osipenko, Dynamical Systems, Graphs, and Algorithms (2006)

Vol. 1890: M. Bunge, J. Funk, Singular Coverings of Toposes (2006)

Vol. 1891: J.B. Friedlander, D.R. Heath-Brown, H. Iwaniec, J. Kaczorowski, Analytic Number Theory, Cetraro, Italy, 2002. Editors: A. Perelli, C. Viola (2006)

Vol. 1892: A. Baddeley, I. Bárány, R. Schneider, W. Weil, Stochastic Geometry, Martina Franca, Italy, 2004. Editor: W. Weil (2007)

Vol. 1893: H. Hanßmann, Local and Semi-Local Bifurcations in Hamiltonian Dynamical Systems, Results and Examples (2007)

Vol. 1894: C.W. Groetsch, Stable Approximate Evaluation of Unbounded Operators (2007)

Vol. 1895: L. Molnár, Selected Preserver Problems on Algebraic Structures of Linear Operators and on Function Spaces (2007)

Vol. 1896: P. Massart, Concentration Inequalities and Model Selection, Ecole d'Eté de Probabilités de Saint-Flour XXXIII-2003. Editor: J. Picard (2007)

Vol. 1897: R. Doney, Fluctuation Theory for Lévy Processes, Ecole d'Eté de Probabilités de Saint-Flour XXXV-2005. Editor: J. Picard (2007)

Vol. 1898: H.R. Beyer, Beyond Partial Differential Equations, On linear and Quasi-Linear Abstract Hyperbolic Evolution Equations (2007)

Vol. 1899: Séminaire de Probabilités XL. Editors: C. Donati-Martin, M. Émery, A. Rouault, C. Stricker (2007)

Vol. 1900: E. Bolthausen, A. Bovier (Eds.), Spin Glasses (2007)

Vol. 1901: O. Wittenberg, Intersections de deux quadriques et pinceaux de courbes de genre 1, Intersections of Two Quadrics and Pencils of Curves of Genus 1 (2007)

Vol. 1902: A. Isaev, Lectures on the Automorphism Groups of Kobayashi-Hyperbolic Manifolds (2007)

Vol. 1903: G. Kresin, V. Maz'ya, Sharp Real-Part Theorems (2007)

Recent Reprints and New Editions

Vol. 1618: G. Pisier, Similarity Problems and Completely Bounded Maps. 1995 – 2nd exp. edition (2001)

Vol. 1629: J.D. Moore, Lectures on Seiberg-Witten Invariants. 1997 – 2nd edition (2001)

Vol. 1638: P. Vanhaecke, Integrable Systems in the realm of Algebraic Geometry. 1996 – 2nd edition (2001)

Vol. 1702: J. Ma, J. Yong, Forward-Backward Stochastic Differential Equations and their Applications. 1999 – Corr. 3rd printing (2005)

Vol. 830: J.A. Green, Polynomial Representations of GL_n, with an Appendix on Schensted Correspondence and Littelmann Paths by K. Erdmann, J.A. Green and M. Schocker 1980 – 2nd corr. and augmented edition (2007)